SolidWorks® for AutoCAD® Users

SECOND EDITION

Gregory Jankowski and David Murray

OnWord Press
Thomson Learning™

Africa • Australia • Canada • Denmark • Japan • Mexico • New Zealand
Phillipines • Puerto Rico • Singapore • United Kingdom • United States

NOTICE TO THE READER

Publisher does not warrant or guarantee any of the products described herein or perform any independent analysis in connection with any of the product information contained herein. Publisher does not assume, and expressly disclaims, any obligation to obtain and include information other than that provided to it by the manufacturer.

The reader is expressly warned to consider and adopt all safety precautions that might be indicated by the activities herein and to avoid all potential hazards. By following the instructions contained herein, the reader willingly assumes all risks in connection with such instructions.

The publisher makes no representation or warranties of any kind, including but not limited to, the warranties of fitness for particular purpose or merchantability, nor are any such representations implied with respect to the material set forth herein, and the publisher takes no responsibility with respect to such material. The publisher shall not be liable for any special, consequential, or exemplary damages resulting, in whole or part, from the readers' use of, or reliance upon, this material.

Trademarks

SolidWorks is a registered trademark and FeatureManager and PhotoWorks are trademarks of SolidWorks Corporation. AutoCAD is a registered trademark of AutoDesk Corporation. Windows, Windows 95/98, and Windows NT are registered trademarks of Microsoft Corporation.

OnWord Press Staff

Publisher: Alar Elken

Executive Editor: Sandy Clark

Managing Editor: Carol Leyba

Development Editor: Daril Bentley

Editorial Assistant: Allyson Powell

Executive Marketing Manager: Maura Theriault

Executive Production Manager: Mary Ellen Black

Production and Art & Design Coordinator: Cynthia Welch

Manufacturing Director: Andrew Crouth

Technology Project Manager: Tom Smith

Cover Design by Lynne Egensteiner

Copyright © 2000 by Gregory Jankowski and David Murray

SAN 694-0269

Second Edition, 2000

10 9 8 7 6 5 4 3 2 1

Printed in Canada

Library of Congress Cataloging-in-Publication Data

Jankowski, Greg, 1959-

SolidWorks for AutoCAD users / Greg Jankowski and David Murray.

-- 2nd ed.

p. cm.

ISBN 1-56690-191-X

1. Computer graphics. 2. SolidWorks. 3. Engineering models. 4.

Computer-aided design. 5. AutoCAD (Computer file) I. Murray, David,

1962- II. Title.

T385 J37 1999

604.2'0285'5369--dc21

99-39244

CIP

For more information, contact

OnWord Press   An imprint of Thomson Learning

Box 15-015   Albany, New York USA 12212-15015

You can request permission to use material from this text through the following phone and fax numbers.

Phone: 1-800-730-2214; Fax: 1-800-730-2215; or visit our Web site at www.thomsonrights.com

About the Authors

Gregory Jankowski is principal officer of CIMCo, a company that develops training materials for the CAD industry, and that provides application development and consulting. Greg has been a mechanical designer and application engineer within the field of computer-aided design for more than 15 years, and has experience with SolidWorks, Pro/ENGINEER, and ComputerVision CAD systems. He is also the author of the computer-based training product *Exploring SolidWorks*.

David Murray works in technical support and is the training manager for CADimensions, a CAD consulting company in East Syracuse, NY. David has had many years of experience with SolidWorks, AutoCAD, and the CAD environment. He is the author of *INSIDE SolidWorks* and is a contributing writer for *SOLID Solutions* magazine.

Acknowledgments

I would like to express my gratitude for the support shown by my wife, Sandy, and my daughter, Alexis. Without their patience and understanding, this work would not have been possible. I would also like to thank everyone at SolidWorks. They have been very supportive of all my efforts since SolidWorks 95.

Thanks also go to Jeff Nauman, Roger Killian, Gopal Shenoy, and Mark Gibson for reviewing the text. Finally, my thanks go to editor Daril Bentley at OnWord Press.

Gregory Jankowski

I would like to thank the folks at SolidWorks Corporation for their phenomenal efforts, who don't seem to get nearly enough respect for all the hard work they put into the software. Also, a hardy pat on the back to Daril Bentley and the rest of the crew at OnWord Press for seeing this work through to its completion and maintaining a semblance of order through their company's transition.

David Murray

Contents

Introduction
Audience, Structure, and Content vi
AutoCAD/SolidWorks Comparative
Discussion ix
Text Conventions and Book Features .. ix

CHAPTER 1
Overview and the SolidWorks Interface   1
Introduction 1
Starting a SolidWorks Editing Session ..2
Opening an Existing Document 6
Types of Documents 9
Creating a New Document 12
Saving and Closing Documents 13
The SolidWorks User Interface 16
Context-sensitive Help 23
Summary 25

CHAPTER 2: Implementation   27
Introduction 27
Implementation Plan 29
Training Plan 29
Translating Legacy Data 32
Hardware Considerations 35
Determining a Reasonable
Pilot Project 42
Summary 45

CHAPTER 3: Getting Started   47
Introduction 47
Startup 48
Sketch Planes 52
Sketch Profiles 57
Design Intent 58
Preference Settings 59
Summary 68

CHAPTER 4: Interacting with SolidWorks   69
Introduction 69
Selecting Objects 70
An Introduction to System Feedback .72
The Right Mouse Button 73
Display Functions 75
Setting Grid and Snap Properties 90
Summary 95

CHAPTER 5: Sketching   97
Introduction 97
Sketch Basics 98
Constraining Sketch Geometry 103
Creating a Sketch 108
Cursor Inferencing 110
Sketch Entities 113
Construction Entities 123
Sketch Tools and Mirroring 124
An Introduction to Dragging Geometry .127
Additional Sketch Entity Tools 128
Sketch Dimensions 142
A Universal Dimensioning Tool 143
An Introduction to Sketch Color Codes 152
Editing a Sketch 152
Editing Sketch Entities and Dimensions .153
An Introduction to the Undo Command .155
Sketch and Relations Functions 156
An Interactive Sketching Exercise ...168
Summary 174

CHAPTER 6: Parts   175
Introduction 175
Part Planning 177
General Feature Creation Order 178
Types of Features 179
Solid Features 181

Feature Names 196
Parent/Child Relationships 198
Other Feature Functions 200
Pattern Features 220
Reference Geometry 225
Modifying a Part 244
Advanced Part Features 252
Thin Features 272
Summary 279

CHAPTER 7: Assemblies   281
Introduction 281
Assembly File Management 283
Assembly Modeling Methodology .. 285
Creating an Assembly 289
Modifying Assemblies 307
Assembly Structure Editing 316
Advanced Assembly Features 318
Example Assembly 336
Summary 340

CHAPTER 8: Drawings   341
Introduction 341
About SolidWorks Drawings 343
Creating a New Drawing 345
Creating Views 358
Modifying and Aligning Views 370
Adding Dimensions 377
Drawing Symbols 394
Bill of Materials 412
Summary 421

CHAPTER 9: Rendering   423
Introduction 423
Lighting Sources 424
Colors 430

Assembly Parts 435
Saving Images 436
PhotoWorks 436
Summary 437

CHAPTER 10: Printing   439
Introduction 439
Printing Basics 440
Page Setup 445
Printing a Document 454
Summary 455

CHAPTER 11: Import/Export   457
Introduction 457
Basics 458
DXF and DWG 462
IGES 465
Stereolithography 469
Miscellaneous Formats 472
Summary 476

CHAPTER 12: Customizing SolidWorks   477
Introduction 477
User Preferences 478
Copy Options Wizard 483
Customize Command 483
Keyboard Shortcuts 487
Programming Languages 489
Summary 493

Appendix   495
Command Cross-reference 495
Sketch Entities Cross-reference 509

Illustrated Glossary   511

Index   519

Introduction

This book was developed to aid users in their transition from a 2D or 3D AutoCAD environment to a 3D solid modeling system. AutoCAD users who have worked in a 3D wireframe environment will find that solid feature-based modeling requires a different mind-set than they are used to. SolidWorks is a native Windows application that takes full advantage of the standards (e.g., user interface, use of right mouse button) and technology (e.g., Object Linking and Embedding) that have made Microsoft Windows and Windows-based products such a success.

There have been over 150 improvements and enhancements made to SolidWorks 99. Many of these improvements or enhancements have been incorporated into this new revision of *SolidWorks for AutoCAD Users*. This includes areas involving sketching, design tables, bill of materials, sheet metal functionality, color editing, lighting, assembly component structuring, assembly mating relationships, drawing views, layers and line formatting, and many other topics of interest. In many cases, references to AutoCAD functionality have been added as well to aid the AutoCAD user in making the transition to SolidWorks.

Audience, Structure, and Content

This book deals with the processes involved in using the SolidWorks product. It also addresses the problems a 2D CAD user faces when moving to a parametric, solid modeling CAD system. The methodology and ways of thinking used to produce designs change when you move to SolidWorks. You will find yourself focusing more on design intent rather than spending a lot of time and effort on maintaining 2D or 3D geometry throughout a drawing. You may also find that with SolidWorks, more time can be devoted to design work than trying to learn the software or carrying out involved commands.

SolidWorks for AutoCAD Users is also meant to guide a new user, or AutoCAD user making a transition, through the phases of implementing SolidWorks. Also covered are topics related to—and issues that arise before, during, and after—the installation of a new CAD system. Many of the chapters in this book include step-by-step processes on how to carry out a SolidWorks command or function. It is an excellent reference guide for anyone who wants to learn the SolidWorks program, or for current users who need a reference source. This book can also be used by those wanting to make an informed and intelligent decision on whether or not SolidWorks is the CAD program for them.

Chapters can be, for the most part, covered in any order for those using the book as a reference guide. However, those individuals attempting to learn SolidWorks software should work from Chapter 1 and proceed sequentially through the rest of the chapters. The chapters in this book are functionally grouped based on topis and are presented in logical order from first to last.

Chapters get increasingly more involved as you move from one to the next. Therefore, you can make a decision as to "how far" you should continue, based on your needs. The most rudimentary basics of SolidWorks are covered up front so that even the newest of users can start from the beginning and work their way up the scale of difficulty. Chapters contain the following standard sections.

Introduction	Each chapter has an introductory section that discusses the main theme of the chapter, as well as prerequisites, content, and chapter objectives.
Prerequisite	If applicable, chapter introductions contain a prerequisite section that details what you should be familiar with before attempting to read the chapter.
Content	This subheading to introductory sections discusses the main topics covered within the chapter.
Objectives	This subheading to introductory sections discusses the objectives for the chapter topics. These objectives consist of the information, skills, or procedures you should walk away with after reading the chapter.
Summary	Summery sections draw general conclusions about the content of a chapter as a whole, and briefly reinforce chapter objectives.

AutoCAD/ SolidWorks Comparative Dsicussion

Each chapter contains notes scattered throughout denoting important differences between the philosophy, procedures, and techniques between AutoCAD and SolidWorks. These sections are denoted by the AutoCAD/SolidWorks comparative text icon, shown at left. In many cases, direct comparisons cannot be made due to the inherent differences between the programs. These differences are also pointed out in the text.

Text Conventions and Book Features

The following text conventions are intended to present material in a consistent and readable form. Formatting conventions such as the previously mentioned AutoCAD/SolidWorks comparative discussion icon, the **NOTE** feature described here, lists, and tables, and numbered steps—along with the illustrated glossary, appendix, and index—are intended to aid you in finding what you are looking for efficiently.

Names of menus, dialog boxes, commands, and similar items appear with initial capital letters. Where names of such items, in keeping with the software's convention, combine initial capitals and lowercase, quote marks (" ") enclose the name. (for example, the "Name feature on creation" field). Colons that appear with names in the software are dropped when such names are referred to in the text. For example, "Line Count:" would appear as Line Count.

File names, Web site names, words used as words, and emphasized terms and phrases appear in italics.

Special notes appear periodically throughout the text that contain information that makes using SolidWorks easier, more efficient, and more productive. An example of these notes follows.

❧ **NOTE:** *When reporting problems to a vendor, customer, or SolidWorks, include a copy of this report file for reference. This can help pinpoint problems.*

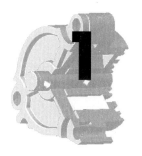

1

Overview and the SolidWorks Interface

Introduction

In this chapter, you will explore the basic look and feel of the SolidWorks user interface and how it compares to AutoCAD. Much of the terminology used in this book will be new to AutoCAD users. These new terms will be explained in this chapter, so that the SolidWorks software will feel more comfortable to you. There are many aspects of the SolidWorks user interface that are similar to AutoCAD, but many that are not. In many cases, it is difficult to unlearn old habits; therefore, the primary focus of this chapter is on helping you establish the correct mind-set for working with SolidWorks.

Prerequisite

At this point, it should be safe to assume that you have a good working knowledge of the Windows 98 or Windows NT 4.0 mechanics. (Windows 95 and Windows 98 have negligible differences when discussing SolidWorks. Therefore, Windows 95 or 98 will be referred to as simply Windows 98 hereafter.) If you do not feel at ease in the Windows environment, it would be to your benefit to run through a Windows tutorial of some sort so that you will feel more comfortable. In addition, some familiarity with drafting would help, although this becomes much more important when discussing SolidWorks Drawing documents in Chapter 8.

Content   This chapter consists of the following major topics. The "Objectives" section that follows informs you, in general terms, of what you should expect to gain as a result of reading and studying the chapter topics.

- Starting a SolidWorks editing session
- Opening an existing document
- Types of documents
- Creating a new document
- Saving and closing documents
- The SolidWorks user interface
- Context-sensitive help

Objectives   At the end of this chapter, you should be able to successfully start a new SolidWorks session, save your work, and work with both left and right mouse buttons to select entities and open menus. Many of the terms used by SolidWorks and solid modelers in general will be familiar to you, as will the SolidWorks interface. You should also have a good feel for what the SolidWorks program in general is meant to accomplish, and how it can make the engineer's or designer's desktop a friendlier place from which to work.

Starting a SolidWorks Editing Session

If you own a copy of SolidWorks, you will be able to follow along with the examples presented in this chapter and elsewhere. If not, the screen shots included in this book will help you along for the time being. Those of you who can follow along should go ahead and fire up the SolidWorks program by double clicking on the SolidWorks icon, or by selecting Solid-Works from your Programs menu under the Start button. Before you explore the SolidWorks interface in depth, a note on "Tip of the Day."

Tip of the Day   This feature does not require much explanation. This is a feature that displays a technical tip when a SolidWorks session is begun. These tips may be beneficial to new users. It can be turned off by unchecking the "Show tips at startup" option, located in the Tip of the Day dialog box. If you do turn off the

Tip of the Day, you can turn it back on later by clicking on the Help pull-down menu and selecting Tip of the Day.

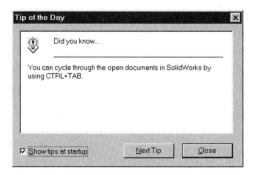

Tip of the Day dialog box.

Windows Standardization

There is a certain amount of standardization that many Windows programs possess. SolidWorks is no exception to this. For example, the pull-down menu structure follows a basic pattern, with File being the first menu selection, and Help being the last. This is no different from AutoCAD, and the options contained within some of the menus are also similar.

Take, for example, the File pull-down menu. There are the standard options File/New and File/Open, to name just a couple. Before a new file is actually started, or before an existing file is opened, you will notice that the pull-down menus are greatly simplified. This follows a common theme that SolidWorks uses to full effect. For instance, if a particular command is inaccessible at a particular moment, it may be grayed out in the menu, or not present at all. This makes selecting the appropriate command all that much easier because there is less clutter.

AutoCAD achieves this to some extent, but not nearly to the extent that SolidWorks does. The AutoCAD menu may have options that are grayed out from time to time, but most of the time the menu choices are simply there, with nothing stopping the inexperienced user from selecting incorrect or inappropriate commands.

Another good example would be the AutoCAD screen menu. When a command is initiated, the screen menu often updates to show you a number of related options for that command. Once the command is completed, the screen menu remains, even if the options for the completed command are no longer valid. Of course, the reasoning behind this is that if you want to repeat the command, it is easily accessed. However, this is not really necessary, considering a command can be repeated in AutoCAD by clicking the right mouse button anyway (to name one option).

A word or two should be said about the standard toolbar icons at this point. Once again, many native Windows programs have adopted a standard set of icons for such basic tasks as saving or opening files. SolidWorks is no exception. Specifically, the first three icons on the toolbar are counterparts to the pull-down menu commands File/New, File/Open, and File/Save, respectively. Whether you use the pull-down menus or the icons is up to you. Generally speaking, it is better to get familiar with the pull-downs first. If you later want to start using the icons, go ahead; they will usually save you at least one mouse click.

A SolidWorks Overview

Before moving to an in-depth discussion of the mechanics of the SolidWorks program, it is important to understand the philosophy behind the software and what its programmers are striving to achieve. If SolidWorks could be summed up in one phrase, it would best be described as a *feature-based, parametric, solid modeling design tool*. Parse this statement one segment at a time and you will better understand the nature of the SolidWorks program. The following sections explain the ideas contained in the phrase.

Feature based

This is a term used to describe the elements that constitute a part. Just as an assembly consists of individual parts, a part consists of individual features. There are two types of SolidWorks features: sketched features and applied features. The differences between them are as follows.

Sketched features: These are based on 2D sketch geometry. Creating a sketch is covered at length in this book. It is the basis of much that is done in the SolidWorks program, and therefore requires some elaboration. Certain steps should be followed when creating sketch geometry, but once those steps are understood, almost everything else will begin to fall into place. Sketched features can be extrusions, rotations, sweeps, or lofts.

Applied features: Features that are applied do not require a sketch. They are applied directly to a model. Examples of this type of feature are chamfers, fillets, and shells, to name a few.

Parametric

With respect to dimensions, the term *parametric* refers to the ability to make changes to dimensions and drive the shape of a part with those changes. In AutoCAD, changing a dimension would result in breaking the associativity with the model. Anything other than the default value for the dimension would keep the dimension from updating if the model were scaled or stretched. In Solid-Works, changing the dimension changes the part.

This concept is an extremely important one and deserves a little more attention. To drive the aspect of parametrics home, consider a simple example. In AutoCAD, you would draw a line, being as accurate as possible. Then you would add a dimension and the value of the dimension would reflect the length of the line. To change the length of the line, you would use Lengthen or Change, Grips, or some other similar command or function. If you did things right, the dimension value would update accordingly. The point is this: *the geometry drives the dimension.*

Take the same example in SolidWorks. First, you sketch a line. The term *sketch* is used because it implies that you do not need to be 100-percent accurate, which is true. Then, you add a dimension to the line. SolidWorks asks you for a value for the length and you type one in. In SolidWorks, *the dimension drives the geometry.* This, in a nutshell, is parametrics.

➥ **NOTE:** *Parametrics is expanded on in Chapter 5, where you learn how to dimension a sketch.*

Parametrics; a fundamental difference between AutoCAD and SolidWorks.

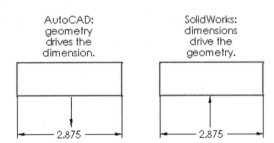

Solid model

There are 2D drawings, there are wireframe models, there are surface models, and there are solid models. Solid modeling packages are the most complete, but are also the most system intensive. This is why when an upgrade to a solid modeling program is made from a 2D drafting tool, the hardware must also usually get upgraded (because the system requirements are more demanding). It is the nature of the beast, and not the program itself, that requires the upgrades to the hardware.

The downstream benefits are very much worth the investment in almost all cases. Surface modeling can be very system intensive also, but is better suited to a different range of disciplines than solid modelers. For instance, mechanical elements are better suited to solid modelers. Anything from minuscule springs and gears to tractor trailers can be easily modeled in Solid-Works, but organic shapes such as the dinosaurs in the Jurassic Park movie would be better suited to surface modelers.

To be fair, surface modelers can be much more demanding than solid modelers in such a situation, because real-time animation of organic shapes with natural textures applied to them is beyond the scope of any solid modeler. For applications such as that, high-powered multiprocessor computers are run in parallel in order to achieve the high degree of computations involved. The number of processors involved can number in the hundreds, and the necessary memory measured in gigabytes.

Opening an Existing Document

An existing document can be opened by clicking on File/Open. This will give you a familiar-looking dialog box, shown in the following illustration, that will allow you to browse your directories, or "folders" as Windows likes to call them. Navigate to the desired folder by double clicking on the folder, or move out to the parent folder by clicking on the Up One Level icon. If you are not sure what icons are what, hold the cursor over the icon in question, and a yellow pop-up "tip" will appear to tell you what the icon is for. SolidWorks conforms to this standard as well, so the cursor can always be held over an icon if you need to see what the icon's function is.

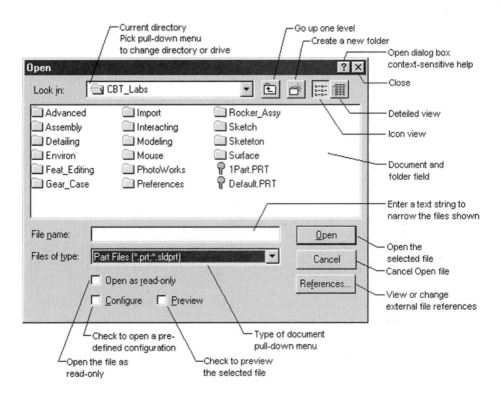

Open dialog box.

When the File/Open dialog box is first opened, it looks for part files. This is fine if you are looking to open a part file. If you are looking for something else, however, you must tell SolidWorks what type of file you are looking for. Hence, the "Files of type" section of this dialog box. Notice that SolidWorks is looking for files with a .prt or .sldprt extension. These are the extensions for SolidWorks part files. Changing "Files of type" allows you to open assemblies or drawings, or to import other file types.

❧ **NOTE 1:** *See Chapter 11 for importing files.*

❧ **NOTE 2:** *Earlier versions of SolidWorks originally used a .prt extension. This file extension was also being used by other software programs, so SolidWorks Corporation changed the extension to .sldprt to keep SolidWorks files from being confused with other file types with the .prt extension. To main-*

tain consistency within the software, all SolidWorks file extensions adopted the additional "sld" lettering.

Take a look at the following table to see what the actual file extensions are for various SolidWorks file types. What these file types are and what functions they serve are discussed later in this chapter. Pay attention to the first three, as they are considered SolidWorks' main file types. You will see other file types listed in the "Files of Type" drop-down list not contained in this chart. These are file types SolidWorks can translate (import) into its program.

Document Type	SolidWorks 99 Extensions
Part	<filename>.sldprt
Assembly	<filename>.sldasm
Drawing	<filename>.slddrw
Drawing template	<filename>.slddrt
Library feature part file	<filename>.sldlfp
Curve file	<filename>.sldcrv
Custom symbol	<filename>.sldsym
SolidWorks Basic macro	<filename>.sldswb

Use "Files of type" to browse for specific file types. It is the proper procedure you would use to open an assembly or drawing. The other file types you will see in the "Files of type" drop-down list not mentioned here are file types not native to SolidWorks. An in-depth explanation of these file types is found in Chapter 11. What is important at present is a general knowledge of the purpose the native SolidWorks file types serve.

The following illustration shows Windows Explorer. A SolidWorks document can be opened by double clicking on the desired icon. Not all native SolidWorks documents will open in this fashion, so this method is best used for opening a file of the basic three file types (part, assembly, and drawing).

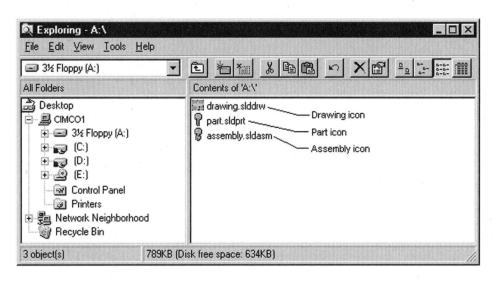

Windows Explorer.

Types of Documents

The following sections explain the three major types of Solid-Works documents (file types). These are part, drawing, and assembly documents.

Part Documents

As touched on in previous paragraphs, SolidWorks has multiple file types. AutoCAD uses a .dwg extension whether you are creating a 2D engineering layout, an assembly, or a solid model. SolidWorks has a separate file format for each of these. A part file is a collection of features (discussed earlier in this chapter) that are added to shape a model. Normally, this is the first file created. More than one solid model cannot exist in the same part file.

Drawing Documents

After a part is created, typically an engineering layout is generated. This generally consists of top, front, and right-side views. It may also consist of section or detailed views, notes and dimensions, a bill of materials, and other items generally found in a typical drawing layout. Drawings can be of one or more parts and can contain multiple sheets, which are the paper pages used for each drawing. Drawings can be of parts or assemblies, in any combination.

You can think of a drawing as the folder that contains your drawing layouts. Drawing sheets are analogous to the pieces of paper containing your drawing formats and views. Each drawing format resides on a different sheet, and you can have as many sheets as you want to in each drawing.

↦ **NOTE:** *Working with sheets is explored in Chapter 8.*

Assembly Documents

Assemblies are, in their most basic form, collections of part files. An assembly can consist of two parts or two hundred parts. Actually, an assembly can consist of just one part, but that sort of defeats the purpose. An assembly can also contain other assemblies, which are generally considered subassemblies in that type of situation. It is possible to have assemblies that contain other assemblies, which in themselves contain assemblies, and on and on. The only limiting factor on the size of an assembly (or a part, for that matter) is the physical hardware in your computer, such as the amount of hard drive space and the amount of memory.

A common annoyance to AutoCAD users is the fact that only one document can be opened at a time. This continues to be the case with AutoCAD release 14. In order to open up two documents at once, you must actually open up two AutoCAD editing sessions. This obviously is a greater drain on system resources. This is not the case with SolidWorks. It is possible to open up multiple parts, drawings, and assemblies—all within the same session and in any combination. This capability is known as MDI, or *Multiple Document Interface.*

All documents related to a design task can be opened at the same time. When changes are made, the related documents can be activated to view the changes immediately. Multiple windows also allow you to cut, paste, or "drag and drop" objects from one window to another.

Another important topic is the term *bidirectional associativity.* This is an impressive term to describe a basic principle, which is: the three basic SolidWorks file types (parts, drawings, and assemblies) are dependent on one another. For example, say you have just finished creating your latest solid part, the infamous widget. It is then time to create your layout, which is literally a drag-and-drop operation, which will be explored later.

The head honcho on the dreaded widget project decides some serious design changes need to be made. Therefore, you open the part file and make the changes. Now what happens to the drawing? SolidWorks takes care of it, because the very next time you open up the layout (probably appropriately named widget.slddrw), SolidWorks automatically updates the drawing to match the part. Any dimensions will change to match the true values, views will get updated to reflect design changes made to the part, and features that have been removed or added will show up correctly.

Now say that Mr. Bossman is looking over your shoulder as you are reviewing the drawing layout. He decides that maybe the dimension changes to the overall length of the widget should not have been altered so drastically, and he recommends another change. You alter the dimension on the drawing, and the part automatically updates to reflect the change. Furthermore, if said widget is being used in an assembly, the assembly will update the next time you open it. Again, the assembly can be edited and the change will reflect back to the part. This is where the "bidirectional" of the term *bidirectional associativity* originates.

Bidirectional associativity among the three SolidWorks main document types.

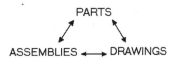

What does all of this mean? Well, it will cut down on your editing time, because it is all being done for you. That much is obvious, but this topic raises some other issues. For instance, maybe you do not want the assembly to update because an earlier revision of the widget is being used in it, and the changes being made to the new widget are only going to affect future assemblies. If the new assembly used a different version of the widget, the file used in the assembly could be easily changed by accessing the assembly component's properties. You could press the right mouse button over the desired component in FeatureManager and select Component Properties. The referenced part file could then be changed by selecting Browse in the Model Document Path field and selecting the new widget version.

⟿ **NOTE:** *You will read more on component properties in Chapter 7.*

Creating a New Document

This is where a new SolidWorks document is born. You are almost ready to start some serious solid modeling. However, there are a few things to get out of the way first. There are two ways to start a new SolidWorks document: click on the appropriate icon or use the pull-down menus. You should be noticing a common theme here (toolbars or pull-downs); therefore, icons will from now on be mentioned less and less, except to tell you what they look like. The use of icons is up to you.

Getting back to the process of creating a new document, the New icon looks like a white sheet of paper with its top right corner folded over. This, once again, adheres to the Windows convention. Clicking on File/New will accomplish the same thing, which is to open the New dialog box, shown in the following illustration.

New dialog box.

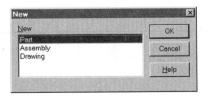

As you can see, it is possible to start any one of the three document types from this point, whether it be part, drawing, or assembly. As is typical of working in SolidWorks, your choice will be the Part option, followed by clicking on OK. This option is typical because drawings are normally generated from parts, and assemblies consist of parts. Therefore, a part is normally the first document you would create on a typical project. This would not matter in AutoCAD, or at least not to the same extent. The closest analogy would probably be to create a solid (or wireframe) in tiled model space, then create your layout in paper space. Again, in AutoCAD's case, this would all be done in the same file.

As soon as the new or existing file is opened, the first thing you notice is the addition of some new toolbars and that the pull-down menus increase in number. This goes along with Solid-Works' idea that fewer choices make the program easier to learn and use.

Some would argue that "ease of use" would translate to "not as powerful" or "less functionality." This is not the case at all with SolidWorks. SolidWorks' use of the Parasolids kernel combined with Windows programming techniques and in-depth use of functions such as object linking and embedding (OLE) makes for a feature-rich and well-rounded software package. In the case of SolidWorks, "ease of use" translates directly to "getting the job done."

Getting started building a solid model is discussed in an upcoming section. However, you first need to know how to save that solid model.

Saving and Closing Documents

It really does not matter if you save your model right now, before anything has really been created, or if you save it in five minutes, or ten. What *does* matter is that you save your work on a regular basis. For those of you who have been working computers for some time, you realize this. However, it is surprising how often even a seasoned veteran forgets to save. *The point is to save your work, and save it often.* To save your work, perform the following steps.

1. Click on the File pull-down menu.

2. Click on Save.

Use the icon that looks like the floppy disk, if you want. It does the same thing. The first time a file is saved, the Save As dialog box appears. This is because the file has not yet been named.

When you first save a file in SolidWorks, and accept the file name it assigns on its own, the name of the file (assume a part file here) will be Part1.sldprt. If you started another new part and saved it, it would be Part2.sldprt, and so on. The next time you start a SolidWorks session, meaning when you next start the SolidWorks program, the naming convention starts all over again. This means that if you do not give the file a new name, it gets saved as Part1.sldprt again, and you will be prompted to see if you really want to overwrite the file of the same name left on your hard drive during the previous session.

This is no big deal. Normally, you would go ahead and type in a unique name for the file anyway, as you would in any program. What is convenient about this is that if you are doing some experimenting and saving your work, but do not really care about permanently saving the file, you do not have to type in a file name. This also cuts down on clutter, because, for example, half a dozen files titled Part1, Part2, Part3, and so on simply keep getting overwritten. If at some point you decide you want to keep one of these generically named part files, just use the Save As option.

Saving Documents with Save As

You will find the Save As option under the File pull-down menu. It is a very simple procedure. After clicking on File/Save As, a dialog box opens that lets you name your current file with a new name or location. This is no different than any other software program, including AutoCAD. What actually happens is that the file you were working on is saved to the hard drive under its original name, and the newly named file becomes the file you are currently editing.

Saving Documents with Save As/Save As Copy Toggle

This is identical to Save As, with some minor differences. When you check the Save As Copy check box while performing a Save As, the newly named file is the one that gets saved to the hard drive, not your current file. In other words, your current file remains your current file, and the copy goes to the hard drive.

Another aspect of using the Save As Copy check box has to do with assembly references. Without getting too technical at this early stage, suffice it to say checking Save As Copy will keep an assembly from referencing the newly named (copied) file. If Save As Copy is not checked, performing a Save As will replace the referenced file in an assembly with the newly named file.

This can be somewhat confusing, so let's look at this one other way. An assembly document references all the files (components) in the assembly. Imagine for a moment that a file "widget" is being used in an assembly. If the widget is opened and renamed to "widget revision 2" by using Save As, and if the Save As Copy option is not checked, then the assembly will begin referencing "widget revision 2" instead of the original "widget" document. You will learn a great deal more about assemblies in Chapter 7.

Closing Documents

Now that you have learned how to save, it is time to learn how to shut down and pack it in for the day. Closing your documents is as easy as clicking on File/Close. Whatever document is active will be closed. If you do not save it first, SolidWorks will ask you if you want to save it before it shuts the file down. The following illustration shows the SolidWorks prompt when closing a document with unfiled changes.

Result of the File/Close command.

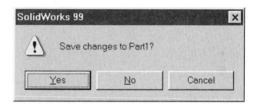

Keep in mind that the program will not ask you if you have already saved a file. Occasionally, a new user will shut down a file and not remember if she saved her work. Have no fear. One of two things happened. Either no changes were made to the file, in which case there is nothing to save, or the work was indeed saved. It is actually difficult to lose your work. You have to try pretty hard. Rest assured that as long as you are reading the warnings SolidWorks is giving you on the screen, your hard work is successfully saved on the hard drive.

Exit Windows shortcut.

Exiting out of the program will give you the same warnings if you have not saved your documents. It does not matter if you have opened and edited two dozen documents, you will get prompted for each of them. To exit out of SolidWorks, either click on the X button to the right of the title bar or use the pull-down menus and click on File/Exit. The illustration at left shows the Exit Windows shortcut.

The SolidWorks User Interface

Most of the SolidWorks interface should be Windows components somewhat familiar to you—at least from a terminology standpoint, if not in practice. For instance, there are the toolbars, whose icons you will get more accustomed to as time goes by. Likewise, there are the pull-down menus, which are referred to throughout this book. There is also the title bar, which resides at the top of every Windows application, whether it be Windows 98, Windows NT, or even Windows 3.1 (although SolidWorks does not run on Windows 3.1). However, there are other elements of the SolidWorks interface that will be new to you. These elements are discussed in the upcoming sections.

If you are sitting in front of a computer and have SolidWorks running, open a new part file using the methods described earlier in this chapter. If not, refer to the following illustration, which shows a new part file.

New part file.

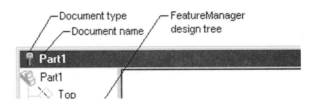

Feature Manager

What is displayed along the left-hand side of the screen is known as the SolidWorks FeatureManager™. As its name implies, this feature allows you to track and manage the creation of features. The features are displayed chronologically from top to bottom. Features require elaboration because AutoCAD does not have a counterpart to a SolidWorks feature (nor does AutoCAD have anything even vaguely resembling a feature manager).

A feature can be thought of as a component to a part. It might be what is known as a boss, such as when material is being added to a part, or it might be a cut, when material is being removed from the part. As features are being added, the part is built up and gains complexity until the final model is complete.

SolidWorks features contain intelligence. This does not mean that a feature can beat you at a game of chess. It means that the part file database contains information that makes the feature more than just a set of spatial coordinates connected by lines and arcs. A solid model contains much more information in its database than does a wireframe model. There is the parametric aspect of the model, along with topological information.

Boolean Versus Parametric Modelers

Another important difference should be drawn between Boolean modelers such as AutoCAD and parametric modelers such as SolidWorks. Consider what you would have to go through in order to create a simple rectangular block with a hole in it in AutoCAD. You would first create a rectangle of the proper size. Then you would have to turn the rectangle into a region, which is a prerequisite to turning AutoCAD profiles into a solid. The region could then be extruded into a solid object. The alternative would be to create a solid by specifying the appropriate primitive command, which would allow you to create basic solid building blocks.

Once the block had been created, you would use approximately the same process to create a cylinder. The easiest method would be to create a cylindrical primitive, located in the correct position relative to the block. If it were not already properly located, that task would also have to be completed. This scenario might exist because it is sometimes more simple to create an object off to the side and then move it to the proper location.

Once the location of the cylinder is correct, a Boolean operation must be performed. A Boolean operation can be thought of as a logical operation between solids, at least in solid modeling terms. This might be adding two solids together, such as a union, or subtracting one or more solids from an existing solid. It might also mean the intersection of two solids to form a separate solid piece of geometry, such as an interference section.

Assume for a minute that the block has now been built and has the hole cut out of it. In other words, the cylinder has been subtracted from the block. If the hole is in the wrong spot, what do you do? This goes to the heart of the matter. First, you would have to create another solid to fill the hole. Then you would have to recreate another cylinder of the correct size in the new location. Last, you would perform another Boolean operation.

Just hope that the hole is in the right spot this time. Needless to say, this greatly impedes the design process.

The SolidWorks Alternative to Boolean Operations

Now look at this operation from a SolidWorks standpoint. It starts out much the same way, by creating a rectangle. Instead of worrying about accuracy, however, the rectangle is "sketched," and then dimensions are added that drive the shape of the rectangle. It is like adding dimensions in AutoCAD, only easier, and the dimensions literally control the shape of the sketch. This is what the term *parametric* means. AutoCAD's dimensions will change if certain editing tasks are performed, such as stretching or scaling a model, but that is as good as it gets. With a parametric modeler such as SolidWorks, you can change a model by modifying its dimensions.

Next, the rectangle is extruded, much the same way you would in AutoCAD. Specify an extrusion height and you are done with the block. Now, sketch a circle, dimension it, and cut it through the part. The ability to "cut" the circle through the part reduces the two steps performed in AutoCAD to one in SolidWorks because there are no Boolean operations.

This reduces the design process in itself, but it is only the beginning. Where the real fun starts is when a design change needs to be accomplished. Take, for example, the task of moving the hole in the block. In SolidWorks, the dimensions placed on the hole can be edited to move the hole. (The mechanics are explored later, but the process is basically nothing more than a dimension change and a click on the Rebuild icon, with the part editing then completed.)

Graphics Window

To the right of FeatureManager, and taking up most of the screen, is the area in which you will see most of your work come together. If you are used to AutoCAD's typical black background, try to adjust to the white background of SolidWorks. It is possible to change it to black (or pink, for that matter), but it is not recommended.

Consider this: in AutoCAD, most of what you are probably used to creating are 2D layouts. Light lines show up nicely on a black background. AutoCAD's wireframe representations of its solid models normally take this form. However, in SolidWorks, you will be creating solid models that can be dynamically rotated in shaded mode; therefore, you do not need that black background.

There are other technical reasons you will not want to alter the white background, which have to do with the way certain entity types are displayed and the color cues SolidWorks gives you. Give it a chance, and you will probably grow to like it.

Between FeatureManager and the graphics window is a vertical bar that acts as a separator. This vertical bar can be adjusted with your left mouse button. If the cursor is moved over this bar, it will change into a double arrow shape. This allows you to either move the bar to the right, which lets you see all of the names in FeatureManager, or to the left to increase the size of the sketch area.

It should be noted that the position of the vertical separator bar is saved with your SolidWorks document. This means that if you open your SolidWorks document in the future, the position of the vertical bar will be remembered.

Origin Point

In AutoCAD, there is the User Coordinate System, or UCS. Solid-Works has the Origin Point. They are used very differently. Gone are the days when the UCS is always down there in the left-hand corner. The origin point (referred to hereafter as the "origin") still represents your *x-y* axis, and it still relates to 0,0 Cartesian coordinates, but that is where the similarity ends. The origin is never moved (as the UCS was) in order to specify planes that will be worked on. In SolidWorks, a plane or planar face is selected, and then the sketch is begun. At that point, the origin is shown and is in alignment with whatever plane or planar face you happen to be sketching on.

Another important aspect of the origin is that it acts as an anchor. In other words, a sketch can be anchored to the origin so that it will not float around in space. Think of the origin as a clamp used to hold a sketch in place so that it can be worked on. Whenever a new sketch is started, another origin is created for that sketch. In contrast, there is never more than one UCS.

When a part is first begun, the origin will appear gray in the work area, as shown at left. Once you start a sketch, a red origin

Original Origin
(Gray)

Sketch Origin
(Red)

Sketch area showing origin in gray.

will appear to indicate that you are in an active sketch. The origin also helps to orient your perspective with regard to where the x and y axes are, as well as the positioning of the current sketch plane. The UCS in AutoCAD behaves much the same way when it appears at different angles on the screen.

➥ **NOTE:** *The origin point is discussed further in Chapter 5.*

Pull-down Menus— A Quick Comparison

If you are used to the DOS version of AutoCAD, you will be accustomed to the pull-down menus remembering the last command selected in that particular menu. AutoCAD is among a few programs that offers this function. Windows programs do not work that way, including SolidWorks.

Customizing SolidWorks pull-down menus cannot be done as easily as editing a text file, as is possible in AutoCAD. Third-party programmers that integrate their menus directly into the SolidWorks menu are usually known as Gold Solution partners by SolidWorks Corporation. There are many of these Gold Solution partners that have brought higher and specialized functionality to the SolidWorks program by seamlessly integrating their software directly into SolidWorks, almost as if it were part of the same program.

These third-party programs range from CAD/CAM tool path software, to photorendering (for example, PhotoWorks, shown in the following illustration), to Finite Element Analysis (see FEM in the following illustration), to many others. In fact, SolidWorks is likely to rival AutoCAD in the sheer number of third-party developers creating software for the SolidWorks program. The number is already quite high.

SolidWorks menu with third-party add-on menus.

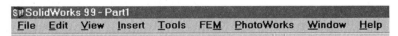

Whether you are working on a part, drawing, or assembly, the titles of the pull-down menus will not change. However, what is underneath the menus will change, depending on what document type is being currently worked on. This is referred to as a context-sensitive command and menu structure. If all three document types are open at the same time, the document active at the time determines what the pull-down menus will contain.

For example, the Insert pull-down menu will allow you to insert features if editing a part, but will allow you to insert mating relationships if working on an assembly.

Incidentally, you may have noticed ellipses (three dots) after certain options in the pull-down menus. This lets you know that a dialog box will be opened if this option is selected, which follows suit with AutoCAD.

Toolbars

AutoCAD release 13 had over 50 toolbars. You would never want to have them all open at once, or it would be unmanageable. Release 14 improved on this by reorganizing the menus and toolbars, with a total of 17 toolbars (if you count the Bonus toolbars). SolidWorks has a total of 17 toolbars. They can be turned on or off by clicking on View/Toolbars and selecting the appropriate toolbar name. The View/Toolbars menu is shown in the at left illustration.

View/Toolbars menu.

Typically, you will have four to six toolbars displayed at any given time. You might have more displayed, depending on your screen size, resolution, and work preferences. The Solid-Works toolbar icons are straightforward and easy to interpret. The toolbars can be dragged to different floating locations by placing the cursor over the edge of the toolbar and dragging it while holding down the left mouse button, which is the same functionality as AutoCAD. They can also be docked to various sides of the screen, as in AutoCAD. This is standard Windows functionality. Most Windows programs that implement toolbars operate the same way.

Toolbars are also context sensitive. By leaving only standard, sketch, and view toolbars checked, the other major toolbars (Sketch tools, Features, Drawings, and Assemblies) will appear when the appropriate document type is activated. The default state of the toolbars is determined by the toolbars active at the end of the last SolidWorks session.

SolidWorks toolbars also have small yellow tool tips that pop up if the cursor is placed over them. To try this, hold the mouse cursor over an icon without moving or clicking the mouse. After a second, the box will appear. A more lengthy description usually appears in the status bar at the bottom left-hand corner of the screen. This may vary slightly, depending on the version of Windows being used, but the implementation is always the same.

Toolbars can also be customized, but you will have to try that at your own risk. It is not recommended that you try this at this point. Wait until you are much more comfortable with the program. The following illustrations show the various SolidWorks toolbars. Their functions are discussed through the course of this book.

Font toolbar.

Sketch toolbar.

Macro toolbar.

Sketch Relations toolbar.

Standard toolbar.

Assembly toolbar.

View toolbar.

Line Format toolbar.

Features toolbar.

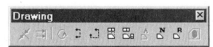

Drawing toolbar.

Web toolbar.

Standard Views toolbar.

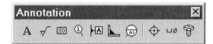

Selection Filter toolbar.

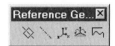

Reference Geometry toolbar.

Annotation toolbar.

Tools toolbar.

Sketch Tools toolbar.

Status Bar — A Quick Comparison

The AutoCAD status bar has a couple of useful functions, such as allowing you to easily turn your grid or running object snaps on or off, and to display information. But for the most part though, the status bar is really just that: a bar to display the status of one function or another at any given point.

SolidWorks does not make much use of the status bar. It displays items such as the current x-y coordinates of the cursor during the sketch process (like AutoCAD), whether or not the current sketch is fully defined, and other information. The status bar also displays a more complete description of the tool tips. If you would like to turn off the status bar, click on View/Status Bar. It is a toggle switch, so this method can also be used to turn the status bar back on. In the case of either AutoCAD or Solid-Works, you will find you can easily live without the status bar, and turning it off increases your screen area. This is strictly a personal preference.

Context-sensitive Help

If you have used Windows help before, you know how to use SolidWorks help. If the Help menu is accessed, there is an option for SolidWorks 99 Help Topics. This will open the Help Topics dialog box, shown in the following illustration.

*Help Topics
dialog box.*

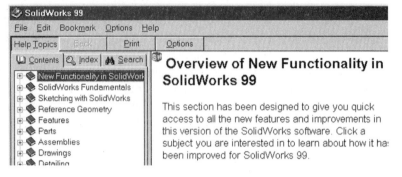

Help is there for a reason. Use it if you need it. It contains step-by-step examples of how to carry out specific functions. Accessing context-sensitive help in AutoCAD is easy, because all one has to do is punch the F1 function key (or F2, depending on the version) during a command and the help dialog box for that command appears on the screen. This is what is meant by the term *context sensitive*. Depending on what is being done at any given time determines what is displayed on the help menu.

SolidWorks Help Functionality

SolidWorks also has context-sensitive help, but it is implemented in a slightly different manner. If you find yourself in need of assistance during a SolidWorks command, click on the help button on whatever dialog box happens to be open at the time. This will bring up a step-by-step explanation of how to carry out the command in question, along with the options available.

SolidWorks makes very good use of dialog boxes. Unlike AutoCAD, there are no commands to type in at the command line, and there is no "Command" prompt. Everything is a user-friendly dialog box and visually interactive. There are not half a dozen command line options you have to know in order to complete a simple command, and most commands are laid out logically and intuitively.

In contrast to AutoCAD, commands in SolidWorks do not vary depending on the approach to entering a command. This is because in SolidWorks there are logical courses of action you must follow in order to implement the commands, as opposed to many ways in which to complete any one task.

SolidWorks 99

This is actually a menu pick that resides under the Help menu. This option allows you to determine the date code of the Solid-Works version being run. The year is stated first, followed by the numerical day of the year. This can be helpful for technical support reasons. SolidWorks Corporation currently provides updates to maintenance subscribers on the Internet so that they can receive enhancements and patches to the software. It also allows for accurate tracking of software glitches that may arise. The following illustration shows the About SolidWorks 99 menu pick.

About SolidWorks 99 menu pick in the Help menu.

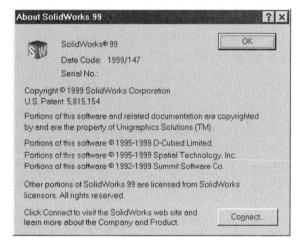

The same dialog box contains a button for connecting to the SolidWorks web site. Clicking on the button will attempt to connect to the SolidWorks web site automatically. However, there are too many configurations possible to guess how this will react on your machine. Clicking on Connect certainly will not hurt anything, but if you do not have a modem or Internet connection, do not expect miracles.

Summary

As similar as AutoCAD and SolidWorks seem in some respects, they are fundamentally different. There is a different mind set involved in creating parametric solid models. Think in terms of adding features when using SolidWorks. A feature can be a sketched feature, such as an extrusion or revolved feature, or it may be an applied feature, such as a fillet or chamfer. But they are always features, and they are always being added, as opposed to carving pieces of material from a block of wood. Because of this parametric feature geometry, elements in the model can be easily edited and manipulated. Many documents can be opened at one time, which might be parts, assemblies, or drawings, in any combination.

Chapter 2, Implementation, deals with the requirements needed to physically get the SolidWorks program up and running in your company. The chapter also deals with many of the hardware questions you might have. If there is a system administrator in your company that takes care of such things, he or she may want to read this. Chapter 3 discusses more about actually using the program, including navigation and selection techniques, and how to begin creating your first solid model.

2 Implementation

Introduction

This chapter discusses issues that arise when implementing a new CAD system. It reviews the concepts behind planning, preparation, training, and management issues you need to address to ensure the maximum return on investment for a new SolidWorks installation. A good understanding of implementation issues will facilitate a smoother transition and help ensure the success of a new system. Two important objectives during the implementation phase are to minimize the time required to become effective with SolidWorks and to develop those skills that will allow you to reduce time to market and produce better-quality designs.

Developing an implementation plan can provide a company with a clear idea of how the system will be installed, how the transition process will be handled, how existing legacy data will be handled, and what hardware is required for the new system. An implementation plan describes the issues a new SolidWorks installation should address. The term *legacy data* refers to existing documents from older software programs. Legacy may also refer to hardware that is obsolete, or that is being replaced by new and improved technology.

Prerequisite You should know how to open files in SolidWorks, so that existing AutoCAD or other files can be brought into the program (see "Opening an Existing Document" in Chapter 1). This chapter deals with networking and hardware issues as well; therefore, a basic understanding of these topics is required.

Content This chapter contains the following major topics.

- Implementation plan
- Training plan
- Translating legacy data
- Hardware considerations
- Determining a reasonable pilot project

Two key sections in this chapter concern legacy data and hardware considerations. The section on legacy data describes methods for handling the legacy data of an existing CAD system. There are a number of approaches that can be applied. This section describes these options so that you can make informed decisions concerning legacy data. This section also explores techniques that make the reuse of legacy data possible. The "Hardware Considerations" section reviews the items that need to be considered when purchasing a system capable of efficiently running SolidWorks.

Objectives With the completion of the section that discusses an implementation plan, you should be able to successfully implement Solid-Works on a project level. When you have finished the section on maintaining and reusing legacy data, you should understand the options available to handle existing CAD data, be able to arrive at an informed opinion as to how the existing data will be handled, and understand how existing 2D legacy data can be used to produce 3D models. You will also understand the requirements for a SolidWorks computer system and how these requirements can affect the productivity and performance of your system. The possibility of a pilot project will be presented to you, including whether or not such a project would be feasible for your company.

Implementation Plan

When implementing a new CAD system, a plan should be developed to determine the requirements and implementation schedule. Goals of the implementation plan should be a smooth transition from the existing system, development of a method for handling existing legacy data, training of new users, and piloting and full implementation of the new system. The following are issues you should address in an implementation plan.

- Development of a training plan
- How to translate legacy data, if needed
- Upgrading hardware to accommodate new software
- Identification of a pilot project

Training Plan

A training plan should provide complete, consistent training for all users. Providing training for users helps ensure that new concepts associated with a new implementation are understood. Even though SolidWorks is easy to use, training helps ensure that new concepts and principles are applied correctly and effectively. Additionally, ideal training addresses old habits and methods of operation carried over from a previous CAD system.

Training is most effective just prior to a user's hands-on use of SolidWorks on the job. If the user does not have a chance to use the CAD system immediately after training, thereby reinforcing the learning, much of the material and information will be forgotten. Training should be coordinated with the timetable of employees' use of the software.

Benefits of Training

The following are benefits you should derive from a successful training plan.

- Better-trained users will use available tools more efficiently and effectively than users who are not trained or who are undertrained. Effective training can also help ensure that users understand solid modeling technology and apply it effectively.
- An effective training program produces users who require less internal and external support. This helps reduce the burden on your support personnel, or cost associated with external support.

- Solid modeling systems can provide more complete information more quickly when applied properly. Training provides the knowledge of available tools and how to apply these tools efficiently.

- A user can become frustrated with trying to implement new technology if he or she is still bound by old habits. Attention should be given by the instructor to mold old habits into a pattern of new habits that will better suit the new software.

Elements of Effective Training

Complete, effective training is important to your company's long-term success. Training needs to be viewed as an ongoing process to ensure that available tools and technology are fully applied. Effective training requires a broad, long-term vision. There are several elements that can be combined to provide a comprehensive training solution. Developing a training plan can help identify the tools, personnel, timetable, and budget required to train existing and new users. The following are elements of an effective training plan.

Training Schedule

A schedule for training users should be based on a few criteria. Try to schedule training so that, upon completion, students will be able to go back to the workplace and begin using the program. It does no good if SolidWorks is taught, only to have the students go back to the office and not touch the software for two months. A comprehensive SolidWorks training course should be at least five days long. Other specialized training for certain aspects of the software may also be needed. Examples of specialized training topics would be mold making, sheet metal, advanced assembly file management, and so on.

Tutorials

Computer-based tutorials can be used as a prerequisite for classroom training. Many of the basic SolidWorks skills can be reviewed and practices introduced using this type of training tool. SolidWorks 99 (along with previous versions) has included tutorials to get the new user warmed up to the software. CIMCo's *Exploring SolidWorks 99* computer-based training CD-ROM also provides an excellent introduction to training. Other publications, such as *INSIDE SolidWorks* (OnWord Press), allow users to teach themselves the software and provide a valuable reference guide.

Classroom Training

Classroom courses provide valuable interaction with an instructor. Having an instructor well versed in the program and understanding of a new user's common mistakes can greatly increase the level of productivity for students. It is very important that each student have his or her own machine with which to explore the program and to complete exercises.

Management should take responsibility for ensuring that the maximum benefits are derived from classroom training before it happens. Having a chance to get oriented within SolidWorks before instruction (using the tutorial included with SolidWorks or using computer-based training tools or books) can help ensure that the user has a general idea of the principles and terms used during the class.

Consulting

Consulting can be used to show a customer how to apply new technology to specific needs, and how to develop implementation and training plans. A customer can also be assisted in an actual project. Use case studies and customer references that are fact and figure based (i.e., time to market savings, cost reduction, measurable quality improvement, and so on).

Mentoring

Customer use of internal one-on-one mentoring can be an effective training mechanism. These internal resources understand the company's application and help reduce external support requirements. Power users of a technology exist within every organization, and a company can benefit by recognizing and encouraging the effective use of this type of user for training and support. Medium to large sized companies may want to employ one user that is very well versed with the software and can answer questions other employees might have.

Internet and Intranet Support

Using a company intranet is a very effective mechanism for providing and disseminating useful information. For those not familiar with the term, an *intranet* is the same as an the Internet, but is localized within a company. The Internet, or World Wide Web, is another excellent way to keep up with changes or obtain information and tips on how to use software. Discussion and user groups at Internet sites can provide an outstanding source of knowledge and diverse ways of doing things.

Translating Legacy Data

One question that arises when a new CAD system is installed is what should be done with existing (legacy) CAD data. The ability to effectively handle existing CAD data can be a key factor in determining the success of a new system. It is rarely practical to take all existing drawings and designs and immediately turn them into solid models and assemblies. The ability to integrate existing data cooperatively with a new system and to reuse existing data can help minimize the impact of the new system and maximize the utility of existing information in producing new designs.

SolidWorks offers features that provide an alternative to redrawing or converting legacy data by allowing the import of existing 2D or 3D geometry. The techniques involved are covered in later chapters.

↝ **NOTE:** *If you have a project that needs to get done right away, turn to Chapter 11 for specific instructions on importing files.*

Strategy Determined by Application

Address the issue of legacy data management early to determine the strategy that should be used for your application. The various methods used will be determined by the existing CAD program. In the case of AutoCAD, .dwg and .dxf files can be directly imported into SolidWorks by clicking on File/Open, and specifying the appropriate extension in "Files of type."

It is rare that when a new CAD system is installed the old system is removed and all existing data is converted to or redrawn on the new CAD system. This would not be a productive use of personnel time. An alternative to this is to maintain two systems, which allows for more gradual transition to the new CAD system. The existing legacy data can be modified on the old system and new designs can be made on the new system. If needed, existing data can be imported into SolidWorks and edited from there.

The advantage of this method is that transfer work is minimized because files do not have to be saved within the new system. The disadvantage is that the existing CAD system might be incompatible with a new computer system. For example, users converting from a DOS-based system may not be able to run the old CAD system on a Windows-based computer. This issue should be recognized and addressed from the onset. It will play an important role in determining which operating system is used.

Legacy Data Maintenance Considerations

The following sections discuss considerations to think about at the implementation planning stage, and then to follow through with as the process continues. A key issue to be weighed is the time required to reuse and complete a design with SolidWorks. Simple changes to an existing product that does not require the advantages of a solid modeling system would not be as good a candidate as a design that has more changes or requires more analysis of the fit or function of the design.

Minimal Changes Required for Existing Drawings

If a couple of 2D drafting changes are all that is needed, it may very well be easier to open the drawing in AutoCAD and make the changes. Print the drawing out if necessary. This requires little time and effort, and no translation is needed.

Major Changes Required for Existing Drawings

This is a different story. It should be considered that more time spent on the AutoCAD system means less time spent on Solid-Works. If many design changes are needed, there are a few options, depending on whether the existing data is 2D or 3D.

Two-dimensional Layouts

A 2D drawing layout can be imported and edited as a Solid-Works drawing. In SolidWorks, it can be edited in the usual fashion as a typical 2D layout. However, this does not take full advantage of the most powerful SolidWorks tools, and has no downstream benefits to speak of.

The preferred method would be to import the 2D data, and then use that data to build a solid model. Once this is done, any design changes that need to be made can be performed efficiently and quickly. The time spent creating the model is well spent in the long run because of the time saved when making future design changes and in creating the new 2D layout.

Three-dimensional Wireframe

This poses a bit of a dilemma. It is possible to import 3D wireframe geometry into SolidWorks, such as via IGES translation. However, there is no way to transform the wireframe geometry into a solid model. It has to be recreated either from scratch or by building the solid up from 2D geometry. This is due to the nature of solid modelers, not the SolidWorks program itself. It would be possible to build up the solid within AutoCAD, little by little, using the existing wireframe, but it would have no intelligence. You are better off recreating the model in Solid-Works. Not only will it be much easier, you will also reap all the rewards of having a parametric solid model when you are done.

One method of reusing 3D wireframe geometry is to break up individual features into separate 2D sketches. These individual sketches can be used to create sketches within SolidWorks using the "Sketch from drawing" function.

Three-dimensional
Solids

Three-dimensional solid geometry can be exported from AutoCAD as what is known as an ACIS solid. AutoCAD uses the ACIS kernel, whereas SolidWorks uses the Parasolids kernel. It is believed by SolidWorks and other high-performance CAD system programmers that the Parasolids kernel is superior. Take for example that there are many versions of the ACIS kernel (1.5, 1.6, 1.7, 2.0, 2.1, 3.0, and so on). The kernel keeps changing, because ACIS keeps trying to improve it. Also take for example that each one of these variations *are not intercompatible*. A built-in translator is needed to move between programs using different versions of the same ACIS kernel.

Another way of exporting solids from AutoCAD is through the IGES format. If you have AutoCAD release 13 or 14, you will have had to purchase the IGES translator separately because it was no longer included with the main program. IGES translators can usually export geometry as wireframe, typically known as 3D curves, or as trimmed surfaces.

A solid model contains topology information, as discussed in Chapter 1. Think of a simple cube with six sides. Each side can be thought of as a surface. Additionally, each surface has an outside and an inside. Last, each surface is "trimmed" to its adjacent sides. The cube has a complete boundary that can be read by SolidWorks and "knitted" to create the solid model. The SolidWorks IGES translator is included with the program.

When a solid or surface model is imported and knitted into a solid model, the imported geometry is displayed as a single feature within SolidWorks. Features can be added to modify the original feature, but the imported geometry cannot be redefined.

New Designs

This is a no-brainer. There is absolutely no reason you would want to create a new part in AutoCAD and translate it over to SolidWorks. The benefits of creating the part in SolidWorks far outweigh any reasons one might have for creating it in AutoCAD. Parametric design, component assembly, real-time rotation of shaded parts, and exploded assembly drawings are just a few of the features available to a SolidWorks part as opposed to an AutoCAD model.

Hardware Considerations

The computer requirements of a solid modeling system differ from those of a 2D CAD system. Solid modeling is more computationally intensive due to the calculations required to build and display geometry. This type of system can benefit from a fast CPU (central processing unit), memory, video cards, and hard drives. The trade-off between saving money and improving a system's performance can make for difficult decisions. There are some system components that can provide more "bang for the buck." This section describes these components but does not recommend a specific brand or manufacturer.

One difficulty when considering hardware is determining how to measure the effective increase in productivity of one system over another. The most effective methods of sorting out the possibilities in hardware are based on a macro file, or benchmark, that runs programs that represent the type of work your company performs on a regular basis. However, most people do not have access to the types of machines they might consider buying for running these types of tests.

Consider a part typical of the parts designed by your company, and use it in profiling various systems to see how the systems will handle the part. For instance, test a part with many complicated features, or a large assembly, which can be used to judge how various machines will dynamically shade, rebuild, or modify the part in some way during demo sessions.

Operating Systems and Network Software

SolidWorks uses Microsoft's Windows 98 or Windows NT operating systems, either on the Intel or Alpha platforms. The choice of operating systems could be based on company standards, support requirements, or availability of operating systems from your hardware vendor. The Windows NT operating system provides a more robust, stable platform. Windows NT provides a multitasking system that handles engineering applications better in terms of memory usage, security, networking, multiple CPUs, and misbehaving applications. Windows 98 better supports old 16-bit applications (i.e., DOS-based programs) much better than Windows NT. Windows 98 is also less expensive than Windows NT.

The general rule of thumb here is that if your company is running older software or DOS programs, go with Windows 98. If networking security is of primary concern and you are looking to squeeze every last ounce of performance out of your hardware, go with Windows NT. Windows NT runs faster because it does not use any of the old 16-bit code, but this is also the reason it lacks backward compatibility. NT is a more stable and robust operating system for full-time CAD use.

Networking software is also an issue, but the out-of-the-box networking functionality of the Windows operating system is very easy to set up and administer. In addition, because networking is included with the operating system software, why buy something extra that will not add functionality? Microsoft's networking software does the job, and does it well.

Processors

With the advent of Intel Pentium II processors and the DEC (Digital Equipment Corporation) Alpha chip, CPUs are processing faster than ever before. The days when an expensive UNIX workstation was required to effectively run a solid modeling CAD system are gone. Many NT workstations rival or surpass the performance of UNIX workstations for a fraction of the price.

The latest CPUs are typically priced at a premium when they first arrive on the market. The trade-off between additional performance and the added cost must be weighed to determine the value of the faster processor. Do not choose a slow processor to save a small amount of money. The increase in productivity over the course of a few months using a faster processor will far outweigh such savings.

Windows NT also supports multiple CPUs. SolidWorks supports multithreading for some (e.g., display) functions. A multithreaded application means that the tasks for a process can be split up so that two or more processors can handle the task. An additional CPU might add to the system performance, but not all functions within your applications, or within SolidWorks, can take advantage of the multiple CPUs. This is why servers benefit the most from having more than one CPU. Many applications are being run at any given time, and the applications can share the CPUs present in the server.

➥ **NOTE:** *Although SolidWorks for the most part does not take advantage of multiple processors (sometimes referred to as multithreading), some of SolidWorks functions do. Particularly, there are some graphics functions that will benefit from multiple processors. You may want to run some benchmarks (mentioned earlier in this section) to determine if multiple processors are worth your investment.*

The Intel Xeon processors (like the original Pentium Pro processors) can be run with up to four processors in the same machine, and are very well equipped to handle heavy server loads due to the second-level cache (or L2 cache), which runs at the same speed as the CPU. Pentium II and Pentium III processors both have an L2 cache that runs at half the speed of the CPU, like all other Pentium processors, and only a maximum of two processors can be implemented at one time. Nevertheless, the Pentium III can currently run at 500 MHz, which easily makes it a chip worthy of SolidWorks.

It would be unfair to not mention some of the other CPU chip manufacturers giving Intel a run for its money as of late. Cyrix and AMD both have good products on the market. AMD is currently offering the stiffest competition to Intel. If you purchase a system with an AMD processor, chances are SolidWorks will run fine. SolidWorks Corporation's stance on this matter, however, is that the software is not tested on these other chips. For this reason, it is best that you start with one system and put it through its paces. If the software runs well and you are happy, consider the trial period a success and make your additional purchases.

Memory   Memory or RAM (random access memory) is used by a system to place data in a temporary, quick-access area for operating system or application use. Memory prices have dropped significantly over the past couple of years, making the addition of sufficient memory a cheap option for improving system performance. Even though SolidWorks can run on a system with 32 Mb of RAM, a full-time SolidWorks design system should be configured with a minimum of 64 Mb of RAM.

To determine the amount of memory required for a system, consider how many applications will be running at the same time and the size of the files. To run Microsoft Word, Excel, and SolidWorks at the same time, a system with 64 Mb of RAM would be acceptable, but the common catch phrase in the industry is "you can never have too much memory."

If the system runs out of RAM, the hard disk will be used to cache the application. This is considerably slower than using RAM and should be avoided. Consider that memory access time is measured in nanoseconds (billionths of a second) and that hard drive access is measured in milliseconds (thousandths of a second). This should give you some idea of the speed difference.

You may need more memory, depending on the sort of parts or assemblies you are creating in SolidWorks. Sixty-four Mb of RAM is acceptable for average parts with dozens of features, and for assemblies with one or two dozen parts. If you are going to be creating parts with many complex features—such as complex spline-type surfaces, swept or lofted features, variable radius fillets, or large patterns (known to AutoCAD users as arrays)—128 Mb or more may be required.

Large assemblies with hundreds of components may require 256 to 512 Mb or more. With extremely large parts, containing in excess of 1,000 features, or assemblies with more than 500 parts, consider the DEC Alpha machines or Intel Xeon equipped computers with very large memory capacities. SGI (Silicon Graphics, Inc.) has recently introduced some very speedy Intel systems you may want to consider. SolidWorks will do the job, but the only limitation is your hardware.

Hard Drives

Hard disks come in a variety of bus styles, such as IDE, SCSI, and Ultra Wide SCSI. The access time (milliseconds) and the transfer rate (megabytes per second) measure the speed of a hard disk. The quicker the access time and greater the transfer rate, the faster a hard disk will perform. The fastest types of hard disks are the Ultra Wide SCSI types. These disks are more expensive than the IDE variety, but offer increased performance.

The current standard are the Ultra DMA type hard drives with a spindle speed of 7,200 rpm, which support throughput rates of 33 Mb/sec. You may also want to check the access times of a hard drive, which is how fast it can access data. This is usually anywhere from 9 to 12 milliseconds. Really though, as long as you purchase a hard drive that is the most recent model from a major manufacturer (such as Seagate or IBM), you should be safe. Make sure you purchase a brand name hard drive from a reputable dealer. After all, your data is going to be stored on it.

The SCSI drives offer a greater degree of performance for more than one reason. The disk rotations on these drive types are faster (up to 10,000 rpm), which accelerates seek times when searching for data. Second, data throughput is higher when using Ultra Wide SCSI drives (40 to 80 Mb/sec). Most importantly, SCSI drives contain their own processor, which makes for less draw on the main CPU, and allows for multitasking on the hard drive.

SCSI drives can process more than one command at a time, whereas an IDE drive must finish the first command before processing the next one. This makes SCSI drives an ideal solution for applications such as video editing and animations, and for file servers. For SolidWorks and CAD applications, an investment in an SCSI drive is probably not warranted unless you want to minimize file loading times for very large assemblies.

Video Cards and Monitors

Video cards and drivers are a key factor when determining the performance of a system. There are graphic cards that work well in a 2D wireframe environment, but do not perform as well in a 3D solid modeling system. A good video card will utilize the OpenGL graphics mode with sufficient memory for running at a resolution acceptable for your monitor size. Resolution is measured in pixels, and is typically anywhere from 640 x 480 (width by height) to 1,280 x 1,024. Even higher resolutions are being reached now, but not all monitors will be able to view the higher resolutions. Generally speaking, more memory on a video card does not increase its speed, but increases the color depth.

Color depth can be anywhere from 256 to 16.8 million colors. For everyday applications, such as word processing or spread-sheets, 256 colors is fine. If you are working with graphic images or CAD software with shaded parts or renderings, a step up from 256 is the least you will want to go. Intermediate settings are typically 32 or 64 thousand colors. Sometimes the color depth is referred to as "high" color (64 thousand) or "true" color (16.8 million). If you are a power user, once you get used to true color, you will never want to go back to anything else. True color makes for very smooth and nicely shaded parts.

Most CAD operators will have at least a 17-inch monitor. Larger (19-, 20-, and 21-inch) models are available today in a wide range of capabilities and prices. With respect to resolution, the larger the monitor, the higher you can go with the resolution settings, assuming the graphics card can support them. At higher resolutions, text and icons (and everything else) get smaller if they have to fit into the same screen space. On a 21-inch monitor, a setting of 1,280 x 1,024 would not be uncommon because there is enough screen space to see everything. If the same resolution were viewed on a 14-inch monitor, everything would be too small to see.

Another thing to consider when purchasing a graphics card is texture memory and 3D accelerator chips. Some cards contain both, and these cards contain varying amounts of texture memory, some as high as 64 Mb. Without going into too much detail, SolidWorks has no use for texture memory. Neither does AutoCAD. Texture memory is important in applications where texture-mapped surfaces are being rotated or animated on your screen.

The Accelerated Graphics Port (AGP) is a fairly recent specification developed to increase graphics performance. This technology is still evolving and has not yet reached its full potential. AGP allows the graphics card access to system memory via the PCI bus, thereby giving the graphics card a much larger memory pool from which to draw. Yet there are many other aspects of this technology that must be considered. For example, the PCI bus is still only 32 bits wide. Some graphics cards use a 128-bit-wide path to access memory already on the graphics card!

There are different implementations of the AGP spec. The multiplication factor of the AGP spec shows the rate of data over the AGP bus. There are graphics cards that use AGP 1x, 2x, and another technology called side banding. AGP 2x, for instance, doubles the data rate at which the PCI bus is transmitting. Side banding increases this flow of data even more, and if you are shopping for an AGP card, you should look for one capable of AGP 2x with side banding. AGP 4x will be the next implementation of the AGP spec, and Silicon Graphics develops its own proprietary motherboards that surpass even the AGP 4x spec.

There are a number of medium and high-end graphics cards on the market. Many of the larger PC manufacturers now have a line of cards they refer to as workstation-caliber graphics accelerators. Keep in mind the main reason you are buying the card and what you want to do with it, and remember that more expensive does not always mean faster graphics processing.

One more note regarding 3D accelerator cards. The market has been inundated with 3D accelerator cards in the past few years. You can buy 3D accelerators for as little as $99 that promise to take your PC to new heights. In contrast to this you can buy "high end" cards for $2,000 or more. What is right for you? Generally speaking, the high-end, expensive graphics cards incorporate a great deal of fast texture memory, which will not benefit the SolidWorks user. The 3D accelerator cards are great for games because they accelerate the rendering of polygons needed to run the game.

For SolidWorks, you will want a good 2D accelerator with a fast RAMDAC and a decent amount of memory on the graphics card. SolidWorks is used for creating solid 3D models, but those models are still being rendered in 2D (albeit very quickly). A card capable of using the OpenGL graphics mode works best with SolidWorks. The RAMDAC frequency establishes your refresh rate for your monitor. 250 MHz or higher works well. The onboard memory limits your resolution and color depth. A card with 16 meg of RAM should be able to do well beyond 1,280 x 1,024 resolution at 16 million colors. You will want the high color depth because your SolidWorks parts will look much more natural.

When you reduce it to essentials, your own eyes are the best judge, just like your ears are the best judge when buying audio speakers. Talk to your hardware vendor or VAR to discuss the applications that will be used and to determine which graphics card would be best suited for running your SolidWorks application.

Determining a Reasonable Pilot Project

Instead of rushing into a complete conversion upon development of an implementation plan, you should consider a pilot implementation project. This allows you to work through known and unforeseen implementation issues on a small scale before you undertake a large-scale project. However, at times a small-scale preview is not possible, probably because a critical project has come up that needs to be undertaken right away.

If this is the case, which is very common, the critical project can serve as your pilot project. Most companies do not have the luxury of picking and choosing a pilot project. If it is feasible, take some of the following thoughts into consideration for your first SolidWorks project.

A pilot project should be limited, if possible, in terms of amount of work and complexity so that new users can focus on the techniques and principles required when developing good SolidWorks work skills. The ultimate goal of a pilot project should be to learn how to apply effective solid modeling skills. A pilot project helps to determine how solid modeling technology can be best applied to your company's products, and to identify additional training that may be required to reach this goal. The following are items to address within a pilot project.

Setting SolidWorks Preferences

SolidWorks contains a number of preference settings that can change the way the program operates, many of which are similar to AutoCAD. Most of the customization options should not be used at all until you get more adept with the program. Unlike AutoCAD, there are no alias files to modify, pull-down menus cannot be edited, and the function of toolbar icons cannot be changed or created from scratch.

The ability to customize AutoCAD is one of its strengths, but it can also be a weakness. For example, it might be difficult or impossible to use someone else's AutoCAD interface if they have modified it completely differently than yours. Another example is conflicts with menus because the menu on the machine that created the drawing is different than the one on the machine currently viewing the drawing. Even something as simple as AutoCAD layer names have always been an issue when trying to standardize throughout a company.

Most of these problems do not exist in SolidWorks. However, there are a few settings that should be agreed upon, such as the size and type of note and dimension fonts, use of ANSI or ISO, the form of drawing templates, and so on. These are essentially the parameters that would need to be standardized whatever the software being used. A meeting among appropriate personnel and departments for resolving these issues is recommended.

Copy Options Wizard

One possibility you have when attempting to standardize your SolidWorks preference settings throughout a company is the Copy Options Wizard. This tool can be found in your Windows Start menu Programs listing, along with all of your other Solid-Works shortcuts. Just click Start/Programs/SolidWorks 99/Copy Options Wizard.

The Copy Options Wizard allows you to export the settings on one machine so that those settings can be imported to other machines. You can quite easily make every machine in your company have the exact same SolidWorks settings by implementing this function. The Copy Options Wizard will guide you through a series of dialog boxes. When you are done, you will have a Windows registry file (with an .reg extension) that you can run on any other computer with the same version of Solid-Works.

To merge a registry file with another computer's Windows registry, simply double click on the registry file from the computer you want to merge the registry settings with. Usually this entails copying the exported SolidWorks registry file to the computer whose settings you want to change. Make sure SolidWorks is not running when you merge the registry file. Once finished, you should be able to open SolidWorks on the secondary computer and it will have all of the SolidWorks settings the original computer had.

➙   **NOTE:** *The optional Preference settings are discussed in detail in Chapter 3.*

Directory Structure and Naming Conventions

This is another issue that exists no matter what program a company is using. As far as SolidWorks is concerned, it is best to set up specific directories for individual projects. Because of the associativity between parts, drawings, and assemblies, it is necessary to keep track of what parts are being used in which assemblies, and what revisions of what parts are being used in those same assemblies.

Depending on the circumstances, this can turn out to be quite a headache. Rest assured that if your company requires it, there are program applications specifically designed for your needs— everything from workflow management to project tracking to automated revision incrementing. Talk to your vendor or VAR for recommendations.

If you do not require anything quite as extravagant as the aforementioned strategies, it is still necessary to keep files organized by creating logically established directories for storing files. This holds true for any software that generates large amounts of data. In addition, now that file names no longer have to be a mere eight characters in length, naming conventions can be much more descriptive. What your company decides on is a matter of internal policy.

Employee Management

When planning a pilot project, identify users who have strong CAD skills and who are receptive to new technology. Selecting users for a pilot project who are motivated to achieve the goal without bias as to how it is achieved can help ensure that the maximum benefit will be derived from the pilot project. This type of user is also useful when the full implementation occurs, serving as an internal resource for other users. The benefit to an internal resource of this type is that the person is familiar with your company's organization and products. Many companies identify and nurture this type of user within their organization.

Lessons Learned from a Pilot Project

When the pilot project is complete, a full implementation can proceed. Lessons learned from the pilot project can be applied to ensure a smooth transition for all users. A training plan should address the training of new users at the transition stage, as well as ongoing training and support.

Summary

Proper planning and training helps ensure a smooth transition to a new software product. This up-front planning can save work in the long run and reduce the time required to get users up to speed using SolidWorks. Any successful project, including the implementation of a new CAD system, will benefit from a well-conceived project plan.

The ability to maintain and reuse legacy data is a strong feature of SolidWorks. SolidWorks effectively handles legacy data and does not necessarily require that this information be recreated. When selecting a new CAD system, the ability to maintain and reuse existing data should be a key consideration.

The type of hardware, network, and operating system that incorporates an effective SolidWorks document and project management system will be based on the size and complexity of your designs and the overall size of your company. Small gains due to a faster system can add up to larger savings when looked at long-term. There are always newer, faster, and cheaper computer systems coming out on the market. Select a system that is adequate for your needs, does not limit your productivity, and allows for room to grow and expand.

3 Getting Started

Introduction

This chapter is concerned with the concepts you will need when considering how a model will be created in a SolidWorks session. Topics covered in the following sections examine basic information required to begin a SolidWorks session and the thought processes involved. Differences regarding basic philosophy of starting a model with AutoCAD and with SolidWorks are addressed.

Prerequisite

At this point, you should have a good working knowledge of the Windows operating system; specifically, how to open and close windows, how to minimize and resize windows, and related functions considered standard to most Windows users. Along with opening, closing, and starting a new SolidWorks document, basic terms used by the SolidWorks program should be familiar to you. If it has been awhile since you opened this book, it would be beneficial to review Chapter 1.

Content

The main goal of this chapter is to convey the fundamental thought process behind starting a part in SolidWorks, as well as how this differs from the same functionality in the AutoCAD environment. Additional terms used by SolidWorks are introduced. This chapter contains the following major topics.

- Getting started
- Sketch planes
- Sketch profiles
- Design intent
- Preference settings

The section on preference settings describes methods used to define and change a few of the more common SolidWorks preferences. These user-definable properties are used to set attributes and characteristics used to create a document and define interaction properties for SolidWorks.

Objectives

With completion of this chapter, you should be able to start a SolidWorks editing session preliminary to actually sketching entities. Completion of the section on preference settings should give you an understanding of how to define and change some of the more common SolidWorks settings. You should know how to distinguish proper sketch profiles to start a part, and be able to determine the best plane to sketch on. You should also understand the concept of design intent.

Startup

SolidWorks can be started using the Windows Start menu, or by double clicking on its shortcut. With a default SolidWorks installation, a program group is created in the Start menu, but no icons are placed on the Windows desktop. The steps in the following section show you how to create a shortcut.

→ **NOTE:** *Whenever the word* click *is used in this book, it refers to clicking once with the left mouse button. With regard to Windows and SolidWorks, this usually selects what is clicked on.*

Creating a SolidWorks Shortcut Icon

To create a shortcut icon for SolidWorks, perform the following steps.

1. Click on the Start button in the lower left corner of your desktop.

2. Click on Settings.

3. Click on Taskbar to open the Taskbar Properties dialog box.

4. Click on the Start Menu Programs tab at the top of the dialog box.

5. Click on the Add button.

6. Click on the Browse button.

7. Navigate to the SolidWorks folder (otherwise known as a directory) and select sldworks.exe.

8. Click on the Open button.

9. Click on the Next> button.

10. When asked to select a folder to place the shortcut in, select Desktop from the list (which should be at the top of the list).

11. Click the Next> button again.

12. Type in a name for the shortcut, such as *SolidWorks*. The default name will be sldworks.exe, but you will not want to use that.

13. Click on the Finish button.

At this point, you can click on OK to exit out of the Taskbar Properties window. You should see your new shortcut on the Windows desktop. This shortcut creation method works in the Windows 95, Windows 98, and Windows NT 4.0 operating systems. It will also work for any other program in your Program group for which you want to create a shortcut.

A simplified method of creating a shortcut on your desktop exists if you are using Windows 98. In your Programs listing you can drag any shortcut with your right mouse button onto the desktop to create a copy of the shortcut there. This is much easier than the 13 steps listed previously, but only works in Windows 98.

➦ **NOTE:** *The properties of a shortcut (e.g., icon graphics, name, default start-up location) can be customized by right clicking on the shortcuts icon and selecting Properties. The shortcut attributes can be modified for the user's needs. Selecting a new default start-up location can be used to start SolidWorks in the directory used to store your parts. This will work for any Windows shortcut.*

Starting a SolidWorks Session

Start the program by double clicking on the SolidWorks icon. Your icon probably says SolidWorks 99, or possibly another version, but for the sake of simplicity, herein it will simply be called SolidWorks. Close the Tip of the Day screen (see Chapter 1), and begin a new part by clicking on File/New and selecting Part. The following illustration shows the New dialog box, used for starting a session.

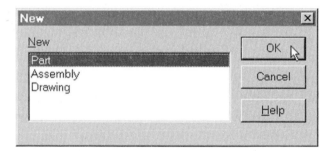

Starting a SolidWorks part using the New dialog box.

Drawings and assemblies are discussed in upcoming chapters. For the time being, parts will be the focus, because that is where the basics of SolidWorks should be developed. The Cancel button should be self-explanatory. It simply cancels you out of the command. The Cancel button's function is the same in any SolidWorks dialog box, and is standard throughout all Windows programs.

The same Windows standardization holds true for the Help button. In SolidWorks, the Help button is context sensitive, and will give you help for whatever dialog box you happen to be in at the time. As is par for the course with SolidWorks, a step-by-step description is usually given, which describes how to carry out the command. To begin, click on OK.

Saving a New Part

When a new part is started, there are always certain components that will appear in the SolidWorks interface. Along the left-hand side of the screen and to the left of the document area—which takes up most of the screen and in which you will be doing all sketching and editing—is FeatureManager, as discussed in Chapter 1. In the middle of the sketch area is the origin point, which should be grayed out at this time. If you have been tweaking settings or experimenting prior to this stage, some aspects of your screen may be slightly different. Do not

FeatureManager.

Name change result in the FeatureManager design tree.

worry about this. Everything should become clear as you proceed. In FeatureManager you should see a few items listed, as shown in the illustration at left.

You will start at the top and work your way down. First, there is the name of the file itself. This should be Part1 on your screen. After saving the file, the name at the top of FeatureManager changes to reflect the new file name. Save the part with the name *Widget* by performing the following steps. The illustration that follows shows the result of this name change.

1. Click on File/Save.

2. Where Part1.SLDPRT is highlighted, type in the name of the part (use the name *Widget*).

3. Click on the Save button.

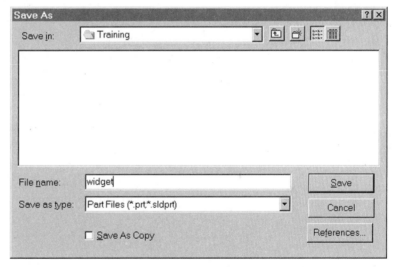

Notice that the name of the part has changed in FeatureManager. Immediately below the part name there are three planes listed, followed by the origin point. Everything listed in FeatureManager will have an icon associated with it. Different features have different icons. This makes finding what you are looking for a lot easier, especially when the feature list begins growing.

At the bottom of FeatureManager are three small tabs, as shown in the following illustration. They allow you to go from FeatureManager to the ConfigurationManager and back again. PropertyManager is accessed with the eye icon in the middle and is used for editing the properties of sketch geometry. Property-Manager's other functionality is discussed in upcoming chapters. If you purchase third-party programs, they may be represented here as well, and will appear as additional tabs.

⇨ NOTE: *ConfigurationManager is discussed in Chapter 6.*

FeatureManager and Configuration Manager tabs.

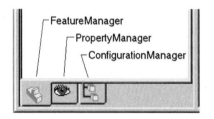

Sketch Planes

Just like a piece of paper is needed before a pencil drawing can be started, so must a sketch plane be selected before a Solid-Works sketch can begin. The sketch plane can be a plane feature or a part face. For this reason, SolidWorks gives you three planes to start with. After a first-time installation, the planes will automatically be named Plane1, Plane2, and Plane3. These names are not very descriptive, so give them new names. Steps explaining how to do this follow.

Many options and parameters can be set from the Options dialog box, which can be very intimidating to new users. Therefore, the entire dialog box will not be discussed here. What works best is to modify a setting here and there as needed, so as not to overwhelm first-time SolidWorks users. To see these options firsthand, click on Tools/Options. The following is a step-by-step procedure for opening the Options dialog box and renaming planes. The illustration that follows shows the Options dialog box.

1. Click on the Tools pull-down menu; then click on Options.

2. Click on the Reference Geometry tab.

3. Rename Plane1, Plane2, and Plane3 to Front, Top, and Right, respectively.

4. Click on OK.

Options dialog box.

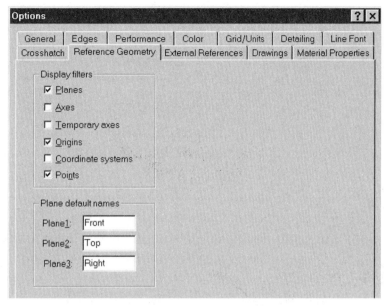

If you currently have a part open, you will expect to see the new plane names in FeatureManager, but that will not happen. This is because the default names were changed, which means that the next time you start a new part, you will see the change.

Renaming Features

Perhaps you would like to rename the planes in the current part. No problem. If you are familiar with renaming files in Windows 98 or NT, this will be easy for you. Do what is referred to as a "slow double click" on the plane you want to rename. This can be accomplished by clicking on an item in FeatureManager, waiting a second, and clicking a second time. A cursor will be displayed at the end of the object being renamed, and the name will be outlined with a rectangle. For practice, go ahead and type in the new names for each of the planes. Remember, Plane1 is Front, Plane2 is Top, and Plane3 is Right.

Features and sketches can also be renamed using the slow double click method. An alternative method you can use is to right click an item in FeatureManager, select Properties, change the Name field, and select OK to close the dialog box. However, the slow double click method is much easier and more straightforward.

Keep in mind that renaming can be accomplished for almost anything listed in FeatureManager. During the creation process, it usually helps to name features as you go. This makes it very easy to find features later if they need editing.

Viewing Versus Selecting Planes

New users have a tendency to think that viewing a plane and selecting a plane are one in the same. This is not the case. Solid-Works will give you visual cues for many things done within the program. This is especially true when selecting objects or sketching. Explore this a little further, with regard to planes.

First, click on one of the planes in FeatureManager. If the front plane is selected, it will appear as a rectangle on the screen. The other two will appear as lines, because they are being viewed from the side. The point, however, is that the plane is green. Anytime something is selected, it turns green (with few exceptions). File this bit of information away for future reference: if something is selected, it turns green.

Now try this: click anywhere in the sketch area. The plane disappears, right? This is because it is being deselected. The Escape button on your computer keyboard will accomplish the same thing. Give it a try. The planes have not actually been turned on. One of them, depending on the one you selected, was merely temporarily selected. To turn on the planes, or Show them, a different method is required.

Introduction to the Right Mouse Button

The right mouse button performs many tasks in Windows 98 or NT. It also plays a big role in SolidWorks. At the moment, you will use it to show the three default planes, which should be named Front, Top, and Right. Use your right mouse button to click on the Front plane in FeatureManager and you should see what is known as a context-sensitive menu. This means that what is in the menu is dependent on what you right click on. To show a plane, you would perform the following steps. The illustration that follows shows a Front plane display.

1. Right click on the Front plane in FeatureManager.

2. Select Show.

Front plane display.

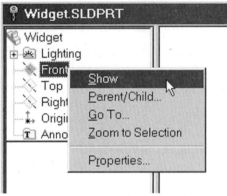

That is all there is to it. Notice that the plane is still selected. Deselect the plane by using either of the two methods previously described. It is usually easiest to press the Escape key on the keyboard, but do whatever is easiest for you. The plane should still be visible, even though it has been deselected. This is because it has been shown. Now repeat this process for the other two planes.

Before moving to discussion of view orientation manipulation, some basic differences between AutoCAD and SolidWorks should be pointed out. What SolidWorks refers to as a sketch plane is similar to what a user would have if he or she manipulated the User Coordinate System (UCS) to a specific orientation. The UCS must be changed every time a new surface needs to be drawn upon. Constantly having to modify the UCS from place to place can get tedious.

In addition, it is not necessarily a surface the UCS is placed on, but a theoretical plane extending in all directions and defined by the user in some fashion, such as with three points. SolidWorks allows you to draw on a plane or a planar face of existing geometry simply by selecting it and clicking on the Sketch icon. Planes can be created for a wide range of functions in SolidWorks, but AutoCAD does not have anything analogous to this, as "plane" entity types do not exist.

Introduction to View Orientation

AutoCAD has its View toolbar, and SolidWorks has the Orientation dialog box. They are very similar in what they achieve, but different in appearance and operation. Access the Orientation dialog box by performing the following steps.

1. Click on the View pull-down menu.

2. Click on Orientation.

3. Click on the "pushpin" icon in the top left-hand corner of the Orientation dialog box.

Alternatively, you can press the spacebar whenever you want to access the Orientation window. Note that the Orientation window appears wherever your cursor happens to be at the time. This is similar to AutoCAD's pop-up snap menu, accessible through the middle mouse button.

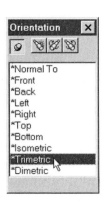

The pushpin keeps the Orientation window on your desktop. Otherwise, the dialog box would disappear as soon as the next mouse click was performed outside the Orientation window. The thought behind this is that you could change a view and be happily on your way, with the Orientation window automatically going back into hiding. The illustration at left shows the Orientation window.

View Orientation window.

AutoCAD's View icons allow you to change views to top, bottom, right, left, and so on. There are also four isometric views: southeast, southwest, and so on. SolidWorks approaches this slightly differently. There are the views Top, Bottom, Right, Left, and so on, like AutoCAD. SolidWorks does not offer four different isometric views. Instead, SolidWorks offers Isometric, Trimetric, and Dimetric views. All system views are preceded by an asterisk, and cannot be deleted. Other views can be added and deleted as desired.

Double click on the desired view in order to activate it (using your left mouse button). There is one other system view Solid-Works offers, and that is the Normal To view. Before Normal To is selected, a plane or planar face must be chosen, unless you are in a sketch. If you are in sketch mode, SolidWorks assumes you want to view "normal to" the current sketch plane.

Without going into too much detail regarding what "normals" are at this point, suffice it to say that they give you a plan view. A plan view is a plan view no matter what software program you are using, which is to say the view will be parallel to your screen. In other words, your line of sight is perpendicular to the view plane. Normal To accomplishes the same thing as

AutoCAD's Plan command but without AutoCAD's additional command line options.

An alternative to the Orientation window is the Standard Views toolbar. Turn on the Standard Views toolbar by selecting the View menu and clicking on Toolbars/Standard Views. You may actually decide you like the toolbar better because it does not take up as much space.

Standard Views toolbar.

The Standard Views toolbar, shown in the previous illustration, does not have a Trimetric or Dimetric view icon, but you can always press your spacebar to access those options in the Orientation window. The following illustration shows a simple part using an isometric, trimetric, and dimetric view.

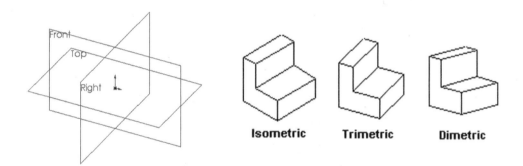

Example of a simple L-shaped part using the Isometric, Trimetric, and Dimetric system views.

If you have not saved your work in awhile, you may want to do so now. Click on File/Save, or just use the icon that looks like a floppy disk. Recall from Chapter 1 that you do not want to use Save As, because that will allow you to rename the file. For now, stick with the name Widget.

Sketch Profiles

An important aspect of creating a solid model is to first mentally visualize how the model should be built. SolidWorks parts are built up from features, and the order in which those features are created is important. Sometimes, feature order can be determined by how it would be easiest to build the part. Other times the features must be created in such a fashion because they cannot physically be created any other way. An example of the latter would be a simple block with fillets and a hole. The fillets

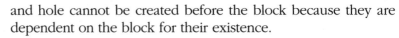

and hole cannot be created before the block because they are dependent on the block for their existence.

All parts, whether working in AutoCAD or SolidWorks, require some sort of base feature. AutoCAD allows the creation of multiple solids without regard to where those solid components exist in space. Solid shapes and primitives are moved to desired locations and then added or subtracted from the base component in order to achieve a desired shape. Because a SolidWorks part file is a contiguous solid model, only one solid can exist at one time within a single part file. This is not a limitation, however, because in assemblies many parts can coexist in any configuration.

Because of SolidWorks' feature-based parametric nature, it is desirable to select a profile that best describes the part when the first feature is created. For instance, if you were creating a new keyboard for a computer company, it would be possible to create a rectangle on the top plane and extrude the shape downward, or sketch a cross-sectional shape on the right plane and extrude it to the left. The second scenario would result in a shape that would more fully represent the actual shape of the part, and would be the preferred way to create the keyboard's initial base feature.

AutoCAD's ability to create numerous shapes in multiple locations is not necessarily a strength. Initially, SolidWorks may seem more rigid in comparison to AutoCAD, but in practice it is actually extremely flexible. In SolidWorks, you decide what would be the best profile to start with, then select a plane with which to begin.

Design Intent

Design intent is a topic more important when working with a parametric modeler than with a Boolean modeler. The term *design intent* is defined as how a part will change if a dimension or constraint is modified. Constraints are a bit more complicated than dimensions. Dimensions, however, are easy enough to understand.

➥ **NOTE:** *Constraints are discussed in detail in Chapter 5.*

AutoCAD certainly has dimensions, but they do not alter the model when changed. AutoCAD dimensions will actually lose their associativity if they are modified to be anything other than their default value. SolidWorks, on the other hand, will rebuild the model for you after changing a dimension. This is one reason SolidWorks is considered a design tool. Likewise, it involves a slightly different way of thinking.

Because AutoCAD does not use parametric dimensions, a great deal of care must go into creating a drawing. When a line is drawn from point A to point B, it must be accurate. With Solid-Works, if it is in the ballpark, that is good enough. When the dimension is added and a value is put on the dimension, it will modify the line to match the dimension value. This holds true for sketch or feature dimensions, and the dimensions can be modified at any stage in the design process. Just keep in mind your design intent.

For example, picture in your mind a simple block with a hole in it. Ask yourself what is more important. Should the hole start on the front plane, or the back? Should it be dimensioned to the upper left corner of the block or the lower right? If the overall dimensions of the block are changed, should the hole remain in the middle of the block? The answers to these questions depend on your design intent.

Preference Settings

User preferences define standard default values and system settings, such as how the system interacts with the user (i.e., single pick per command, naming features or dimensions upon creation, and so on). They also set default attributes and values, such as units of measure or number of decimal places. These preferences are used to define default properties and characteristics for sketches, parts, assemblies, and drawings.

Default values should be defined prior to starting your first design project. Setting consistent, standard configuration values for multiple-seat sites will make sharing documents easier. Carefully reviewing and setting default user preferences can help eliminate the need to redefine properties after documents have been created. Standard configuration values will make it easier to produce documents that look consistent and do not require users to change standard values when working on a document created by someone else. The mechanics involved in getting to the Options dialog box are easy. To do this, perform the following steps.

1. Click on the Tools pull-down menu.

2. Click on Options.

3. Make the desired changes.

But wait! You should read on before proceeding.

What Is Affected by Option Settings

Most option settings will affect the part that is open. However, some tabs have an option box that changes this condition. There are a few tabs that allow you to select whether the changes made will affect the current document, future documents, or both. What you select depends on whether the changes made should become the system defaults or not. The following is a list of the three option setting types and their individual scope.

System Defaults	Settings will affect all new documents.
Active Document	Settings will affect only the document being edited.
All Possible	Settings will affect the current document and all new documents.

To date, the tabs that have the Apply To drop-down dialog box (see the following illustration) are the following.

- Color
- Grid/Units
- Detailing
- Line Font
- Crosshatch
- Material Properties

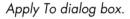

Apply To dialog box.

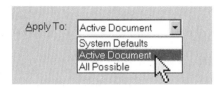

In the previous illustration, only the current drawing will be affected. It should be noted that even if option settings are set to affect All Possible, they will not retroactively affect previously made documents, whether they are parts, assemblies, or drawings. Previous documents already created in SolidWorks have had their individual settings established and must be changed independently. For this reason, it is all the more important to decide on company-wide standards before embarking on a large, company-wide project.

Realistically speaking, there just are not that many settings that need to be decided upon when attempting to implement standards. Decide on ANSI or ISO, for example, and on other options, such as English or metric, tolerance types, and maybe plane names. That pretty much covers it. Templates and title blocks are defined elsewhere and are covered under a separate topic.

One more word on setting options: if any options are set before a document is actually opened or begun, all changes will automatically affect only future documents. This is why the Apply To drop-down box is grayed out. When you think about it, the reason should be obvious.

Copy Options Wizard

One utility that will make implementing standardized options throughout your workplace is the Copy Options Wizard. This utility is found in your SolidWorks start menu listing (shown in the following illustration). Once you have decided on a particular set of user preferences in your Options dialog box, you can export those settings to save them or import them to another machine.

Copy Options Wizard shortcut.

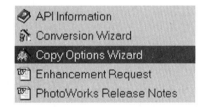

The interface for the Copy Options Wizard is very straightforward and easy to use. By default, SolidWorks will save a file called *swOptions.reg* that will contain all of your Options settings. This registry file can easily be merged into another computer's registry by double clicking on it. Simply copy the swOptions registry file to another computer. Actually, you can also run the registry file over a network. The registry settings will be merged into whatever computer you run it from.

The Options dialog box is divided into functionally grouped tabbed sections. Each tab covers a different group of option settings, as shown in the following illustration. Changes to preference settings can be made to as many of the categories as you desire. Click on a tab to view the settings for that section.

Options dialog box tabs.

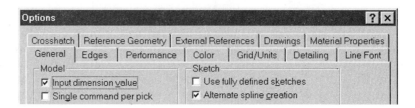

↦ **NOTE:** *The following section is for reference only. It is not intended that new users begin tweaking all of the options available in the Options dialog box. It is highly recommended that settings be changed on an as-needed basis. SolidWorks is a very easy program to learn, but attempting to customize the interface at this point could result in making your interface different than that used in this book. Use caution when changing settings you are not familiar with.*

General Options Tab

This tab defines default settings for various commands, and for sketching, document backup, FeatureManager, and rotational behavior. The following characteristics can be defined within this tab.

Model	Defines model commands and input defaults.
Sketch	Defines sketching preference values for constraining sketches, model reference properties, and default settings when adding defining dimensions.
FeatureManager Design Tree	Defines the default characteristics of FeatureManager.

View Rotation	Defines the increment/decrement values for view rotation when using keyboard shortcuts. Defines the speed for rotation using a mouse.
General	Defines whether to use the last open document when starting a SolidWorks session, to maximize documents by default, and whether to create a backup copy when a document is saved. Also used for modifying other general functions, such as whether or not thumbnail graphics or shaded face highlighting is used.

Edges Options Tab

Edges preferences define the default display mode for edges within documents. The following characteristics can be defined within this tab.

Hidden Edges	Defines the display method for hidden lines and the ability to select hidden objects.
Part/Assembly Tangent Edge Display	Defines the display method for tangent lines.
Shaded Display	Controls how edges are displayed in shaded display mode.

Performance Options Tab

The Performance tab defines the default display quality for documents. The following characteristics can be defined within this tab.

Display Quality (Shaded)	Defines the smoothness for shaded display. A smaller deviation will produce smoother cylindrical surfaces.
Display Quality (Wireframe)	Defines the display quality for wireframe display.
Rebuild	Check this option to provide a higher level of error checking when creating or modifying features. Slower model rebuilds will occur with this option checked, because all surfaces will be verified. If you encounter problems rebuilding a model, check this option.
Windows 98 Zooming	This allows zooming in on small features or parts very closely, but may slow down the display. Not active in Windows NT.
Transparency Quality	Defines the level of transparency quality. If your computer has a graphics card that can perform alpha blending in hardware, this option works very well.

Curvature Generation	Controls when curvature data is generated for shaded models.
Assemblies	Controls whether parts are loaded lightweight or resolved when opening assemblies, and controls prompting for rebuilding and resolving assemblies and components.

Color Options Tab

Color defines the default color for parts displayed in shaded, wireframe, and hidden modes. The following characteristics can be defined within this tab.

System	Different color attributes can be assigned to the items listed. Double click with the left mouse button on the item to be changed. Select or redefine the desired color.
Features	Different types of features can have different color attributes assigned to them. Double click with the left mouse button on the feature type to change it. Select or redefine the desired color.
Reset All	Restores the default system colors.
Edit	Edits the color of the selected object type. A preview of the color is shown below the Edit button.
Advanced	Used to set shading characteristics. Shading must be selected from the System field for this button to be active.
Curvature	Sets curvature display radius values.
Reset Feature	Resets the feature color settings to their default values.

Grid/Units Options Tab

Grid/Units defines the default unit of measure, angular units, and grid characteristics. The following characteristics can be defined within this tab.

Grid (Properties)	Defines the default grid characteristics. These include grid display, snap, and major and minor grid spacing.
Grid (Snap behavior)	Defines snap-to-grid characteristics.
Units (Length unit)	Defines the default linear unit of measure (inch, mm, cm, feet, or feet/inches) and decimal precision used for linear dimensions.

Units (Angular unit)	Defines the default unit of measure (degrees, degrees/minutes, degrees/minutes/seconds, or radians) and decimal precision used for angular dimension.
Spin Box Increment	Defines the increment/decrement values used for length and angular spin boxes.

Detailing Options Tab

The Detailing tab defines default dimension characteristics for documents. The following characteristics can be defined within this tab.

Dimensioning Standard	Select ANSI, ISO, JIS, DIN, or BSI drafting standards. Dual dimensions and datum display are also defined within this field.
Notes	Defines the font, balloon, and leader properties for notes.
Break Lines	Defines the default value for the distance between break lines on broken views.
Center Marks	Sets the default size and display for center marks.
Witness Lines	Defines the default value from the object to the start of the extension line, and the default value from the dimension line to the end of the extension line.
Datum Feature Symbols	Enters the letter or number for the next datum label.
Dimensions	Defines the font, tolerance, and precision for dimensions.
Virtual Sharp	Controls the appearance of virtual sharps.
Arrows	Defines the default characteristics for dimension and leader arrowheads.
Annotations	Filters the display of annotations and sets annotation display characteristics.
Leaders	Allows for controlling the alignment of dimension text with respect to the leaders, thereby overriding the default alignment.
Section Font	Sets the font used for section line and section view labels.
Detail Font	Sets the font used for detail circle and detail view labels.

Line Font Options Tab

The Line Font tab defines the default line font display for drawings. The following characteristics can be defined within this tab.

Type of Edge	Selects the line style to be modified.
Line Style	Selects various styles of lines.
Line Weight	Selects various line weights.
Preview	Previews image of the selected line font.

Crosshatch Options Tab

The Crosshatch tab defines the default display characteristics for crosshatch entities. Crosshatching is used to denote the cutaway area of a section view. The following characteristics can be defined within this tab.

Type	Defines the currently selected crosshatch pattern.
Properties	Defines the default crosshatch pattern, angle, and spacing.

Reference Geometry Options Tab

Reference Geometry options allow setting the default display characteristics of reference geometry and stipulating the default names of the initial three planes.

Display Filter	Sets whether or not various reference geometry is displayed when a new part or SolidWorks document is begun.
Plane Default Names	Sets the default names for the planes.

External References Options Tab

The External References tab defines default read-write access and file locations. This option's setting can be used to define documents that are read-only or to create a search list for files not within the same directory. A read-only file allows access to a document (part, assembly, or drawing) for which changes cannot be saved.

You can also use External References to create a list of search directories. In this way, if a file referenced by an assembly is moved, SolidWorks will search the Folder list before asking you if you would like to browse for the file yourself.

Folders	Allows specifying the path for specific SolidWorks functions (primarily Palette items).
Assemblies	Controls how names are generated for referenced entities or replaced components.

Drawings Options Tab

Drawing options are used to define the default sheet characteristics for drawings, such as the default view scale, method of projecting views (first or third), display characteristics, and others.

Default Sheet	Defines the default sheet scale.
Type of Projection	Sets whether first- or third-angle projection is used for typical engineering layouts.
Default Display for New Drawing Views	Defines the default display mode for new views and how tangent lines are displayed. Typically, hidden lines are shown in gray.
Detail Item Snapping	Allows for inferencing when dragging dimensions.

Material Properties Options Tab

The Material Properties options tab defines the density of the current part.

Properties	Defines the density of a part.

→ **NOTE:** *In previous versions of SolidWorks there were tabs in the Options dialog box for Import and Export file translation options. These options have been incorporated into the File/Open and File/Save dialog boxes. You can read more about file import and export options in Chapter 11.*

Summary

An important aspect of SolidWorks is selecting the appropriate sketch plane. Consider what the best profile should be when creating the base feature of the part, then take into consideration what plane the profile should be placed on. That plane should be your initial sketch plane.

Use the right mouse button to your advantage. It serves many functions in Windows 98 and Windows NT, and SolidWorks follows this principle. The right mouse button opens what is known as a context-sensitive menu, providing options relevant to what was selected with the right mouse button.

Design intent refers to how the solid model will change if dimensions or constraints are modified. Consider design intent before applying dimensions or constraints, as this will affect your model if design changes are made at a later time.

Defining consistent preference settings within your company will produce consistent patterns between SolidWorks documents that are easier to share and understand. These values should be carefully reviewed during the implementation process to determine which settings should be defined.

4 Interacting with SolidWorks

Introduction

This chapter is an interactive introduction to the concepts you will need when beginning a SolidWorks session. The topics covered in the following sections examine the skills and information required to become familiar with the rudimentary and general operating techniques of SolidWorks. Similarities and differences will be addressed along the way regarding some habits that may have been carried over from working with AutoCAD.

Prerequisite

The SolidWorks interface should be familiar, as well as most of the basic terms used by the SolidWorks program. Concepts such as design intent and using planes to sketch on should also be understood. Being able to select an appropriate profile and plane prior to beginning a part should also be understood.

Content

The majority of this chapter will deal with how to select objects, using the right mouse button, using the zoom and view commands, and comparing SolidWorks grid and snap features to AutoCAD. The following major topics are covered in this chapter.

- Selecting objects
- An introduction to system feedback

- The right mouse button
- Display functions
- Setting grid and snap properties

The section on preference settings describes the methods used to define and change certain SolidWorks preferences. These user-definable properties are used to set the attributes and characteristics that create a document and define the interaction properties for SolidWorks.

Objectives

With completion of this chapter, you will be able to begin a SolidWorks editing session preliminary to actually sketching entities. You will also understand how to navigate using the zoom and pan commands, and the differences between AutoCAD's grid and snap settings and how these settings will affect your work in SolidWorks.

Selecting Objects

Before you begin, you should understand the term *objects*. Objects are just what they sound like. Anything created by the user in SolidWorks is considered an object. These can be sketched entities, dimensions, features, assembly components, planes, views, and anything else you might place into a drawing, part, or assembly. *Entities* is also a common term used to describe objects. Generally speaking, entities may often describe sketched objects, such as in the phrase "Select entities by placing a window around them." However, the two words, *objects* and *entities*, may be used interchangeably.

The reason something is selected depends entirely on what is being accomplished at the time. In AutoCAD, you may select entities in order to erase them, move, copy, stretch, array, or any number of other operations. It is much the same in SolidWorks. The left mouse button is always used to select objects. That is the left mouse button's primary function, and this holds true throughout almost all Windows programs.

There are a couple of default selection methods used by AutoCAD. Clicking on an entity would select it, with AutoCAD highlighting the entity to look like a dashed line. In SolidWorks, selecting is done the same way, with the exception that if you are selecting more than one entity, the Control key must be depressed. When objects are selected in SolidWorks, they turn green. Solid model edges (and sometimes faces) are the exception to this rule, and turn into blue dashed lines when selected.

There is an option known as *shaded face highlighting* that will make a face turn green when selected. This option can be turned on in the Options dialog box (Tools/Options). In the General tab, shown in the following illustration, place a check in the option that states "Use shaded face highlighting."

Turning on shaded face highlighting.

☑ Use shaded face highlighting

Deselecting objects works the same way in SolidWorks as it does in AutoCAD. Click on an object a second time and it will become deselected and lose its green highlight. Just remember to hold that Control key down if there is more than one entity selected already!

Another default selection method used by AutoCAD is the window or crossing box method. This is handy for selecting many objects at once. SolidWorks allows you to use the window method as well. To implement this, click in a blank area on the screen, hold down the left mouse button, and drag the opposite corner to create a "window." It makes no difference if you drag from right to left or from left to right. There is no crossing box in SolidWorks; therefore, entities need to be completely enclosed in the window before they are selected.

There is not as extensive an array of selection options in Solid-Works as there is in AutoCAD (e.g., crossing polygon, fence, and so on). This is not an issue, however, because these optional selection techniques are simply not needed. One reason for this is that SolidWorks breaks a model down into features. This makes the selection process much easier because you can easily pick items from FeatureManager.

An Introduction to System Feedback

SolidWorks is always giving you visual cues as to what is going on in the program. The chapter on sketching (Chapter 5) explores this topic much more fully, but for now, discussion will be limited to system feedback as it pertains to the realm of selecting objects.

By placing the tip of the cursor over certain objects or entities, you are given feedback by SolidWorks that confirms what type of entity the cursor is actually over. At that precise moment, if the left mouse button is clicked, that entity will be selected. This can be very important when trying to select specific entity types, such as edges, faces, vertex points, or dimensions. The cursor is constantly changing to let you know what the cursor is over. These changes take place dynamically as the mouse is moved over the part.

Remember the Widget part? Go ahead and open it now if you have not done so already. You should still have the three default planes turned on and the view set to Trimetric. Now move the cursor slowly over the planes and the origin point. The cursor should change as it encounters the various objects. This is the system feedback just discussed, and it plays a major role in how you interact with SolidWorks.

If you have the sample parts loaded on your system, you can open a sample part to test this system feedback functionality and get a better feel for what is happening. If you are not sure if the sample parts were loaded, use Windows Explorer to check the SolidWorks folder and look for the *Samples* directory. There will be a few directories in the *Samples* folder. Look for the *Examples* subdirectory. You will find some sample part files there. Open one of the sample parts and use it to view the system feedback under discussion. The Sump sample part file is shown in the following illustration.

Sump sample part file.

Just move the cursor over the part and you will see the cursor change, depending on what the cursor is currently positioned over. Experiment with selecting various entities. If the samples were not loaded on your machine, or you do not have Solid-Works, refer to the illustration at left, which shows examples of system feedback.

Examples of system feedback.

Later, when there are more objects on the screen to work with, you will see firsthand how the cursor changes, depending on what it is over. A vertex point is considered an endpoint of an edge. When the cursor is placed over a vertex, a small square representing a point is displayed. If the cursor is placed over an edge on the solid model, a line will be displayed. Faces can be either planar or nonplanar surfaces. If the cursor is over a face, it will change to show a symbol that appears as a wavy surface, almost like a flag blowing in the wind.

There are other examples of system feedback, such as with dimensions, planes, sketch entities, and much more. You will be spending a good deal of time on system feedback in upcoming chapters, particularly in Chapter 5.

➤ **NOTE:** *It is very important to pay close attention to cursor feedback. It cannot be stressed enough that you must be aware of the changes taking place with the cursor and be able to correctly interpret this information.*

The Right Mouse Button

The SolidWorks right mouse button is not used at all the same way that AutoCAD's right mouse button is. For this reason, new SolidWorks users will find it annoying when they right click, expecting to reinstate their last command or enter a return, only to get a pop-up menu.

The general rule of thumb is that the left mouse button is for selecting objects, and the right mouse button for opening a context-sensitive menu. You saw an example of a context-sensitive menu in Chapter 3 when turning on the planes in the Widget part. There are many other tasks that can be accomplished with the right mouse button in addition to just hiding or showing planes.

The point to remember is the term *context sensitive*. Depending on what the cursor is positioned over at the time the right mouse button is clicked will make a difference between what is actually displayed in the menu. The menu is *in context* to what is being selected with the right mouse button. Often, the cursor does not have to be positioned over anything, in which case many of the items in the menu will be nothing more than short-cuts to commonly used functions. The following illustration shows an example of this.

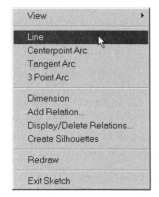

Typical context-sensitive menu.

Context-sensitive menus typically contain sketching shortcuts, dimensioning and relation (constraint) shortcuts, and zoom shortcuts. Of course, it is not much of a shortcut when the respective icons are only an icon away. In terms of mouse clicks, clicking on an icon is only one, but using the right mouse button and selecting from that menu can be either one or two clicks. It is possible to access and select from the right mouse button with only one click. To do this, right click to bring up the menu, continue to hold down the right mouse button, then position the cursor over the item you want to select. When the right mouse button is released, the desired selection will be made. Try it and see for yourself.

With this selection capability, what is the sense in using the context-sensitive menu? If your screen is large and it is more of a chore to slide the mouse over to the toolbar, it may be easier to right mouse click and bring up the menu. The second reason is that sometimes the right mouse menu contains options that cannot be found quite as easily in other areas, such as the pull-down menus. The right mouse button menu is an easy place to find commands that relate to the particular item over which you are right clicking. It can save you a lot of time and effort in the long run.

➥ **NOTE:** *If you cannot remember where a specific command is located, try the right mouse button method of locating options.*

Display Functions

When it comes to ease of use and functionality, SolidWorks is much more convenient than AutoCAD in the area of display functions. For example, take the simple task of manipulating a shaded part. SolidWorks can easily pan, zoom, and rotate the shaded part in real time. AutoCAD comes close to this functionality with its real-time pan and zoom functions in release 14. (Release 13c4 had RTPAN and RTZOOM). AutoCAD also lets the user dynamically rotate a part using the Dynamic View command. However, rotating with Dynamic View is strictly limited to wireframe display.

How about performing dynamic rotation of a part or assembly with hidden lines removed, or when the part is shaded? It is just not going to happen in the AutoCAD environment. As a matter of fact, once you get accustomed to SolidWorks' ability to perform these functions, going back to maneuvering in AutoCAD seems extremely constricting. Take another look at the View toolbar, shown in the following illustration.

View toolbar.

The View toolbar shown in this illustration may look a little different than yours if you are comparing this to what is on your SolidWorks screen. This is because the toolbar shown here has been customized to show all available icons. Do not worry too much about any icons that may be missing on your toolbar. Chances are you will not miss them much anyway. You could add the additional icons through the Customize command in your Tools menu, but read the chapter on customizing before you attempt this (additionally, see the following Note).

↝ **NOTE:** *It is highly recommended that new users to the SolidWorks program do not attempt any customizing of the SolidWorks interface. Doing so can result in the program not working as it was intended to in its default configuration, and your version of SolidWorks may not match the configuration used by this book. In addition, it is best that new users learn the program in its "natural" state.*

Do you still have a sample part open? If not, open one. You are now going to explore some of the view and display commands.

Zoom and Pan Commands

Start at the left of the toolbar and work your way to the right. The first seven icons are the View Orientation, Previous View, Zoom, and Pan icons. These functions can also be found under View/Modify in the pull-down menus.

View Orientation

The View Orientation icon opens your Orientation window (see "Introduction to View Orientation" in Chapter 3). Alternatively, you can press your spacebar to open the Orientation window.

Previous View

AutoCAD has the Previous option in the Zoom command, and SolidWorks has the Previous View icon. They both do exactly the same thing, with one minor exception. SolidWorks Previous View icon will remember your last 10 views, whereas AutoCAD will remember all of your previous views back to the start of your drawing in your current AutoCAD session. This is rarely limiting, if at all, however, because you will probably never need to go back to more than your tenth previous view. If you had to click on the Previous View icon that many times, you would be better off creating a new view.

Zoom To Fit

The magnifying glass with the red rectangle is the Zoom To Fit icon. Its function most closely resembles AutoCAD's Zoom command with the Extents option. Clicking on it zooms the model to fill the screen, leaving some space around the edges. Only the model is taken into consideration when Zoom To Fit is used. Planes or dimensions may not be shown in their entirety if they are larger than or are at some distance from the model.

Zoom To Area

This is represented by the magnifying glass with the red plus sign. It most closely resembles AutoCAD's Zoom/Window operation and is implemented exactly the same way. After clicking

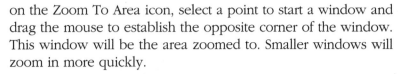

on the Zoom To Area icon, select a point to start a window and drag the mouse to establish the opposite corner of the window. This window will be the area zoomed to. Smaller windows will zoom in more quickly.

AutoCAD's zoom function is very powerful in the respect that it can zoom in or out great distances. This is because AutoCAD was created primarily as a 2D architectural drafting tool. Solid-Works is primarily a 3D mechanical design tool. With AutoCAD, you could theoretically write a book on the head of a pin, or draw the solar system at full scale. The solar system example has actually been done, but AutoCAD has stopped including it in their sample files.

SolidWorks can be used to create trains, boats, or tractor-trailers, or it can be used to create tiny gears and springs. However, if you are looking to design a new solar system or write your next novel on the head of a pin, maybe the SolidWorks program is not for you.

Zoom In/Out

To implement this command, click on the Zoom In/Out icon, then hold the left mouse button down and move the mouse either toward you to zoom out or away from you to zoom in. The RTZOOM command in AutoCAD release 13c4 or the Real-time Zoom function in release 14 is closely analogous to this function.

Zoom To Selection

There is no counterpart in AutoCAD to SolidWorks' Zoom to Selection function. The Zoom to Selection icon is usually grayed out unless you have an object selected. For example, if you select a face on a part you are editing and click on Zoom to Selection, the selected face will fill the screen. This is very convenient, especially during occasions such as sketching on a face of a part. You can select the face to sketch on, click on the sketch icon to enter sketch mode, then click on the Zoom to Selection icon and you are ready to work.

Rotate View

New users sometimes find this to be a little on the tricky side. Click on the Rotate View icon, which looks like the two clockwise circling arrows. Hold the left mouse button down and move the mouse cursor left and right or up and down to obtain the desired view. It may seem tricky at first, but with a little practice, you will master it.

It should be mentioned that there is additional functionality to the Rotate command in SolidWorks. If you find it necessary to rotate about a particular edge, for instance, all you need to do is to click on that edge while in the Rotate command. Once you click on the desired entity, use the Rotate command like you normally would. That is, hold the left mouse button down and rotate the part. You may also select a face and rotate about its centroid, or select a point (vertex) to rotate about.

If you are using this additional functionality of the Rotate command and decide you would rather go back to the traditional way of rotating your model, simply click anywhere in a blank area of the work area (away from the model). This will put you back into the default rotate mode of operation without first having to exit the Rotate command.

There is no counterpart in AutoCAD for this command. You might argue that the Dynamic View command can be used to dynamically rotate an AutoCAD model, but it does not come close. AutoCAD's most recent release of version 14 still does not have any capabilities that match SolidWorks on this level.

Rotate About Screen Center

This is an option that slightly changes the way the Rotate View function works. It is a toggle found under View/Modify/Rotate About Screen Center. When not checked, parts rotate about their center of mass (centroid). When checked, parts rotate about the center of the screen. This is most easily demonstrated if a part is close to the edge of the screen. Try it with one of the SolidWorks example files to see if you can notice a difference.

Pan

This icon has the four arrows, which point north, south, east, and west. It functions similarly to the AutoCAD (R14) Pan command, but not like the original Pan command. Release 13 (and previous) AutoCAD users will generally try to pick two points that define the pan vector, but SolidWorks does not require this. In SolidWorks, you simply hold the left mouse button down and drag the model in the direction you want it to go. As a quick reference, use the following table to compare SolidWorks view commands to their respective AutoCAD counterparts.

SolidWorks Command	*Similar AutoCAD Command*
Previous View	Zoom/Previous
Zoom To Fit	Zoom/Extents
Zoom To Area	Zoom/Window
Zoom In/Out	Zoom/Realtime
Zoom To Selection	None
Rotate View	Dview
Pan	Pan

Keyboard Shortcuts

In addition to the icons and pull-down menus, there are some keyboard shortcuts that can be used. In some cases, the keyboard shortcuts do not actually have counterpart icons that perform the same function. These are pointed out in the following table with the use of italics. Use the following keyboard shortcuts for panning and zooming if you find they work better for you.

Command	*Keyboard Shortcut*
Rotate the model incrementally	Arrow keys
Rotate the model 90 degrees	Shift + Arrow keys
Rotate clockwise/counterclockwise	Alt + left/right arrow keys
Scroll the model (Pan)	Ctrl + Arrow keys
Zoom to Fit	F
Zoom in	Shift + Z
Zoom out	Z

There is an optional setting when rotating the model. For clockwise or counterclockwise rotation, you would use Alt + left or right arrow keys. For rotation about the model's centroid, you would use just the arrow keys. To change the incremental amount the part is rotated, you would perform the following steps. The illustration that follows shows this optional setting.

1. Click on Tools/Options.

2. In the View Rotation section of the General tab, specify the number of degrees in the Arrow Keys box.

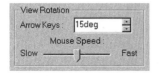

Adjusting the rotation increment.

The default value for this setting is 15 degrees, and that works out quite well. The Mouse Speed slider bar determines the mouse sensitivity when rotating the part using Rotate View. Moving the slider to the right increases sensitivity, and decreases sensitivity if moved to the left. Do not turn up the sensitivity too high until you get the hang of using the Rotate View command.

Display Options

Here is another area where SolidWorks excels over AutoCAD. The next grouping of icons on the right-hand side of the View toolbar are the Display icons. They can also be found in the pull-down menus under View/Display. Some of the Option settings that affect display quality are also explored in this section.

Wireframe

This option will display a simple wireframe view of a part. There is nothing fancy going on here. What is displayed will be very similar to a typical AutoCAD part in its natural display state, which is also wireframe. There is an option that allows you to set the wireframe display quality. The following steps adjust this setting. The illustration that follows shows this operation.

1. Click on Tools/Options in the pull-down menus.

2. Select the Performance tab.

3. In the Wireframe section, select Custom.

4. Adjust the slider bar as desired.

Adjusting wireframe display settings.

What determines where this setting should be set is dependent on the hardware in your computer. Graphics card and processor speed will make a big difference as to how high this setting can be adjusted. If you have a fast machine, move the slider bar all the way to the right. If you start to see that the computer is slowing down, it might be necessary to move the slider bar to the left. Make small adjustments at a time until a good "middle of the road" setting is found. The goal is to establish the highest display quality with the best performance. This is especially true with regard to shaded display quality. The wireframe display quality also affects the next two display types, discussed in material that follows.

One final note: AutoCAD has a setting called VIEWRES and FACETRES. Basically speaking, VIEWRES sets a system variable that allows you to zoom in and still have arcs and circles look like arcs and circles, instead of looking like various sided polygons. Set the VIEWRES too low, and your circles look like octagons. Set VIEWRES too high, and there may be a performance penalty. If working with solids, FACETRES controls how many facets the model will have when hidden lines are removed. The SolidWorks wireframe performance setting is similar to VIEWRES.

Hidden In Gray

This option could actually be expanded to say "Hidden In Gray or Dashed." It all depends on another Options setting, discussed in material that follows. The Hidden In Gray icon will calculate where hidden lines are, depending on the current view, and display those hidden lines in a light gray color.

If the model is rotated, hidden lines are recalculated and redisplayed as required for the new view. If the model has many free-form faces, such as lofted or swept geometry, calculating hidden lines may take more time. This holds true for both hidden line display options, Hidden in Gray and Hidden Lines Removed (discussed in the next section).

To adjust whether or not hidden lines are displayed as gray or as dashed lines when using the Hidden in Gray display setting, you would perform the following steps. The illustration that follows shows this operation.

1. Click on Tools/Options.

2. Select the Edges tab.

3. Select either Gray or Dashed in the Hidden Edges section.

Selecting line display for hidden edges.

The other two check box options in the Hidden Edges section of the Edges tab are Select Hidden for Wireframe/HLG and Select Hidden for HLR and Shaded. Select Hidden for Wireframe/HLG allows you to select edges during the editing or design process that would normally be behind the part (hidden lines) if displayed in Wireframe or if using the Hidden In Gray display mode. (The acronym HLG stands for Hidden Lines Gray.) Checking Select Hidden for HLR and Shaded allows the selection of lines that would be considered hidden while the model is displayed with Hidden Lines Removed or Shaded.

If you select both of these options, it will be very easy to select any edges in the model regardless of the display mode being used. On the other hand, if you are having difficulty narrowing in on edges you want to select due to large numbers of edges in your part, you may want to uncheck one or both of these options. Also note that when the Select Hidden for Wireframe/HLG option is used, it also refers to hidden lines if they are being shown as dashed lines (see the previous three steps).

Hidden Lines Removed

This option does not need much explaining. However, you can draw a comparison with AutoCAD here. AutoCAD's hidden line algorithm has been improving through every major release, but it is still archaic when compared to SolidWorks. AutoCAD's hidden line command will remove hidden lines, but there is no way to display those hidden lines as dashed. In AutoCAD, hidden lines are invisible, and that is that.

Furthermore, if the model is zoomed, panned, or regenerated for any reason, the hidden line display is lost. Hidden line removal will only work with certain AutoCAD entity types. A simple wireframe cannot be displayed with its hidden lines removed; therefore, care must be taken in the creation process to ensure that entities are used that will allow for the hidden line process to be implemented. For SolidWorks, editing a part while its hidden lines are removed is second nature.

Shaded

As previously mentioned, AutoCAD has no capabilities that compare to SolidWorks when dynamically rotating a shaded solid model. The best that AutoCAD can do is render the part, but as soon as the next regeneration takes place, the image reverts back to wireframe, just as with the "hidden lines removed" function. Be it said that AutoCAD is an excellent product, and many of its 2D functional strengths are still superior to SolidWorks. However, in its display capabilities, Solid-Works is superior to AutoCAD.

As with wireframe, shaded display quality can also be adjusted in SolidWorks. Follow these simple steps to change the display quality of the shaded display mode.

1. Click on Tools/Options.

2. Select the Performance tab.

3. In the Shaded Display Quality section, select Coarse, Fine, or Custom.

4. If Custom is selected, adjust the slider bar as necessary.

When adjusting the shaded display quality, as shown in the following illustration, a preview is given that illustrates the "choppiness" of the display. No preview is given for the shading itself. Setting the preview to look like a circle by moving the slider bar from left to right usually makes for a good setting. This gives the best performance while retaining a good shaded display quality. If you will be doing a lot of zooming in close to your models, it may be necessary to move the slider farther to the right.

Shaded Display Quality dialog box.

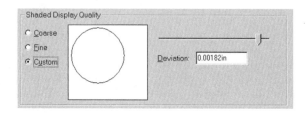

As with adjusting the wireframe display quality, a good middle ground needs to be reached. The more powerful your machine is, the farther to the right you will be able to place the slider bar.

•➔ **NOTE:** *See the hardware section of Chapter 2 for more information regarding shaded display quality.*

Other Display Options

The last two display options are Perspective and Section view icons, shown in the illustration at left. These icons may not be present on your toolbar, but this does not mean that their functions are not present. Do not worry about customizing toolbars at this stage. Just use the View pull-down menu as specified in the instructions that follow.

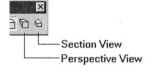

——Section View
——Perspective View

Perspective and Section View icons.

•➔ **NOTE:** *If you must customize your toolbars, read Chapter 12 first.*

Perspective View

To show a model with Perspective mode enabled, click on View/Display/Perspective. There will not be a huge difference with perspective enabled, but you should be able to notice a "vanishing point." The intensity of the perspective can be adjusted, which is measured as the distance the model is from the viewer's eye. Perform the following steps to adjust the degree of perspective. The illustration that follows shows the Perspective View function enabled.

1. Click on the View pull-down menu.

2. Click on Modify.

3. Click on Perspective.

4. Modify the value in the Perspective Information dialog box (a smaller number increases perspective).

5. Click on OK.

Modifying perspective settings.

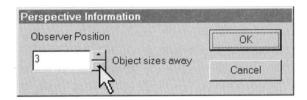

The default value for perspective is 3. As stated previously, a lower value will increase the perspective. The spin box arrows can be used to increment the perspective value up or down. Try setting this to a value of 1 to see if you can perceive a difference. You may need to dynamically rotate the model in order to fully realize the changes. Modify the perspective setting a second time and decrease the value even more by typing in a value of *.25*. With a value of *.25*, there should definitely be a noticeable difference.

AutoCAD's version of a perspective view is enabled through the use of the Dview command. This has always been a chore. For those of you who have worked with Dview, you know that some nice results can be achieved, but obtaining the correct Target and Camera angle is cumbersome, to put it mildly. Perspective is enabled through the Distance option, and the UCS changes to remind you that you are in Perspective mode. Solid-Works does not have the limitations AutoCAD does when enabling perspective. All of the Zoom and Display options are still available, and there are no limitations imposed on the selection methods used.

When you are finished playing with SolidWorks perspective command, you may want to reset the perspective setting to 3. Make sure you turn off Perspective mode when finished. If you do not, you will find the planes and dimensions look particularly weird. Perspective mode makes for a nice effect, but it is not a mode that should be used when designing or editing parts. To turn off Perspective mode, click on View/Display/Perspective or click on the Perspective View icon. Because Perspective View is a toggled option, it is very easily turned on or off.

Section View

An excellent way to check the design of a model, especially with increasingly complex parts, is to use the Section View function. This allows you to "look inside" the model and check various aspects of the design. It is very useful with parts that have a high degree of inner detail that might otherwise be obscured by other lines and edges of the part. The following are the steps you would use to activate a section view. The illustration that follows shows the Section View dialog box.

1. Click on View/Display/Section View.

2. Click in the Section Plane(s)/Face(s) list box area of the dialog box.

3. Select the desired plane from FeatureManager.

4. The selected plane should be listed in the Section Plane(s)/Face(s) list box. If another plane is also listed, you can select that plane from the list box and press the Delete key on your keyboard.

5. Click on the Display button.

6. If the wrong side is being displayed, click on Flip the Side to View.

7. Click on OK.

Setting up a section view.

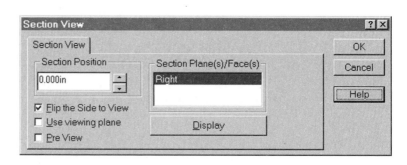

The Pre View check box will automatically update the display every time an option or setting is changed in the Section View dialog box. It is usually easier to make whatever setting changes are necessary, then click on Display to see the effects. The "Use viewing plane" option will use a plane parallel to the current view (in other words, your screen). No selection of a plane is necessary because the plane used by SolidWorks is dependent on whatever view you happen to be using at the moment.

Section Position will offset the section plane selected in the Section Plane(s)/Face(s) list box. If adjusting this setting, look at the screen to see a preview of where the section view plane will be positioned. Click on Display or OK when the desired offset value is achieved.

Section views can be created in AutoCAD through the use of a workaround method. There is no user-friendly, straightforward way to create one. Sections can be created for drawing layouts, but that is a different topic.

•➤ **NOTE 1:** *See Chapter 8 for comparisons regarding 2D section views.*

•➤ **NOTE 2:** *Section views are used for viewing purposes only. The geometry that results from creating a section view does not actually exist in a model and cannot be selected or used in any way to edit or build on the model. Section views can only be used for verification purposes. For this reason, it is best to turn off the section view before continuing the editing or design phase of a model.*

To turn off a section view, click on View/Display/Section View. You should notice that the check mark is then removed from the menu item and the original model view is restored. Section Views and Perspective mode can be used in combination. See the following illustration for an example of this.

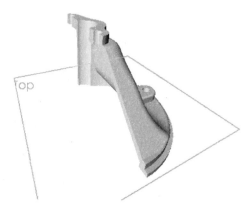

Sump part with Section View and Perspective mode enabled.

↝ **NOTE:** *You may be wondering if the previously discussed method is used in some way to create section views in drawing layouts. Two-dimensional section views can be created for layouts as well, but this procedure is much different and is covered in Chapter 8.*

Curvature

Curvature is a function that does not get used every day. For this reason, Curvature does not have its own icon in Solid-Works. In AutoCAD, there is no command that performs this function. What does the SolidWorks Curvature command accomplish? It shows where the radius of curvature is smallest on a part. A part, such as a bottle, would be a good example to use for this command because many bottles have varying degrees of curvature. If you are running on a slower computer (Pentium 200 or less), you may want to think twice about attempting this. The mathematical calculations will take some time.

To perform this command, click on View/Display/Curvature. The model will be shown using shades of green, yellow, orange, and red. Green represents slight curvature, with red representing the highest degree of curvature (smallest radius). It may surprise AutoCAD users to learn that the part can be freely rotated and zoomed without causing a regeneration. To toggle the Curvature display off, click on View/Display/Curvature. The model will return to its normal state. The following illustration shows an example of Curvature display.

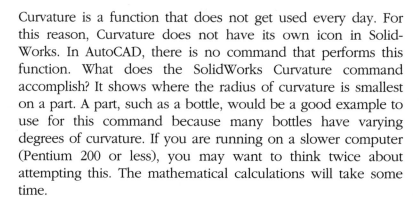

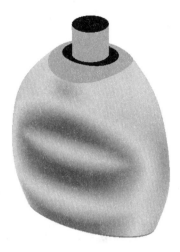

Example of the Curvature display.

The Curvature display also allows you to pinpoint precise areas on the model and gather curvature information. For instance, the following illustration shows the same bottle with Curvature display enabled, and the cursor placed over a position on the bottle. Notice how the cursor displays the radius of curvature. These values will change dynamically as the cursor is moved over the faces of the part. Any areas that are black will have a zero radius curvature.

Extracting curvature data with the cursor.

Redraw

This icon is not on the View toolbar, but this chapter is a good place to introduce the Redraw command. Redraw functions much the same way as AutoCAD's Redraw command. The screen is repainted, or refreshed. It is nothing more than a refresher for the graphics display. Nothing in the part or model database is recalculated; no regenerations are performed. To complete a Redraw, use any one of the following options.

Redraw icon.

- Click on View/Redraw.
- Click on the Redraw icon located on the Standard toolbar (see the illustration at left).
- Use the Ctrl + R hot key combination.

If at any time you notice dots or lines on the graphics display you do not think should be there, perform a redraw to see if the images in question go away. Often, remnants from inferencing lines or highlighted entities remain on the screen, but this is nothing to be concerned with. You can either ignore the left-over images or redraw the screen, whichever you prefer.

Setting Grid and Snap Properties

The following sections discuss grid and snap settings. These settings establish characteristics of grid and snap behavior that determine how a part is sketched or drawn.

Grid Settings

Grid properties can be set to define the characteristics of the grid and snap behavior. The grid can be applied when creating part sketches or drawings. The grid can be used as a visual aid and to set the snap grid characteristics. This can make creating a sketch easier, but also has some drawbacks, which are explored in this section.

The Grid Function

The grid can be left visible without the grid snap active, and vice versa. The grid can be defined with system default values or changed as required during sketching. Aligning the grid to a model edge can make the spacing easier to work with if the model edges do not necessarily conform to the horizontal and vertical nature of the grid. Accessing the grid can be done in two ways. The easiest method is to click on the Grid icon on the Sketch toolbar, as shown in the illustration at left.

Accessing the Grid/ Units tab from the Grid icon.

The second method is to perform the following steps.

1. Click on Tools/Options.

2. Select the Grid/Units tab.

The following list is a quick rundown of what the various settings will accomplish in the Grid Properties section of the Grid/Units tab, which is shown in the illustration that follows.

Display Grid	Toggles the grid display on and off.
Dash	Displays the grid as dashed lines as opposed to light gray.
Automatic Scaling	Redefines grid parameters if zooming in close or far away in order to achieve a more reasonable grid spacing. Otherwise, zooming out too far would result in the grid being too dense and therefore not displayed at all.
Major Grid Spacing	Sets the spacing for major grid lines.
Minor Lines Per Major	Specifies how many divisions the major grid spacing will have.

Grid Properties section of the Grid/Units tab.

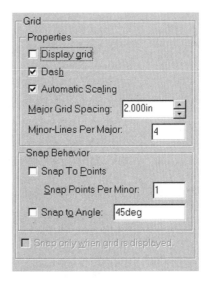

Most CAD programs have some sort of grid function. They are implemented differently, and look slightly different, but the functionality remains basically the same. Using a grid for creating a mechanical part is usually not necessary. However, sometimes a grid has a tendency to help new users get their bearing on what plane they are sketching on. The grid will not be visible unless sketching; therefore, in order to see the grid, you have to enter sketch mode. Without worrying about the mechanics involved with entering a sketch, discussed in Chapter 5, follow these steps so that you can experiment with the grid settings.

1. Make sure you have a part file open (it can be a new or existing part).

2. Select a plane from FeatureManager (these may be called Plane1, Plane2, or Plane3 on your system). This will be your sketch plane.

3. Click on Insert/Sketch.

If the grid is not visible, complete the process previously mentioned in order to turn it on. Feel free to rotate the part and make some modifications to the grid settings to get a feel for how they operate.

Aligning a Grid to a Model Edge

If it is your preference to use the grid, there may be an occasion when the horizontal alignment of the grid does not lend itself very well to the current situation. Take for example a typical AutoCAD drawing in which most of the lines are orthogonal, but at an angle other than horizontal.

In this scenario, the AutoCAD operator would typically enter the Drawing Aids dialog box and set the snap angle to the desired value. This rotates the grid and crosshairs so that entities can be created with minimal effort. Aligning the grid to a model edge in SolidWorks has much the same effect. Align a grid to a model edge using the following procedure. The illustration that follows shows the Align Grid dialog box.

1. Make sure you are in an active sketch and that Grid is turned on.

2. Click on Tools/Sketch Tools/Align Grid.

3. Select a linear edge with which to align the grid.

4. Select Apply.

5. If the alignment is not satisfactory, select a different edge and reapply.

6. Click on OK when done.

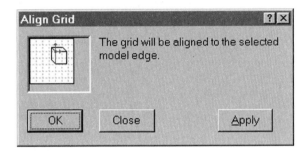

Align Grid dialog box.

Snap Settings

SolidWorks' snap settings functionality is similar to AutoCAD's. However, as with grid settings, snap setting implementation is different between the two environments. For those not familiar with what a snap grid is, the snap grid can be thought of as an invisible grid that controls where the cursor can draw. Endpoints of lines and arcs will always "snap" to the closest node on the snap grid.

The Snap Function

If the snap grid is set to the same size as the display grid, all snap points will appear to be on the visible display grid. An example of setting snap behavior is shown in the following illustration. The list that follows describes the various Solid-Works snap settings and the functions they perform.

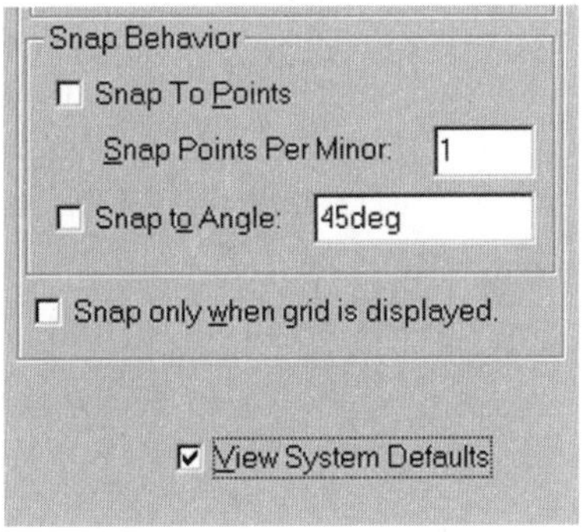

Setting snap behavior.

Snap To Points	Toggles the snap grid on and off.
Snap Points Per Minor	Sets the number of snap points per minor grid line.
Snap To Angle	Toggles angle snap on and off and sets the value of the angle snap.
Snap Only When Grid Is Displayed	Enables snapping when grid is on only. If the grid is off, snap mode is disabled. View System Defaults must be checked in order to activate this option.
View System Defaults	Enables you to see what the current system defaults are. These may be different from settings for the current document.

↦ **NOTE:** *The Snap To Points option will only be available when the "Snap only when grid is displayed" option is not checked. As mentioned in the preceding table, you must be viewing the system's default settings before access to "Snap only when grid is displayed" is available.*

AutoCAD does not have a Snap To Angle function. Snapping to an angle is convenient if all or most of the angled lines in a sketch have the same angle. For instance, a setting of 45 degrees would make drawing an octagon much easier because all of the angled lines would be at 45 degrees.

Using a snap grid and snapping to geometric points are two completely different functions. SolidWorks contains the capabilities that allow you to create a snap grid, much like AutoCAD. AutoCAD also has Object Snaps and Running Object Snaps.

↦ **NOTE:** *SolidWorks has a version of these geometric snap functions, the similarities and differences of which are discussed in Chapter 5.*

A Final Word on Using Snap

Whether or not you use snap is ultimately up to you. If it works for you, and you find it a useful and convenient tool, use it to your advantage. It should be noted, however, that implementing the snap function is something normally just not needed in SolidWorks. This goes back to the basic underlying differences between AutoCAD and SolidWorks.

When creating wireframe or solid geometry in a nonparametric program such as AutoCAD, accuracy is the name of the game. If lines are not of precise length, the model is no good and will have to be edited or rebuilt. In SolidWorks, lines are roughly sketched in, then driven by dimension values. It is not necessary to be 100 percent accurate when drawing the sketch, because SolidWorks takes care of the accuracy when the dimensions are added. For this reason, a snap grid is just not necessary while using SolidWorks.

One other point should be added. The Snap function can actually be detrimental to your productivity in SolidWorks. For example, when sketching very small entities, you may find that the snap points are actually outside the zoom area. This can make it appear as though your sketch function is not working properly. Do not be alarmed. Simply turn off the Snap function and all will be well.

Summary

Selecting objects in SolidWorks is done with the left mouse button. Clicking on an entity will select it. If selecting more than one item, hold down the Control key. When selecting sketch geometry, use the window method to select more than one entity at a time. The window drag direction makes no difference in SolidWorks, as it does in AutoCAD.

Always pay attention to the cursor! It changes to let you know what is going to be selected. This may be a vertex point, edge, face or plane, dimension, or some other entity, such as a line or arc. The right mouse button displays a context-sensitive menu. What is displayed in the menu that pops up as a result of a right mouse click depends on what the cursor is positioned over when the right mouse button is clicked.

Display functions in SolidWorks are much more user friendly and powerful than AutoCAD's display functions. The ability to dynamically rotate a shaded part aids in editing a part and finding any aspects of the part that may need correcting. Most of SolidWorks' zooming and panning commands are similar to commands found in AutoCAD.

Use the grid and snap functions if they will offer some benefit to the design process of the part being created, but keep in mind that SolidWorks parametric capabilities render the snap function superfluous. When creating a sketch, dimension values shape the model. This is opposite to an AutoCAD user's way of thinking.

In Chapter 5 you will get down to business with creating sketch geometry. All of the sketch entity types will be covered, from lines and arcs to elliptical arcs and splines. System feedback will be explored in much greater depth, and you will delve into feature geometry and how to create your first base feature.

5 Sketching

Introduction

Sketches are a collection of 2D entities (e.g., lines, arcs, circles, and splines) used to define a profile of a solid feature. These 2D sketches are the foundation of the solid modeling process. Planes and model faces are used to define sketching planes. A sketch profile is created on a 2D plane and then used to define a solid feature (e.g., an extruded boss or cut).

Mastering sketching skills is fundamental to becoming a proficient solid modeler. Sketches are the building blocks of solid features. Much of the parametric intelligence and flexibility of a model is defined during this stage of the design process.

Content

Because sketch geometry is so fundamental to the SolidWorks design process, much importance needs to be placed on the techniques used when creating a sketch. The "Sketch Basics" section describes some of the basic sketching concepts and fundamental principles used in creating a sketch. The "Sketch Entities" section discusses how individual SolidWorks sketch entities are created, with steps to guide you through the process. The section "Sketch Tools" introduces other tools used to draw various sketch entities and shows you how to extract existing sketch geometry from features.

The "Sketch Dimensions and Constraints" section shows you how to create sketch dimensions and add intelligent relationships between sketch entities. The "Modifying a Sketch" section discusses various aspects of how an existing sketch can be edited. SolidWorks allows for an existing sketch feature to be redefined, thereby modifying the shape of the part. The sketch can be redefined to add or delete sketch entities, dimensions, or constraints. The sketch plane for the sketch can also be redefined, and this section contains process steps that show you how this is accomplished.

The chapter ends with an example sketch that guides you through an actual design of a part. This example demonstrates how a part is created using sketch geometry to create solid features.

Objectives

With completion of the "Sketch Basics" section, you should understand how to select a sketch plane and how the complexity of a sketch affects a part. You should also be able to create dimensions and add geometric constraints. Completion of the "Sketch Entities" section will enable you to create all of the various sketch entities. You should also be able to modify an existing sketch, dimension, or constraint, or change the sketch plane itself. In addition, you should understand how to reuse an existing sketch to create a new feature.

Sketch Basics

The skills required for sketching include the ability to identify features and sketch planes, as well as the methods necessary for constraining sketch entities geometrically and dimensionally. The methods used to define these features determine how a part will behave when edited. There are some basic steps that need to be performed repeatedly in order to build a new solid model. There are also some basic rules that need to be followed. These rules and steps are examined first. The following illustration is an example of a very simple sketch. It contains simple line segments and dimensions. Also notice the origin point in the center of the sketch.

Sample sketch.

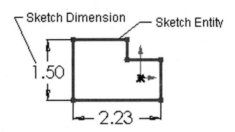

Feature-based Modeling

Feature-based modeling is a term used to describe the various functional components of a part. The ability to identify features is key to intelligently creating parts and assemblies. For example, a sketch might contain a single circle to define a cut, or a group of holes that defines a hole pattern. If a group of sketch entities are related to one another, and they perform a common function (e.g., mount hole pattern), they can be created within the same sketch.

The following illustration shows an example of feature-based modeling. Take a moment to look at the various feature types used to create this solid model. Pay more attention to the actual feature type and not the names of the specific features.

Many separate features constitute a complete model.

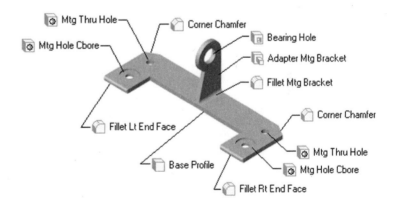

This sample part consists of a series of boss and cut extrusions, fillets, and chamfers. The holes can be added by performing a cut and extrusion, or by using the hole wizard. You will learn both of these methods in Chapter 6.

Sketched Features Versus Applied Features

Features can be divided into two groups. These are sketched features and applied features. The main difference between these two feature types is that a sketched feature requires a sketch, and an applied feature does not. Examples of sketched features are extrusions, revolved features, or swept or lofted features. Applied features are applied directly to the part and do not require sketch geometry. For this reason, they are easier to create. Examples of applied features are fillets, chamfers, domes, and shells.

↦ **NOTE:** *All of these types of features are explored in depth in Chapter 6.*

An important distinction that should be made between AutoCAD and SolidWorks is that SolidWorks uses feature-based modeling techniques, whereas AutoCAD does not. AutoCAD's equivalent, Boolean operations, requires adding and subtracting solid geometry to shape a part. In SolidWorks, features are always added, whether or not said features are actually adding or removing material. In addition, SolidWorks features can be changed later, whereas Boolean features cannot.

The fact that a SolidWorks feature contains intelligence is the true distinction between AutoCAD's simple Boolean operations and SolidWorks feature-based approach. Without this intelligence behind each feature, there would be no way to edit those features. Try to think in these terms, as it will help you to understand the different mind-set needed to work well in SolidWorks. Remember that SolidWorks features contain definitions (intelligence) and AutoCAD's geometric shapes do not.

Sketch Guidelines

There are a few simple rules that should be followed when sketching. There are always exceptions to rules, but generally speaking, a sketch should be a *closed non-self-intersecting profile.* Take a look at the following example to see what a sketch should *not* look like. The following illustration shows a self-intersecting profile, an open profile, and a profile with intersecting islands. Islands can be defined as independent profiles.

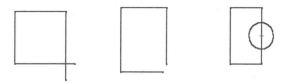

*Examples of what a sketch
should not look like.*

The process of creating a sketch can be broken down into five steps that are repeated numerous times to create a solid model. It would be to your advantage to memorize these five steps, which follow.

1. Select a sketch plane.

2. Enter sketch mode.

3. Create the sketch.

4. Add dimensions and constraints.

5. Create the feature.

There are a few points that should be made regarding the previous five steps for creating a feature. The first point concerns steps 1 and 2. If you are very new to SolidWorks and have little to no experience sketching, you should make sure you click on the sketch icon (step 2) *immediately* after selecting your sketch plane. The reason for this is that SolidWorks will put you on the Front sketch plane (named Plane1 by default) if you do not select a sketch plane first. This applies to your first base sketch only.

The common mistake for new users is to select their sketch plane, but then get distracted and attempt to perform some other function before they click on the sketch icon. In the process, the original sketch plane that had been selected becomes deselected. Without realizing this, the user clicks on the sketch icon thinking they are on the Top or Right planes, when in reality they are on the Front plane. This can be very confusing. Clicking on the sketch icon as soon as you select your sketch plane will keep this from happening.

The second point is a simple one, which is that you should not have to worry about exiting out of a sketch when creating simple extruded or revolved features. Once you enter a sketch, stay away from the sketch icon. Stay away from the rebuild icon as well, which resembles a green light, because rebuilding will also exit you from a sketch. Once you create your feature (step 5), you will automatically be exited from your sketch.

All of this will begin to make more sense as you progress through the book. For now, file these five steps away somewhere in the back of your mind for later use. Every sketched feature created will be based on them. Applied features (which do not require sketch geometry) do not require the five steps.

✎ **NOTE:** *Sketched and applied features are explored in Chapter 6.*

Sketch Complexity

Another good rule of thumb is that *it is better to have less complicated sketch geometry and more features than complicated sketch geometry and fewer features.* Sketches that are overly complex can be difficult to create, dimension, and maintain. Making many overly simplistic sketches can create too many features in a part, which makes the part difficult to understand and modify. A good rule of thumb is to create sketches with logically grouped sets of features. The following illustration shows an example of a sketch with more geometry than a single sketch should have.

Too much sketch geometry for one feature.

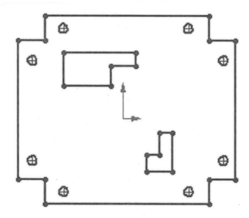

If this sketch were extruded, it would probably create a set of features you do not want. Because of the islands (separate profiles) within the larger main profile, SolidWorks would interpret the smaller islands as cuts.

Creating complex sketches can make feature editing more difficult. Combining too many features into a single sketch can make the part less flexible, and in some cases unusable. It is easier to modify, suppress, reorder, or delete the features of a sketch when the features are contained in a separate sketch. The following illustrations show how the features could have been created by separating the sketch geometry into logical groups.

Same sketch divided into separate features.

The first sketch shows a good sketch for the main feature, otherwise known as the base feature. Sketch 2 illustrates a typical sketch for cutting the eight holes out of the base feature. Sketch 3 contains two shapes used to create bosses on the base feature. The sketch geometry is grouped logically and performs independent functions.

Constraining Sketch Geometry

A constraint is a method used to define the position or size of a sketch entity. A constraint can be either geometric or dimensional. A geometric constraint is a relation added to the sketch to control an attribute (i.e., horizontal, vertical, tangent, and so on) of a sketch entity or between two or more sketch entities. The number and type of dimensions and constraints added to a sketch depend on the design intent of the model. Some constraints are added automatically by SolidWorks during the sketch process, with other constraints added using the Add Relation function.

A dimensional constraint is a sketch dimension used to define the size or angle of a sketch entity. A sketch dimension can be modified to change the size or shape of the dimensioned sketch entity. SolidWorks does not require that dimensions be added to a sketch. For instance, if a dimension value is not known, the dimension does not have to be added. Be aware, though, that in order to modify a dimension, one must be present.

It is good practice and common sense to place dimensions in the sketch. This way, the dimensions can be accessed at a later time and the values of those dimensions can be changed as needed. Additionally, dimensions placed in the sketch will carry over to the drawing layout when that drawing is created. If there are no dimensions in the sketch, there will be no dimensions brought over from the part file into the drawing. As a result, reference dimensions will have to be added.

Design Intent

The term *design intent* is used to express how the rules (in the form of dimensions and geometric constraints) are applied to your model. To phrase this another way, design intent is how your model is going to change if dimensions are altered. Once again, bear in mind that due to the parametric nature of Solid-Works, dimensional location is very important.

You should dimension a sketch in a manner that defines the design intent. By dimensioning the sketch in this manner, the sketch and model can be changed to tweak the design and achieve the desired shape. References to existing feature geometry can be made, resulting in a part that will change shape predictably when modifications are made.

The following illustration shows how a hole pattern was dimensioned to impart the design intent for the feature. The spacing between the holes and the distances from the top left corner of the base feature define the locations for the holes. Additional geometric relations (i.e., horizontal, vertical, and equal radii) were defined to reduce the number of dimensions required to both produce the sketch and maintain the ability to change one value and have both circles update together. The various constraints available are discussed later in this chapter.

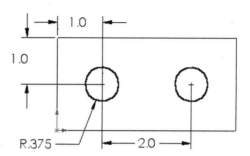

Sketch dimensions placed to create design intent.

In the previous illustration, take a close look at the options available for placing dimensions. If the horizontal dimension of 1 inch were modified, the two circles would move left or right. This is because a horizontal dimension from center to center between the circles has been included. In this model, it was more important in the designer's eyes to directly control the distance between the centerpoints of the circles. Alternatively, the circle on the right could have been dimensioned to the upper right corner of the rectangular base feature. It all depends on design intent, which is what must be considered during the design process and when placing dimensions and constraints.

Design intent is an important topic, and one you should think about when designing your model. To reiterate, design intent is how your model behaves when dimensions are modified.

Constraints

You need to understand what *constraints* (called "relations" in SolidWorks) mean in the SolidWorks environment. If two line segments are drawn in SolidWorks, and one line is constrained parallel to the other, moving either line will result in the other line moving to maintain the parallel relationship. If the same two lines were drawn in AutoCAD, they may very well be parallel to start with, but moving the end of one line will not effect the other line whatsoever. This is a fundamental difference between the two programs.

Constraints are added in one of two ways. They can be added by SolidWorks automatically while sketching, or they can be added by the user afterward. When constraints are added by SolidWorks during a sketch, the cursor changes to let you know exactly what constraints are being added at that time.

➤ **NOTE:** *Just as it is very important for a new AutoCAD user to read the command prompt, it is very important for a new SolidWorks user to watch the cursor. This fact cannot be stressed enough!*

When SolidWorks changes the cursor to let you know what is going on, it is known as *cursor inferencing.* (This concept is covered in detail later in this chapter, and you will have a chance to see inferencing in practice.) The following are the SolidWorks constraints available to you.

- Horizontal
- Vertical
- Perpendicular
- Parallel
- Midpoint
- Coincident
- Collinear
- Coradial

- Concentric
- Intersection
- Tangent
- Symmetric
- Equal
- Merge Points
- Pierce
- Fix

Many of these constraints are self-explanatory. Some should look similar to AutoCAD snap options, and others may be completely new terms. (The section "Modifying a Sketch" later in this chapter discusses exactly what all of these constraints do and how you can add and delete them.)

Sketch Planes

When you create a sketch, you must select a sketch plane on which to place the new 2D sketch. This is step 1 of the five-step process previously mentioned. Similarly, if you were a drafter working on a wooden drawing board, you would need a sheet of paper before you could start drawing.

The SolidWorks sketch plane is the plane that will be used to define the location of the sketch. The sketch plane can be a planar model face or a planar entity. If the plane does not exist, the plane must be created prior to creating the sketch. This can be accomplished in one of seven ways, which are explained in material that follows.

AutoCAD does not require a sketch plane. A 3D line can be created by just typing in its *x-y-z* coordinates. This is possible in

SolidWorks, but it is not important to this discussion. What is important is to realize the similarities and differences between the UCS and a SolidWorks sketch plane. The UCS can be moved by defining its geometric location through a variety of options, such as AutoCAD's 3Point option.

SolidWorks allows the creation of a plane through a similar option. AutoCAD can create wireframe 3D geometry without need of a plane, but SolidWorks does not need or use wireframe geometry. When creating an extrusion in AutoCAD, the extrusion direction is perpendicular to the UCS (using default options). Likewise, in SolidWorks the extrusion direction is perpendicular to the sketch plane.

SolidWorks planes are represented by a rectangular box, but theoretically extend infinitely in all directions. The direction the solid feature should go, such as for an extrusion, determines the desirable orientation for the sketch plane. The example shown in the following illustration shows how a plane or a part face can be used to create a sketch. The sketch can then be used to create a cut (material removed) or boss (material added).

Selecting sketch planes.

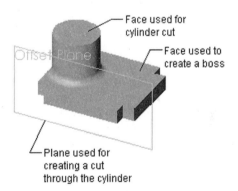

Compare the previous illustration to the following to better understand what features were being added and why the sketch planes were chosen the way they were. A plane can be used when a planar face does not exist on a part, such as with the cut through the side of the cylindrical boss. There is no flat face on a cylinder that can be used to create a 2D sketch. Therefore, a plane (Offset Plane) would need to be defined to create a sketch.

It should be noted that the flat planar face at the front of the part in the previous illustration could be used as a sketch plane. This could still have been the case if the planar face had been angled (drafted, for instance). However, the extrusion direction is always perpendicular to the sketch plane. You must take this fact into consideration when selecting your sketch plane if an extrusion is your intended feature type.

Resulting features are perpendicular to their respective sketch planes.

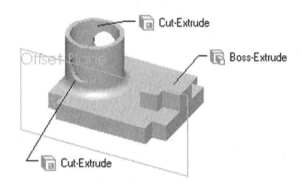

NOTE: *It is usually a good idea to rename a new plane using a meaningful name. This helps document the design, and makes understanding and modifying the design easier. Remember the slow double click technique of Chapter 3. This renaming function works for features and sketch names as well.*

Creating a Sketch

Sketches are used to define 2D geometry, which in turn is used to define features in a solid model. Starting a sketch is a very simple procedure. It involves selecting a plane and clicking on an icon. Do you remember the Widget part started earlier? If you would like to follow along at this point, open up the part *Widget* created earlier. If you did not save your work, start a new part. Right click on the front plane to show it and turn off (hide) any other planes that might be on. If the view needs to be changed, remember to click on View/Orientation and change to the Front view by double clicking on Front. You may want to click on the View Orientation pushpin to keep this dialog box on top.

Entering Sketch Mode

Perform the following steps to enter sketch mode.

1. Click on the Front plane to select it—either from FeatureManager or in the sketch area, it does not matter. It should be green when selected.

2. Click on the Sketch icon.

You should now be in sketch mode. You can verify this a number of ways. The first tell-tale sign is the origin point, which should appear as two small red arrows. The origin represents the *x-y* axes. Another sign is the sketch icon. It will appear depressed, as though it is pushed in. Take a look at the title bar. It should now read *Sketch1 of Widget.sldprt*, or something similar. The title bar lets you know what sketch is currently active and the part being edited.

In addition, FeatureManager will have a new item in it. At the bottom of the FeatureManager list should be the name of the new sketch. SolidWorks starts out naming sketches as Sketch1, then increments the number as more are created. If Sketch1 were deleted, SolidWorks would still name the next sketch Sketch2, even though Sketch1 no longer existed.

To recap, the following are brief versions of the four ways you can tell if you are currently in an active sketch.

* Sketch icon will appear depressed.
* Red axes of the origin will appear on your current sketch plane.
* Title bar will contain *Sketch(x) of part_file_name.sldprt*.
* There will be a new sketch listed at the bottom of FeatureManager.

Common Mistakes— Planes and Views

New users to the SolidWorks software commonly make certain mistakes with the software. It is to your advantage to understand what some of these mistakes are so that you can avoid them. One such mistake is to confuse the Orientation window with the sketch planes listed in FeatureManager.

If you renamed the standard three planes, as suggested in Chapter 3, your planes will be named Front, Top, and Right in FeatureManager. These are the actual planes themselves. Selecting one of these names in FeatureManager will select a plane.

The Orientation window (discussed in chapters 3 and 4) contains the names of the views, not the planes. It just so happens that if you renamed your planes, they would then have the same names as the views in the Orientation window. This is obviously desirable, because you want the system views created by SolidWorks to correspond to the plane names in FeatureManager.

➥ **NOTE:** *Remember this fact: The Orientation window contains system views; FeatureManager contains the actual planes.*

Sketch View Orientation

Sketching can be done in any view orientation. There are times when an orientation normal to the sketch plane may make sketching, constraining, and dimensioning sketch entities easier. Sometimes sketching in a plane view may be more difficult because it is difficult to see certain aspects of the existing solid geometry. The view orientation will not affect the sketch geometry; therefore, use any view that makes your job easier.

Cursor Inferencing

The cursor can offer many visual cues for two main reasons. The first reason for the cursor changing is to provide you with a symbol showing what entity will be selected if the left mouse button is clicked. As previously stated, this entity might be a vertex point, plane, face, edge, and so on. There is no excuse for selecting the wrong entity as long as you are paying attention to the cursor.

A second reason for providing visual feedback through the cursor is to allow constraints to be added during the sketch process and to aid in sketching. When working in AutoCAD, it is very good advice to always read the command prompt. This is especially true for new users. While working with the SolidWorks software, it is especially important to watch the cursor. SolidWorks will give you the information you need while sketching. You just need to know enough to pay attention to this information.

Cursor inferencing is a function that displays the cursor in a different manner when creating a sketch entity, depending on the object being created and the placement of the cursor with respect to other entities in the sketch. Understanding the meaning of the cursor icons can save time and effort, while producing better sketches and models.

The inferencing cursor graphically shows the relationships added automatically during sketching. Inferencing can be used to automatically define geometric relationships. SolidWorks will tell you exactly what constraints are being added as you sketch. It is up to you to recognize them and use these intelligent cursor symbols to your advantage.

↦ **NOTE:** *Good sketching techniques will define the quality and robustness of a solid model.*

The importance of carefully choosing geometric constraints is that the choices made during sketch creation will determine how the model can be modified. Always try to choose meaningful geometric constraints. Adding unnecessary constraints can prove inconvenient when modifying the model and may have to be removed at a later time. They may even overdefine the model to the point that the model is unsolvable. The following illustration shows an example of cursor inferencing.

Cursor inferencing while sketching a line.

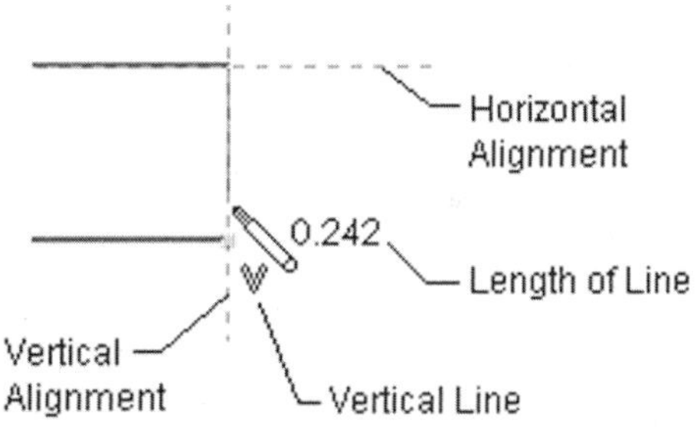

Inferencing Cursor Symbols

The symbols attached to the cursor will change when creating a sketch entity. The cursor will also display additional information (e.g., length or radius) based on the type of entity being created. If the cursor is placed over an entity, the symbol will be displayed for that entity. The following is a list of the various sketched entity symbols you might see.

- Arc

- Centerline
- Centerpoint Ellipse (SolidWorks term for an elliptical arc)
- Circle
- Ellipse
- Line
- Parabola
- Point
- Rectangle
- Spline
- Text (listed as sketch geometry here, not as a note)cursor inferencing;inferencing, cursor

Other Sketch Symbols

In addition to symbols that tell you what type of entity the cursor is over, SolidWorks will also display many other symbols to help while sketching. With AutoCAD, you would use object snaps to control the geometric placement of entities. In SolidWorks, the "snapping" happens on the fly. There is no middle mouse button pop-up menu to access, because it is not needed. The following list contains a good majority of SolidWorks symbols you might see while actually sketching geometry. The symbols that will result in an actual constraint being added by SolidWorks appear in *italics*.

- 0 quadrant point
- 90 quadrant point
- 180 quadrant point
- 270 quadrant point
- Arc included angle = 90
- Arc included angle = 180
- Arc included angle = 270
- *Coincident*
- *Endpoint*

- *Horizontal*
- Horizontal alignment
- *Intersection*
- Midpoint
- Parallel
- Perpendicular
- Tangent
- Vertical
- Vertical alignment

In the next section, "Sketch Entities," you should try to follow along with the book. It is also urged that you pay close attention to the cursor and watch for the various inferencing symbols (otherwise known as system feedback) that will appear.

Sketch Entities

In this section, which explains the various types of sketch entities, a step-by-step process is provided so that you can follow along or refer back to this book at a later time. Open the Widget part or start a new part if you will be following along in Solid-Works. Before you start, review the following points.

- Prior to sketching, a plane or planar face must be selected.
- The sketch icon should be clicked on immediately following plane selection. (This is not a requirement, but benefits new users.)
- As you sketch, the cursor will display symbols that aid in the sketch process.
- Some constraints will be added automatically by SolidWorks as you sketch.
- Always pay close attention to the cursor while sketching or selecting entities.
- Use View/Orientation or the arrow keys to change your view of the part.
- A slow double click on an item in FeatureManager allows you to rename the item.

In AutoCAD, setting the display grid and snap grid would probably take place at this point. If you want to set up the grid, which might help keep you oriented as to what plane you are sketching on, turn the grid on at this point. It is strongly recommended that the snap grid be left off. The reasons for this were explained in Chapter 4.

Right click on the Front plane and show it. Remember, it usually helps to show the plane being sketched on. This serves as a visual guide, but is not required. Next, make sure the Front plane is selected (entities are green when selected), then click on the Sketch icon. This is the icon that looks like a pencil on the Sketch toolbar, as shown in the illustration at left.

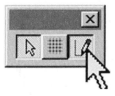

Clicking on the Sketch icon.

If you have successfully entered sketch mode, you will notice that a new sketch has been started in FeatureManager and that the origin has turned red to indicate an active sketch. In addition, the Sketch Tools toolbar should now be active. You will initially be dealing with the first nine icons on this toolbar.

During this section, hold the cursor over an icon if you need to see the yellow tool tip that will tell you what the icon's name is. This information will be displayed at the bottom left-hand corner of your SolidWorks screen as well. You can also reference the following illustration to see what the names of the first eleven Sketch Tool icons are.

The sketch entity icons.

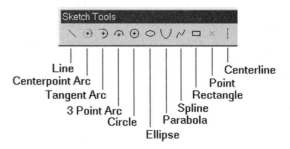

→ **NOTE:** *In the following examples, the terms* pick *and* drag *will be used. When an item is picked, it is being clicked on by the left mouse button. If the term* drag *is used, the left mouse button should be held down and the mouse moved at the same time. In other words, the common phrase "pick and drag" means to select something and move the mouse while keeping the left mouse button depressed.*

In order to begin sketching an entity, you must click on the appropriate icon. To draw a line, you must click on the line icon. Simple enough, right? Check one optional setting before you begin. Click on Tools/Options and look in the top left corner of the General tab. There will be a check box option "Single command per pick." Make sure it is not selected. The reason for this is because if this option is selected, a command (i.e., line or circle command) will exit after using the command. This is inconvenient when you want to draw more than one line or circle, or whatever you happen to be sketching.

Leaving "Single command per pick" unchecked allows you to continue using the selected command until that command is no longer needed. In this case, the command must be exited manually. This can be done in a number of ways. In order to exit a command, you can perform any one of the following.

- Click on the command's icon again to deselect it.
- Click on the Select icon (this looks like a cursor) on the Sketch toolbar.
- Right mouse click in any empty area of the sketch work area and choose Select.
- Press the Escape key on your keyboard.

In the steps contained in the sections that follow, which describe entity types, it is assumed you know enough to click on the appropriate icon in order to start the command. Reference the sketch entity icons in the previous illustration if you need to. It should be noted that when an entity is drawn, it is selected by default. You will notice this if you are working along with this book. It is standard operating procedure for SolidWorks.

Line

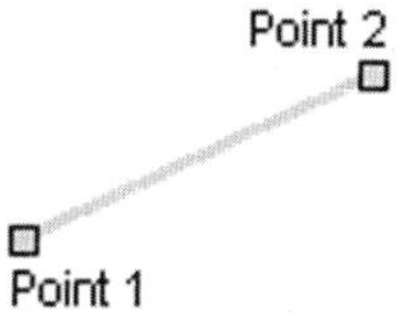

Point 2

Point 1

A simple line segment.

A line is defined as a straight line segment between two points, as shown in the illustration at left. To sketch a line, perform the following steps.

1. Pick a point to start the line.

2. Drag the second endpoint.

Centerpoint Arc

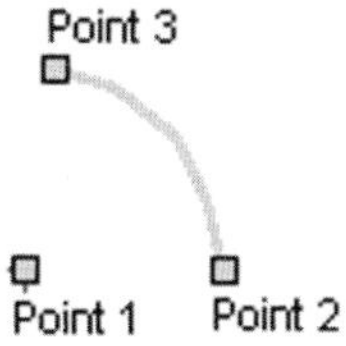

Point 3

Point 1   Point 2

A completed arc using Centerpoint Arc.

A centerpoint arc, shown in the illustration at left, is defined by its centerpoint, radius, start point, and endpoint. There are three ways to define arcs, and Centerpoint Arc is the trickiest. Give it a few tries for practice by performing the following steps.

1. Pick a point to define the centerpoint.

2. Drag the radius. Where you release the left mouse button determines the radius *and* the arc start point.

3. Pick a second time (usually where you let go in step 2, but it does not really matter). Drag to determine the arc's endpoint. Release the button to drop point 3.

Tangent Arc

A tangent arc, shown in the illustration at left, is defined by selecting a sketch entity endpoint and dragging the arc to the

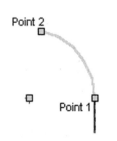

Creating a tangent arc from a small line segment.

desired location. The arc will remain tangent to the selected sketch entity. This arc option is very user friendly, but it is important to remember to select an existing entity endpoint. Whether it is the endpoint of an arc or line makes no difference. To sketch an arc tangent to an existing sketch entity, perform the following steps.

1. Pick an endpoint to draw the arc tangent to.

2. Drag the arc length and radius. Where you let go determines the endpoint of the tangent arc.

3 Point Arc

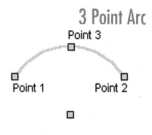

A typical three-point arc.

A simple rule of geometry states that any three points can define an arc. A three-point arc, shown in the illustration at left, is defined by sketching three points the arc must pass through. To sketch a 3 Point Arc, perform the following steps.

1. Select the start point for the arc.

2. Drag the endpoint for the arc.

3. Pick and drag the arc midpoint (the green dot, or point 3 in the previous illustration) to define the radius.

Circle

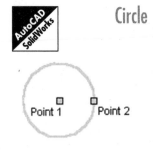

Picking and dragging a circle.

A circle does not need much of an introduction. It is one of the most user-friendly of all sketch entities, and is similar to AutoCAD's Circle command with the Radius option. To sketch a circle, perform the following steps. Picking and dragging a circle is shown in the illustration at left.

1. Pick to place the center of the circle.

2. Drag the radius.

Ellipse

Some sketch entity icons, such as Ellipse, may not be present on the toolbar. You can customize the toolbars to contain icons not present by default. Customizing is not recommended for new users, but if you must, read Chapter 12 first. Otherwise, just use the pull-down menu for now. The ellipse function defines an ellipse, shown in the following illustration, by specifying major and minor axes. To sketch an ellipse, perform the following steps.

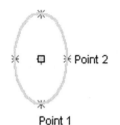

Pick and drag twice to create an ellipse.

Centerpoint Ellipse

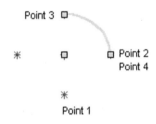

To complete a centerpoint ellipse, you must pick and drag three times.

1. Select Ellipse from the Tools/Sketch Entity pull-down menu (or use the Ellipse icon if present).

2. Pick the center of the ellipse and drag the first axis length. Where you let go of the mouse button will also determine rotation of the ellipse (see point 1 in the illustration at left).

3. Pick and drag the second axis length (usually at a point 90 degrees from the first axis, such as with point 2 in the illustration at left).

The order in which the major or minor axes are created does not matter. If the first axis is longer, it would be considered the major axis, and vice versa.

A centerpoint ellipse, shown in the illustration at left, defines an elliptical arc with a major and minor diameter and a start and endpoint. Do not let the name throw you, as this command could very well have been called Elliptical Arc instead. The centerpoint ellipse is a partial ellipse. The start and endpoints are defined after an elliptical outline is created.

The steps to create this entity start out exactly the same as those for Ellipse, but include an additional step. If you were having trouble with Ellipse, go back and practice before you attempt this procedure. To sketch a centerpoint ellipse, perform the following steps.

1. Select Centerpoint Ellipse from the Tools/Sketch Entity pull-down menu (there is no icon for this function).

2. Pick the center of the ellipse and drag the first axis length. Where you let go of the mouse button will also determine rotation of the ellipse (see point 1 in the illustration at left).

3. Pick and drag the second axis length, usually from a point 90 degrees from the first axis (such as point 2 in the illustration).

4. Pick the start point of the elliptical arc (somewhere on the light blue outline of the ellipse, such as with point 3 in the illustration) and drag the endpoint (point 4).

Parabola

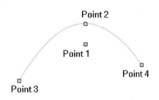

Creating a parabolic curve.

The parabola sketch tool is the newest addition to SolidWorks sketch tools. There is an icon for Parabola, but it may not be present on your toolbar. If this is the case, use the pull-down menus. To create a parabola, perform the following steps.

1. Select Parabola from the Tools/Sketch Entity pull-down menu.

2. Pick to place the focal point of the parabola (point 1 in the illustration at left) and drag to enlarge the parabola (point 2).

3. Pick somewhere on the blue dashed guidelines to start the parabolic curve (point 3 in the illustration) and drag to position the curve's endpoint (point 4).

Spline

A spline is a curve defined by a set of control points. These control points define locations the curve will pass through. There are two methods by which a spline can be created. An alternative spline creation method can be employed by selecting the "Alternate spline creation" check box from the General tab in the Tools/Options menu (see the following illustration).

"Alternate spline creation" check box.

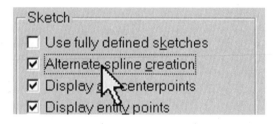

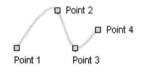

Simple spline with four control points.

Both spline creation methods will be covered so that you can see how each functions. For both methods, it is assumed you have already clicked on the Spline icon. If "Alternate spline creation" is *not* checked, perform the following steps. The illustration at left shows a simple spline.

1. Pick the start point of the spline and drag the second control point.

2. Pick the second control point and drag the third control point.

3. Repeat as necessary for all of the control points.

4. To complete the spline, click once anywhere away from the spline.

If "Alternate spline creation" *is* checked, perform the following steps.

1. Pick wherever a control point is needed (no dragging is necessary).

2. Double click on the final control point to complete the spline.

As you can see, the alternate method for creating splines is much easier. The method used in creating the spline will make no difference to the physical qualities of the spline. Use whichever method is easiest for you.

Rectangle

A rectangle, shown in the illustration that follows, is a set of two vertical and two horizontal lines joined at the ends. This function is quicker and easier than sketching four separate lines, and SolidWorks will add the horizontal and vertical constraints automatically. To create a rectangle, perform the following steps.

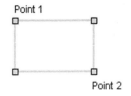

Point 1

Point 2

Rectangle created with the Rectangle command.

1. Pick to place the first corner of the rectangle.

2. Drag to place the opposite corner of the rectangle.

Parallelogram

The Parallelogram icon has two functions. It can be used to create a parallelogram, which is obvious, but it can also be used to create a rectangle at an angle. When used to create a rectangle at an angle, two sides of the rectangle are constrained to be parallel and another two perpendicular constraints are added. The sides of the rectangle are not constrained to be horizontal or vertical, as is the case with the Rectangle command. In the case of a parallelogram, both opposing sides are constrained parallel. Both methods are shown here. First, to create a rectangle at an angle, shown in the following illustration, perform the following steps.

1. Pick to place the first corner of the rectangle.

2. Drag the length of one side of the rectangle.

3. Pick and drag to establish the height of the rectangle.

Creating a rectangle at an angle using the Parallelogram icon.

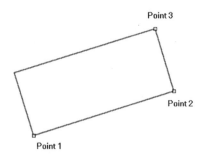

To create a parallelogram, shown in the following illustration, perform the following steps.

1. Pick to place the first corner of the parallelogram.

2. Drag the length of one side of the parallelogram.

3. Pick and drag to establish the height of the parallelogram while simultaneously holding down the Control key.

Creating a parallelogram.

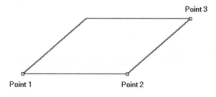

Point

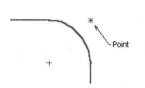

Example of a point entity constrained to an intersection of two lines.

A point is a reference entity only. Points are useful for referencing projected intersections, such as in the illustration at left, and sometimes for constraining purposes. Another good example is the use of points to define a plane. Dimensions can also be attached to sketch points. Unlike AutoCAD, SolidWorks points cannot have their appearance changed. A SolidWorks point will look like an asterisk, or a green dot if selected. To create a point, all you need to do is select a location with the left mouse button.

It should be noted that the Point command can also be used to create an item known as a virtual sharp. Virtual sharps are very useful for dimensioning purposes. The section "Introducing Virtual Sharps" later in this chapter provides more information on this topic.

Centerline

A centerline, shown in the illustration that follows, is created similar to a line and is used for mirroring, revolving, dimensioning references, and establishing geometric relationships. Even though it has all of these functions, it is still considered a reference entity only. Points and centerlines are unique entity types because they do not contribute to the solid geometry of a model. To create a centerline, perform the following steps.

1. Pick to place the centerline start point.

2. Drag the second endpoint.

A simple centerline entity

As far as SolidWorks is concerned, there is absolutely no difference between construction entities and centerlines. Any entity can become a construction entity, which you will discover in the upcoming section "Construction Entities." AutoCAD has specific construction line entities, sometimes known as "X" lines (or simply xlines). This is not required in SolidWorks.

Text

Sketch text can be used to create text on a part face or plane and then used to create a feature. This feature can be a cut, extrusion, or other feature type. Common uses are for engraved or embossed text. When using text in a sketch, the text must be the only entity in the sketch. For instance, if the word *Text* is inserted into the sketch, all of the Sketch Entity icons will become disabled and appear grayed out. Likewise, if any entity is drawn in a sketch, the Text option will no longer be available.

Adding text to a sketch in this fashion is not the same as adding a note. Sketched text has one function, and that is to create a solid feature for your solid model.

↦ **NOTE:** *Adding text as a note is covered in Chapter 8.*

You may want to open another part before attempting this. Sketch text cannot be created as a base feature (the first feature in your part) because it would create islands of geometry. As you might recall from Chapter 1, a part must consist of one contiguous solid. If you feel confident enough to create a basic block shape to use for this example, go ahead. To create a sketched text entity, perform the following steps.

1. As with any sketch, first pick a plane or planar face on which to sketch.

2. Select Text from the Tools/Sketch Entity menu.

3. The cursor will appear as the letter A. Pick an approximate location on your sketch plane where you want the lower left insertion point of the text to be placed. This will open the Text dialog box (see the following illustration).

4. Type in the text you want to use for this sketch.

5. If you want to specify a different font, uncheck Use Document's Font and select a new font. You will have access to all Windows fonts on your computer.

6. If you want to see a preview of what the text will look like, click on the Preview button.

7. To change the size of the text, change the value in the Scale box and click on Preview again.

8. Click on OK when you are satisfied with the results.

Text dialog box.

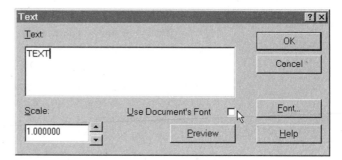

When you are finished, you will see blue text on your Solid-Works part. You will want to locate the text at a precise position. To do this you must dimension the insertion point of the text, which will be at the lower left area of the first line of text. Dimensioning is covered later in this chapter. An example of sketch text is shown in the following illustration.

Sample of sketch text.

A couple of final words on sketch text. SolidWorks has to perform some technical wizardry on the Windows fonts in order to be able to do this. To be precise, the sketch text has to be "vectorized." This means that the text winds up being made of lines, arcs, and splines. More often than not, the text is usually splines, which are computationally intensive. To make a long story short, it will make your computer chug. Turning text into solid geometry is not something that generally needs to be done on a regular basis, but when it does, it is recommended that it be one of the final operations in the design process.

Construction Entities

Any sketched entity can be turned into a construction entity. The construction entity is a reference entity used to constrain other sketch entities or for dimensional reference. Construction lines are not usually necessary in SolidWorks, but they do have their time and place.

AutoCAD has a large need for construction lines, and even created an actual entity type devoted to construction lines in release 13, otherwise known as xlines. Do not view SolidWorks construction lines in the same light. They are usually used as reference entities only. To make a sketch entity a construction entity, first make sure you are currently in an active sketch, then perform the following steps.

1. Right mouse click over the entity you want to convert.

2. Select Properties from the context-sensitive menu. This opens the PropertyManager for the selected entity.

3. Check the For Construction check box (see the dialog box in the illustration at left).

As you can see from the illustration, there are many other aspects of PropertyManager besides converting entities into construction geometry and vice versa.

↝ **NOTE:** *PropertyManager is discussed in more detail later in this chapter.*

It should be noted that construction lines can also be changed into regular sketch entities by using this same procedure. There

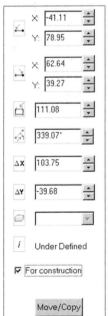

Changing a line to a construction entity in PropertyManager.

really is no difference between a centerline and a construction line. Both perform the same tasks. They are reference entities.

In some cases you may find it necessary to convert many entities into construction entities at once. This can be accomplished by performing the following steps.

1. Select the entities you want to convert to construction entities (remember to hold down the Control key).

2. Right click on any one of the selected entities. This will open the Common Properties dialog box, shown in the following illustration.

3. Click on the Construction Geometries check box.

4. Click on OK.

Common Properties dialog box.

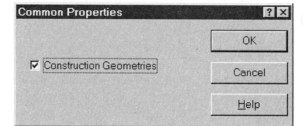

Once again, this method of converting regular sketch entities into construction entities can be used in reverse. As a matter of fact, it is also possible to select a mix of regular sketch and construction entities, in which case the Construction Geometries check box will appear with a gray background, as opposed to the definitive checked or unchecked appearance.

Sketch Tools and Mirroring

Sketch tools are used to manipulate 2D sketch geometry and to create sketch geometry from model geometry. These are accessed from the last six icons on the Sketch Tools toolbar, shown in the following illustration.

Sketch tools are the last six icons on the Sketch Tools toolbar.

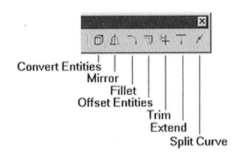

Convert Entities
Mirror
Fillet
Offset Entities
Trim
Extend
Split Curve

Not all of these tools can be demonstrated at this time, although the commands can all be explained. This is due to the fact that solid geometry needs to exist before some of the tools can be used. Specifically, Convert Entities and Offset Entities require solid geometry. These are covered last.

Mirror is used to create a symmetrical (mirrored) copy of selected sketch entities around a centerline. Only one centerline can be selected for mirroring, even though there can be more than one centerline in a sketch. Once the new geometry has been created, it is associated with the parent geometry and the mirror line, but not dependent on it. In other words, the original geometry could be deleted after mirroring the geometry without deleting the mirrored entities. Deleting the mirror line or the original entities, however, would delete the symmetrical relationship.

Mirroring can be done one of two ways while working with sketch geometry. You will examine both of them. First, assume you have already created some sketch geometry you want to mirror. Make sure there is a centerline in the sketch that will serve as a mirror line. To mirror selected sketch features around a centerline, perform the following steps. The illustration that follows shows the selection of entities for the mirror operation.

1. Select the entities to be mirrored. Make sure the Control key is depressed if selecting more than one, or use the window method to select entities.

2. Select the centerline to use as a mirror line while holding down the Control key.

3. Click on the Mirror icon or select Mirror from the Tools/ Sketch Tools menu.

Selecting entities for the mirror operation.

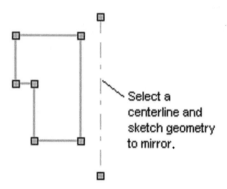

Select a centerline and sketch geometry to mirror.

There is an alternative method of mirroring in which objects can be dynamically mirrored as they are created. Once again, a centerline is needed. Therefore, in order to attempt this operation, draw a centerline down the middle of the screen. Complete the following steps to dynamically mirror sketch geometry. An example of a dynamically mirrored line is shown in the illustration at left.

Example of a dynamic mirror line.

1. Select the centerline to be used as a dynamic mirror line.

2. Click on the Mirror icon. The centerline should look like that shown in the illustration at left.

3. Create sketch geometry as needed. It will be mirrored as you sketch.

4. Click on the Mirror icon a second time to turn off dynamic mirroring when finished.

If you have successfully created some mirrored geometry, leave it on the screen for the discussion on dragging geometry, which follows mirroring.

An Introduction to Dragging Geometry

This is the perfect opportunity to introduce you to dragging geometry and why it can be beneficial. Now that you have an idea of what the term *dragging* means with regard to sketching, you can apply the same phrase to geometry that has already been sketched. If there are no sketch entities on your screen at this time, it would be a good idea to create a few odds and ends for the purpose of this topic.

In SolidWorks, it is possible to move sketch geometry around the screen, much like AutoCAD's use of grips. Clicking on a grip allows AutoCAD users to move endpoints, lines, centerpoints, and so on. This same type of "pick and drag" functionality can be accomplished in SolidWorks. There are no grips, just endpoints, and all you have to do is select a point and drag the mouse while holding down the left mouse button. As a matter of fact, you can actually drag any geometry, not just endpoints. Try this on the mirrored geometry from the last exercise.

Dragging Various Geometry

Depending on what portion of specific sketch geometry is selected makes a difference as to how it is dragged. Dragging an endpoint of a line moves just the endpoint. Picking and dragging a line moves the entire line. Dragging a circle by the edge resizes the circle. Dragging a circle's centerpoint moves the entire circle.

There are a few reasons this is especially important. First, dragging geometry can tell you a lot about a sketch you may not have realized. For instance, dragging might tell you that the line you thought was attached to the other three sides of a rectangle was not attached. Dragging will show you your mistakes and it will show you if the correct constraints have been added to the sketch.

✓ **TIP:** *It is a very good idea to drag sketch geometry for the feedback it will give you. Use this technique often.*

No Solve Move

Dragging sketch geometry, as mentioned in the previous section, can tell you much about what relations are currently associated with your sketch or what relations you may still need to add. That is, unless you have No Solve Move turned on.

No Solve Move allows you to drag geometry without solving for the coincident relationships that exist between, for instance, the endpoints of lines. In the case of a rectangle, usually the rectangle would lengthen if one end of the rectangle were dragged outward. If No Solve Move is enabled, the rectangle is not lengthened. Instead, the line becomes detached from the rectangle.

It is usually not desirable to have entities fly apart as soon as you drag them to a new position. However, if you want to enable No Solve Move, select No Solve Move from the Tools/Sketch Tools menu. It is a toggle switch, and can be toggled on or off in this fashion. You can also find an icon for No Solve Move on the Sketch toolbar.

Additional Sketch Entity Tools

The following sections describe the process for using various sketch entity tools. These include Fillet, Trim, Extend, Convert Entities, Offset Entities, and Split Curve.

Fillet

A fillet is a tangent arc entity attached to two line segments, a line and an arc, or between two arcs. A fillet can actually be a round or a fillet. SolidWorks does not differentiate between the two, and it does not really matter. Typically, small cosmetic fillets are not included within a sketch unless they are a design element. Inserting a sketch fillet is an easy way to place a tangent arc between two existing lines.

NOTE: *This fillet function is a sketch tool only. Adding fillets to feature geometry (applying the fillets as features) is a different procedure and should not be confused with sketch fillets.*

AutoCAD SolidWorks

Similar to AutoCAD, SolidWorks will automatically trim or extend lines as needed. Draw yourself a rectangle or even some arbitrary lines and arcs on the screen and try stepping through the fillet process with this book. The Sketch Fillet dialog box is shown in the following illustration.

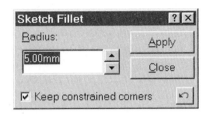

Sketch Fillet dialog box for filleting sketch geometry.

Walk through the following steps to complete a fillet. The illustrations that follow show lines selected in this process and the outcome.

1. Click on the Fillet icon, or select Tools/Sketch Tools/ Fillet from the menu.

2. Type in a value for the radius.

3. Select two entities to fillet. If the items form a perfect corner, you can optionally select a vertex point.

4. Repeat as necessary, changing the radius if desired.

5. Click on Cancel when you are finished.

Selecting lines to fillet, and the result of filleting.

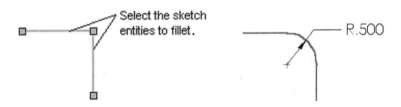

↝ **NOTE:** *When dialog boxes are open, it is not necessary to hold down the Control key when selecting more than one entity.*

The Apply button in the Sketch Fillet dialog box is not usually needed. Simply selecting the entities to be filleted is enough. You will find that if you preselect entities before opening the Sketch Fillet dialog box, the Apply button will be activated. At that time, you may click on Apply to add the fillet.

The Undo Command

You may have noticed that there is an Undo icon located in the Sketch Fillet window. This Undo icon is similar to the undo function incorporated into many AutoCAD commands. Simply click on the Undo icon if you want to reverse the previous sketch fillet operation.

The "Keep constrained corners" option is checked (enabled) by default and is used to add a virtual sharp to sketch geometry when there are dimensions present. If it were not for the automated addition of virtual sharps, dimensions already present in a sketch would be eliminated as the fillets were added. This is because the sketch lines' endpoints are being eliminated as the fillets are added and the dimension extension lines no longer have an endpoint to dimension to. If there are no dimensions on your sketch when you add fillets, you can safely uncheck the "Keep constrained corners" option. Read on for more on virtual sharps.

Introducing Virtual Sharps

Virtual sharps are often used for dimensioning purposes and can be created through the use of the Point command. They can also be created automatically during operations such as when adding sketch fillets, as you learned previously. A typical scenario in which you might need a virtual sharp would be an edge you want to dimension to but no longer exists because a fillet has removed it. An example of such a situation is shown in the following illustration.

Dimensioning to a virtual sharp.

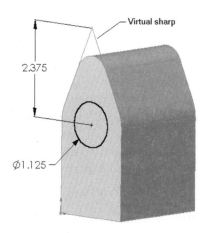

When creating a virtual sharp, it is mandatory that you first select the items you will be using to define the virtual sharp. You must also be in an active sketch. To create a virtual sharp, perform the following steps.

1. Select the lines or edges at whose projected intersection you want to place a virtual sharp.

2. Click on the Point icon.

To control the appearance of virtual sharps, access the Detailing tab from the Tools/Options dialog box. Next, click on the button labeled Virtual Sharps. You will see a small window appear that gives you five options for the appearance of the virtual sharps (see the following illustration). This is a global setting and will affect all sharps in the SolidWorks document. You cannot control the appearance of individual virtual sharps.

Controlling the appearance of virtual sharps.

Trim/Extend

SolidWorks Trim and Extend commands are a great deal easier to use than AutoCAD's similar trim and extend functions. First, there is no need to select cutting or boundary edges, and there are no command line options to go along with the Trim or Extend commands. Second, SolidWorks' Trim command allows for the complete removal of entities. This can be extremely handy, as the Trim command doubles as a Delete command.

Trim

As previously discovered in the section on dragging geometry, it is possible to lengthen or shorten lines and arcs by simply dragging their endpoints. However, there will be times when you will need something a little more powerful than this. Solid-Works' Trim command is similar to AutoCAD's in some ways. As with AutoCAD, you will be clicking on the portion of the entity to be discarded. To trim (shorten) a sketch entity, perform the following steps. The illustration that follows shows implementation of this command.

1. Click on the Trim icon or select Trim from the Tools/Sketch Tools menu.

2. Select the portion of the entity to be removed.

*Implementing
the Trim command.*

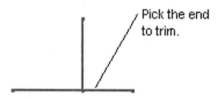

Pick the end
to trim.

This procedure represents the Trim command in its most generic state. SolidWorks trims the entity selected back to the next intersection it encounters. SolidWorks will additionally show you a preview of what will be trimmed. Unlike AutoCAD, SolidWorks can totally eliminate an entity.

⊶ **NOTE:** *Typically, to delete objects in SolidWorks, make sure the object is selected and press the Delete key. This holds true for all SolidWorks objects.*

What if the entity to be trimmed needs to trim back to an entity other than the first one it encounters? The answer is "dragging." Drag the entity to be trimmed to the entity you want to trim it to so that the trimming entity (or "cutting edge") is highlighted. When the mouse button is released, the entity will be trimmed.

Extend   As with the Trim function, SolidWorks' Extend command is much more user friendly than AutoCAD's Extend command and requires no preparation. The illustration that follows shows implementation of the Extend function. To extend (lengthen) a sketch entity, perform the following steps.

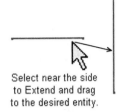

Select near the side
to Extend and drag
to the desired entity.

*Implementing
the Extend command.*

1. Select the Extend icon or select Extend from the Tools/ Sketch Tools menu.

2. Select the end of the entity to be extended (optionally, drag to the entity to be extended to so that the "boundary" entity is highlighted).

Similar to AutoCAD's Extend command, you must use caution when selecting the end of the entity to be selected. Make sure you select on the appropriate side of the entity's midpoint to be extended. Otherwise, SolidWorks will attempt to extend the entity in the wrong direction and will not find anything to extend to. This should be blatantly obvious, because as with the Trim function, SolidWorks will show you a preview of the extended entity before you actually click the left mouse button.

Convert Entities

The Convert Entities function is used to create sketch entities based on model edges. The key word here is *convert*. What is happening is that edges of existing geometry are being converted to sketch geometry. Think back to one of the primary steps needed to create a feature. The following are the five steps.

1. Select a plane or planar face on which to sketch.

2. Enter sketch mode.

3. Create the sketch.

4. Add dimensions and constraints.

5. Create the feature.

The step you should be most concerned with at this point is step 3: Create a sketch. A sketch must be created prior to creating a sketched feature. This may seem obvious to you, but it is being stressed here because it is one reason the Convert Entities function is useful. What Convert Entities does for you is to convert existing *model edges* into sketch geometry for the current sketch. This sketch geometry can then be turned into a feature.

Why are converted entities important? Converted entities have external references from sketch geometry to existing feature geometry. Any changes made to existing geometry will be reflected in the sketch entity made using Convert Entities. Single edges can be converted or offset. By selecting a face, all edges that define the face are converted or offset. (The Offset Entities function is explained in material to follow.)

There is a prerequisite for implementing either of these two sketch tools, and that is to select model edges, a model face, or geometry from another sketch prior to initiating either command. If the Convert Entities or Offset Entities icon is selected without any model geometry selected, the system will display an error message.

Another important aspect of these two commands is the projection of the entities selected. To put this quite simply, any edge selected for conversion or offset will be projected perpendicular to the sketch plane. The illustration that follows shows the conversion of edges. To convert model edges into sketch geometry, perform the following steps.

1. Select the edges or model face to be converted.

2. Click on the Convert Entities icon or select Tools/Sketch Tools/Convert Entities from the menu. The selected edges or model face edges will be projected to the current sketch plane and converted to sketch geometry.

Converting feature edges into sketch geometry.

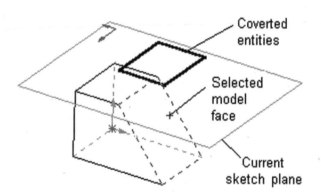

Coverted entities

Selected model face

Current sketch plane

Offset Entities   The Offset Entities function is used to create sketch entities based on model edges offset to a specified distance, as shown in the following illustration. It is very similar to Convert Entities, with the single exception that offset is added to the converted entities.

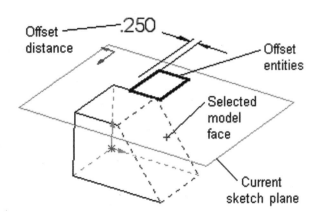

Using the Offset Entities command.

All of the rules that govern Convert Entities apply to Offset Entities. All you need to do is supply an offset and an offset direction. This is easily accomplished through use of the user-friendly Offset Entities dialog box, shown in the following illustration.

Offset Entities dialog box.

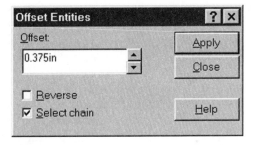

It is possible to select the appropriate edges to be offset either before or after you click on the Offset Entities icon. An offset distance is required, along with a side to be offset. This is very similar to AutoCAD's Offset command, except that in Solid-Works you do not first have to select the entity to be offset. Another difference with SolidWorks is the references established to existing feature edges, which make for an intelligent model.

As with converted entities, anything offset using the Offset Entities command results in a link back to the original feature edges. If the original feature is modified, the dependent sketch will also update. If the entities being offset are simply existing sketch lines or arcs, this association back to the original entities is not as significant. This is due to the fact that the offset items are in the same sketch as the originals. Modifying the sketch will not necessarily have such drastic effects on subsequent dependent features. To offset selected model edges or faces, or entities within the same sketch, perform the following steps.

1. Click on the Offset Entities icon or select Tools/Sketch Tools/Offset Entities from the menu.

2. Select the edges or model face to be offset. The selected edges or model face edges will be projected to the current sketch plane, offset the specified distance. If selecting sketch geometry in the current sketch, offset entities will remain on the current sketch plane.

3. Check the Reverse option to reverse the offset direction.

4. Check the Chain Select option if you want to select all of the entities in a continuous loop (contour).

5. Click on Apply to offset the selected entities.

6. Repeat as necessary, and click on Close when finished.

Split Curve

Using the Split Curve function is in essence creating two entities where before only one existed. It is similar to AutoCAD's Break command. For example, if you were to specify the Break command, select a line you wanted to break, and then press the @ key, the line would be broken at the point where you selected the entity. That is exactly what the Split Curve function does.

All sorts of sketch entities can be split, including lines, arcs, and splines. Even circles can be split, although you must split them twice, thereby creating two arcs. Construction entities, such as centerlines, cannot be split. To use the Split Curve function, perform the following steps.

1. Select the Split Curve icon or select Split Curve from the Tools/Sketch Tools menu.

2. Click on the entity you want to split at the location you want the split to occur.

That is all there is to it. You will notice that you can now drag the split point to a new location, or delete one side or the other of the split. The choice is yours. The following illustration shows the Split

Curve command in operation. In this "before and after" image, the "after" shows dragged geometry after the split was performed.

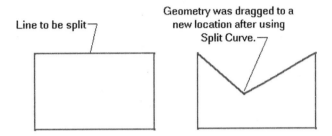

Using the Split Curve command to split a line and then change the shape of the geometry.

Once you have used Split Curve to split an entity, you can remove the split simply by selecting the split point and pressing your Delete key. This only works, however, if you have not dragged the split point to change the shape of the geometry. For example, the split point in the previous illustration has been dragged to a new location. The resulting two lines are no longer collinear. Now the split point is seen by SolidWorks as an entity endpoint and deleting it is not allowed.

Patterning Sketch Entities

Sketched entities can now be patterned. This is known in Solid-Works lingo as a "step and repeat." It is very similar to AutoCAD's Array command. It is possible to create a linear or a circular pattern using the Step and Repeat function.

One thing you must ask yourself when patterning sketch entities is whether or not you would be better off patterning a feature instead. If you pattern a feature, all patterned instances will update if the original is modified. It is also possible to parametrically alter the definition of a feature pattern. This is not the case with a sketch pattern. Sketch patterns have no dimensions associated with them and have no editable definition.

Because there are two types of sketch entity patterns, the steps for each will be listed here separately. The Linear Sketch Step and Repeat function is explained first. To create a linear pattern of sketch entities, perform the following steps.

1. Click on the Linear Sketch Step and Repeat icon, or select Linear Step and Repeat from the Tools/Sketch Tools menu.

2. Select the entities to be patterned.

3. Specify the Number, Spacing, and Angle parameters for the pattern Direction 1.

4. Specify the Number, Spacing, and Angle parameters for the pattern Direction 2 if required.

5. Click on the Reverse Direction button for either direction if necessary.

6. From the Instances list area, you can select instances and press the Delete key. This deletes the instances from the pattern and lists the same instances in the Instances Deleted list area.

7. Click on the Preview button to see a preview of the pattern if you make adjustments to the pattern parameters.

8. Click on OK to create the pattern when you are satisfied with the preview.

The illustration that follows shows the Linear Sketch Step and Repeat dialog box. Notice that there is an undo button built into this dialog box. This makes it easy to back up if you make a mistake. Use the Preview button to your advantage. Remember that sketch patterns are not features. Unlike feature patterns, they have no dimensions associated with them. Therefore, you cannot go back and adjust the pattern after it has been created.

Creating a sketch pattern using Linear Sketch Step and Repeat.

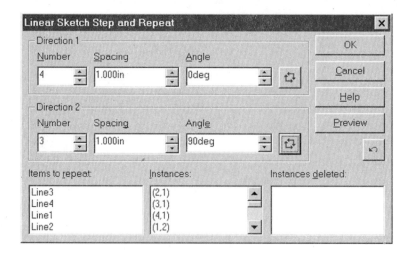

When you are creating a linear pattern of sketch entities, you are not held down to using the dialog box to fill in all of your parameters for the pattern. You will notice small green handles, similar to AutoCAD's Grips. These green handles can be dragged to specify the pattern Spacing and Angle between instances. The preview of the pattern specified in the previous illustration is shown in the following illustration. Notice the handles.

A linear sketch pattern preview.

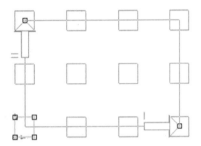

Once you click on OK, to create the pattern, whether it is a linear or a circular pattern, the entities will still need some defining constraints and dimensions. Otherwise, you will have a pattern of underdefined geometry.

The process for creating a circular pattern is similar to that for a linear pattern. You must plug in the parameters that define the pattern, such as the total number of instances in the pattern and the spacing between instances. In the case of a circular pattern, the spacing is defined by an angular measurement in degrees. This is not to be confused with the angle measurement in the linear pattern, which describes the angle of the direction the linear pattern will take. There is an Angle and Radius parameter for circular sketch entity patterns, but before matters get too confusing, take a look at the Circular Sketch Step and Repeat dialog box, shown in the following illustration.

Circular Sketch Step and Repeat dialog box.

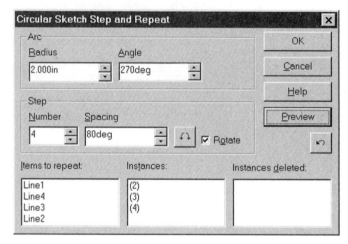

The steps for creating a circular sketch entity pattern follow. Subsequent to the steps, you will take a closer look at what these various parameters are actually accomplishing. To create a Circular Sketch Step and Repeat, perform the following steps.

1. Click on the Circular Sketch Step and Repeat icon, or click on Circular Step and Repeat from the Tools/Sketch Tools menu.

2. Select the entities to be patterned.

3. Specify the total number of entities to be patterned by modifying the Step Number value.

4. Specify the angular distance in degrees between instances by adjusting the Spacing value.

5. Select the Direction button to change between patterning in a CW or CCW direction.

6. Check or uncheck the Rotate option to control whether or not instances are rotated as they are patterned.

7. Modify the Arc Radius as necessary to adjust the radius of the circular pattern.

8. Modify the Arc Angle as necessary to adjust the angle of the circular pattern. This is the angle as measured between the original entity and the center of the pattern circle relative to 0 (0 being east, 90 being north, and so on).

9. Select the Preview button to preview the pattern.

10. Select OK to complete the circular pattern.

Now take a look at one more image. The following illustration shows the resultant circular pattern using the values shown in the previous illustration. The dimensions, construction geometry, and large circle have all been added for illustration purposes. All you would actually see after creating the pattern are the four squares.

A circular sketch entity pattern.

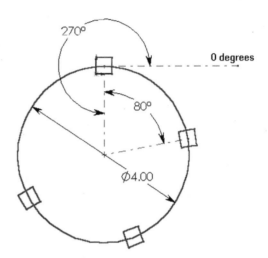

The overall 4-inch diameter of the circle is controlled by the Arc Radius setting of 2 in the Circular Sketch Step and Repeat dialog box. The Arc Angle of 270 controls where the center of the circular pattern will be. Think of zero degrees as always being off to the right (like AutoCAD). This setting makes sense. The 80 value is the Spacing, and controls the distance between instances. There are a total of four instances, and the direction is CW, as controlled by the Direction button (to the left of the Rotate option). The Rotate option rotates the instances as they are patterned, just like the rotate option when performing a Polar Array in AutoCAD.

That is all there is to creating a sketch entity pattern. Remember that it is a good technique for fully defining your sketch geometry. Otherwise, your sketches may behave in an unpredictable fashion.

3D Sketching

Creating a 3D sketch is a bit different than creating a 2D sketch. The first thing you would probably notice upon entering a 3D sketch is how limited your sketch entities appear to be. In fact, the only sketch icons available are lines, centerlines, points, and fillets. On the plus side, the 3D sketch icon can be used to create sweep paths or guide curves quickly and easily.

To create a 3D sketch, either click on the 3D Sketch icon on the Sketch toolbar or select 3D Sketch from the Insert menu. You will probably want to make sure you are in an isometric view if attempting this. Next, try clicking on the Line icon and drawing some lines the way you typically would. Notice the XY attached to the cursor. This means you are sketching on the XY plane.

If you press the Tab key, you can toggle the plane being sketched on. There are only three choices here, and you do not have the ability to "select" a plane the same way you do as when creating a typical 2D sketch. Pressing the Tab key, however, will cycle between the XY, YZ, and ZX planes. As you continue to sketch more lines, your 3D sketch will begin to take shape.

Relations will be added to 3D sketch lines much like 2D lines, such as Horizontal, Vertical, and Along Z. You can even add fillets to 3D lines. If you find it necessary to dimension your 3D sketch, go ahead. As with any sketch, it is always a good idea to fully define the geometry. Add relations and dimensions as needed to define your design intent.

Sketch Dimensions

Dimensions are used to display distances. This is nothing new to anyone reading this book, and is a simple concept. Dimensions in SolidWorks are used for the same reason, but have additional functions. Dimensions are placed on 2D sketch geometry so that the geometry can be shaped and driven with these dimensions. This allows you to edit the values of these dimensions and have the software update the geometry.

The dimensions add intelligence to the part, and dimension placement determines how the part can be modified. This is related to the concept of design intent, previously discussed. It is possible to create a complete model without dimensions, but this is not practical. Again, the following are the two main reasons for putting dimensions on your SolidWorks part.

1. Dimensions allow you to parametrically control a model.

2. Dimensions added to a part can be automatically transferred to a 2D layout.

That is the nuts and bolts of it. Now all that is left is the mechanics of adding dimensions. The following section discusses how this is accomplished.

A Universal Dimensioning Tool

Starting the Dimension command.

The Dimension command can be used to place a dimension between two vertices (endpoints), a single sketch entity, a line and a point, two arcs, radial or diametric dimensions, and so on. All of this can be done with SolidWorks' one dimensioning sketch tool.

The following examples assume that you have already selected the Dimension icon and are ready to apply dimensions. The Dimension icon, shown in the illustration at left, is found on the Sketch Relations toolbar. This toolbar usually consists of three icons, one of which looks like a dimension. This is the icon you want.

To insert a point-to-point sketch dimension, perform the following steps. The illustration that follows shows this function.

1. Select the first point to dimension to.

2. Select the second point to dimension to.

3. Select a location to place the dimension lines. This will determine the distance the dimension lines are from the items being dimensioned.

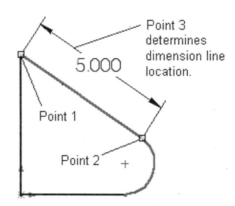

Placing a point-to-point dimension.

There are a couple of optional settings that should be addressed at this point. What is on the screen may differ from what is being displayed in this book; therefore, you need to make sure you are working with the same set of rules. The following are a few optional settings found under Tools/Options, along with what these options do and the recommended settings for each.

Input dimension value. This option is found under the General tab. When checked, this option, shown in the following illustration, allows you to supply a value for the dimension as soon as it is placed on the sketch. Leave this option on so that you can plug in values on the fly.

Leave "Input dimension value" checked.

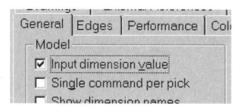

Center text. This option, shown in the following illustration, is found under the Detailing tab. When this option is checked, SolidWorks automatically centers dimension values if the dimension values are positioned within the extension lines. If the dimension value is placed outside the extension lines, the dimension value will stay where you placed it. Leaving this option off allows for greater flexibility when positioning dimension values. Leaving this option on or off is up to you.

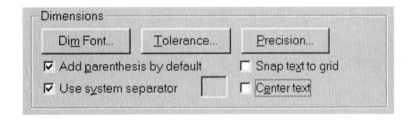

The "Center text" option.

Dim Font. In the same area as "Center text," this button allows for changing the size and font of dimensions. Usually a setting of .125 inches or 12 point works well, but the exact setting is up to you. Monitor size and screen resolution will also play a role in this setting.

Always display text at the same size. To find this option, click on the Annotations button, which is found under the Detailing tab. Click on Annotations to open the Annotation Properties dialog box. If "Always display text at the same size" is checked, text will stay the same size if you zoom in or out while in a part or assembly. Otherwise, text may get extremely small or extremely large. This is a nice feature. Make sure it is checked.

The "Always display text at the same size" option does not affect drawings. Drawings attempt to follow the WYSIWYG (What You See Is What You Get) school of thought; therefore, dimensions and text appear as they would when printed.

This takes care of optional settings for now. The following are other dimension types to be considered.

A point-to-point dimension is really the same as a parallel dimension. In AutoCAD, this dimension type would be considered an Aligned dimension. Also in AutoCAD, it is possible to select an entity and pick a location to place a dimension. This can be done in SolidWorks the same way. The results are the same. To place a parallel (or "aligned") dimension, perform the following steps. The illustration that follows shows this function.

1. Select the line to be dimensioned.

2. Select a location to place the dimension lines.

Placing a parallel dimension.

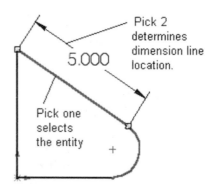

You can see from the previous illustration that there is no difference in the aesthetic value of the dimensions. The next thing to consider is diametric dimensions.

Diametric Dimensions

Basic diametric dimensions, such as when dimensioning a circle, are very easy to create. Use the same universal dimensioning tool you would use for any other dimension. To place radial or diametric dimensions, perform the following steps. The illustration that follows shows both of these types of dimensions.

1. Select the arc or circle to be dimensioned (point 1, following illustration).

2. Select the location for the dimension value (point 2, following illustration).

Create diameter or radial dimensions the same way.

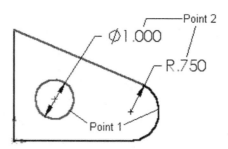

There is another type of diametric dimension that can be created for revolved parts, which is shown in the illustration that follows. The need for this results from bringing dimensions into a drawing from part files that have been revolved and contain only linear-style dimensions in the sketch.

•→ **NOTE:** *Revolved features, and features in general, are covered in Chapter 6.*

In the following example, it would be standard procedure to place a dimension between the centerline and the right-hand edge of the part. When a dimension of this type is seen from a top view of the revolved cylindrical part, it does not look right. To remedy this, perform the following steps, to create a diametric dimension from a linear dimension.

1. Add a dimension from the centerline to the desired point or line.

2. Right click on the dimension value and select Properties. This will open the Dimension Properties dialog box.

3. Check the Diameter dimension box.

4. Click on OK.

This will turn the dimension into a diametric dimension. However, a diametric symbol needs to be supplied because this is still not actually an arc or circle being dimensioned, and you must force SolidWorks to display the appropriate symbol. Add a symbol by performing the following steps.

1. Right click on the dimension value and select Properties. This will open the Dimension Properties dialog box.

2. Select the Modify Text button. This opens the Modify Text dialog box, where text or symbols can be prefixed or appended to an existing dimension.

3. Select the Diameter symbol with the cursor positioned in front of the <DIM> text. This is the default position for the cursor.

4. Select OK on the Modify Text menu to go back to the Dimension Properties menu.

5. Select OK to accept the changes. Your dimension should look similar to that shown in the following illustration.

A diametric dimension for revolved parts.

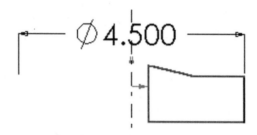

This trick only works if there is a centerline in your sketch. This is because SolidWorks assumes a revolved feature will be created because you are using a centerline. However, simply including a centerline in the sketch is not enough. You must actually dimension to it or the Diameter dimension check box will not be available. Make certain you select the centerline itself, not its endpoint.

Angular Dimensions

Angular dimensions are used to show the angle between two lines, as shown in the illustration that follows. To create an angular dimension, perform the following steps.

1. Select the first line to dimension (point 1, following illustration).

2. Select the second line to dimension (point 2, following illustration).

3. Click to locate the dimension value (point 3, following illustration).

Creating a 68-degree angular dimension.

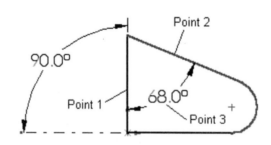

Horizontal Dimensions

Horizontal and vertical dimensions, shown in the illustration that follows, can be placed between a line and a point automatically without any extra user intervention. For example, if a dimension is placed between a vertical line and a point, it will be horizontal by default because the line is already vertical. There is no other possible solution; therefore, SolidWorks acts accordingly.

The same holds true if a dimension is placed between a horizontal line and a point. However, if a dimension is placed between two points, SolidWorks needs to be told that the dimension should be horizontal or vertical. Otherwise, Solid-Works assumes it should be a parallel dimension. The key to performing this task is the position of the mouse. To insert a horizontal or vertical dimension, perform the following steps.

1. Click on the Dimension icon.

2. Select the first vertex point to dimension to (point 1, following illustration).

3. Select the second vertex point (point 2, following illustration).

4. Move the mouse (cursor) around the screen until the proper dimension preview is shown.

5. Click to place the dimension line (point 3, following illustration).

Forcing a dimension to be either horizontal or vertical.

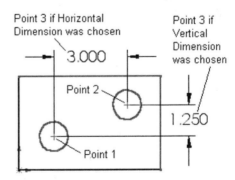

It is sometimes desirable to "lock in" a particular dimension type while moving the mouse and observing the various dimension previews. For example, you may want to use the aligned dimension style, but when you attempt to center the dimension value between the extension lines the dimension style reverts back to horizontal. (This will occur only if you have the "Center text" option turned off, which was described earlier in this chapter.)

When adding dimensions between two points there is the capability to lock in a particular style of dimension. To perform this function, click once with your right mouse button when the preview of the dimension looks correct. This action will lock the preview into that particular dimension style and then allow you to place the dimension value wherever you choose.

Driven Dimensions

Driven dimensions are usually created by accident by new users to the SolidWorks program. When there are too many dimensions or constraints on a sketch, the sketch becomes overdefined. What this means is that there are too many rules applied to the sketch for SolidWorks to interpret, and something has to give way for all criteria to be met. This usually results in a dimension being driven.

Keep in mind that dimensions control the shape of sketch geometry in SolidWorks, which is very different from AutoCAD's way of doing things. If there are too many driving dimensions, SolidWorks' solution to this dilemma is to make the last one added a "driven" dimension. Driven dimensions do not control the shape of geometry.

Driven dimensions are actually much closer in nature to the type of dimensions AutoCAD uses. These dimensions will update accordingly, just like AutoCAD dimensions, if a part changes size or shape. If an overdefining dimension is added, you will see the dialog box shown in the following illustration.

What you see if you have added too many dimensions.

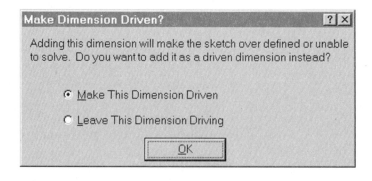

The default is to make the troublesome dimension driven, but you are not obligated to do so. If this situation is encountered by accident, it is recommended that you leave the dimension driving and determine the reason for the overdefinition. The reason for this is that it is very important you understand the interacting relationships taking place within a sketch and why the geometry is being overdefined. In addition, it is simply good practice for new users. Methods of determining what is causing overdefining sketch geometry are discussed later in this chapter.

Adding a few driven dimensions as reference dimensions will not cause any problems in the sketch because driven dimensions cannot overdefine a sketch. Dimensions only cause problems when you leave them as driving dimensions. Driven dimensions will appear gray (and surrounded by parentheses if created in a drawing). A driven dimension is shown in the following illustration.

Driven dimension.

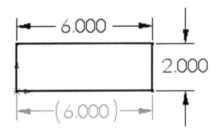

The default operation used when adding overdefining dimensions can be controlled with settings in the Options dialog box. Click on Tools/Options and under the General tab look for the section shown in the following illustration. Make adjustments as needed, but for new users still getting accustomed to the software, leaving both boxes checked is a good idea.

Setting overdefining dimension characteristics under the General tab.

An Introduction to Sketch Color Codes

There are a number of problems that can be encountered when creating a sketch. SolidWorks supplies color codes to aid you in recognizing these problem areas. The following is a list of the various color codes and what they mean.

- *Blue.* Sketch geometry is underdefined. Entities may move about if dragged, and may change position unexpectedly.
- *Black.* Geometry is fully defined. You have full control over sketch geometry.
- *Red.* Geometry is overdefined. There are too many constraints or dimensions.
- *Brown.* Dimensions or geometry are referencing entities that no longer exist. This condition is known as "dangling."
- *Pink.* The geometry's position cannot be determined using current constraints.
- *Yellow.* Sketch would result in invalid geometry if solved.

The last two color codes are not very common. If you are a new user, the first three are the color codes you should concentrate on.

✓ **TIP:** *As a rule of thumb, blue is okay, black is best, and red is an undesirable situation you should always try to remedy right away.*

Editing a Sketch

An existing sketch can be modified at any time. This allows you to change, add, or delete geometry, dimensions, and geometric constraints after the initial sketch has been exited. It is easy and quite common for a new user to accidentally exit out of a sketch. You can exit a sketch by clicking on the Sketch icon or by rebuilding the part, but the point of the matter is that you do not need to exit from a sketch for a simple extruded or revolved feature. SolidWorks does this for you automatically when you create the feature. (By the way, the Rebuild icon is found on the Standard toolbar and looks like a street light with the green light lit.)

The key to editing a sketch is by right clicking on it in FeatureManager. A very common mistake is to click on the Sketch icon to enter back into an existing sketch, but this is not correct! By clicking on the Sketch icon, SolidWorks thinks you are attempting to begin a new sketch. If you want to edit an existing sketch, perform the following steps.

1. Right mouse click on the sketch in FeatureManager to activate the context-sensitive menu.

2. Select Edit Sketch.

3. Edit the sketch geometry as needed.

4. Click on the Rebuild icon (or the Sketch icon) when done to accept the changes made to the sketch.

Alternatively, it is possible to right click on a feature that owns the sketch you want to edit. You will still see the Edit Sketch menu choice. You can also select (highlight) the sketch you want to edit by picking it with your left mouse button and then clicking on the Sketch icon. This will place you in edit mode for the selected sketch. The preferred method, however, is to use your right mouse button, as described in the previously listed steps.

Editing Sketch Entities and Dimensions

Existing sketch entities can be manipulated by dragging them. Dragging sketch entities, as shown in the following illustration, will maintain geometric constraints. These constraints cannot be violated; therefore, these relationships will be maintained as the geometry is dragged and reshaped.

Dragging sketch geometry.

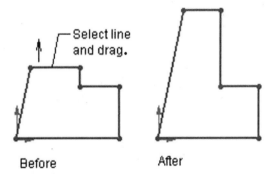

Before   After

If there are too many dimensions on a sketch, dragging will not be possible. In this case, dimension values must be modified in

Modifying a dimension.

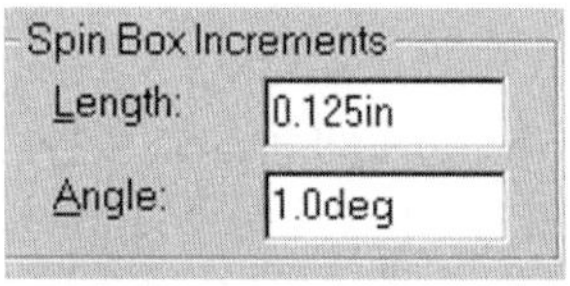

Setting spin box increments.

Enabling PropertyManager for Direct Edit functionality.

Direct Edit

order to resize and reshape the sketch as needed. To modify a dimension value, double click on the dimension value with the left mouse button. The illustration at left shows this function.

When modifying a dimension, either type in a new value or use the spin box arrows to increment the dimension value up or down, as shown in the illustration that follows. When the desired value is obtained, click on the green check to accept it or the red X to exit the box without accepting the change. The spin box increment amount can be adjusted to better suit your particular needs. If the spin box increment is too coarse or too fine for your application, perform the following steps to adjust it.

1. Click on Tools/Options.

2. In the Grid/Units tab, specify the desired spin box increments for Length and Angle.

3. Click on OK.

Another option you have when editing a sketch is the Direct Edit functionality, new to SolidWorks 99. Before you can use Direct Edit, you must first enable this function. To do so, access the Options dialog box (Tools/Options) and check the Enable Sketch PropertyManager option found at the bottom of the General tab, as shown in the following illustration.

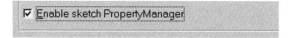

When Direct Edit is enabled, access to PropertyManager becomes available and takes the place of the Properties option typically found by right clicking on an entity. Once you select an entity, PropertyManager appears, giving you control over many aspects of the item selected. An example of the Property-

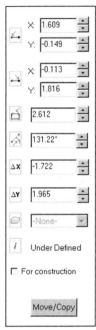

*PropertyManager
for a line.*

An Introduction to the Undo Command

Manager function is shown in the illustration at left. This illustration shows the properties for a line, but different entities have different properties. An arc, for example, would have properties for included angle, radius, and so on.

Direct Edit Versus Property Manager

Old hands at SolidWorks may decide they do not want to use the Direct Edit capabilities of SolidWorks. After all, PropertyManager does not take the place of adding dimensions or constraints. It is largely used for editing sketch geometry, which can quite easily be accomplished simply by dragging. PropertyManager should be looked at as a convenience for those who find it easy to use. It is not a necessity, though, and will not detract from your Solid-Works performance skills if you opt not to use it.

If you decide to leave the Direct Edit functionality disabled by not checking the Enable Sketch PropertyManager option, you can still access PropertyManager. All you do is access the properties of an entity by right clicking on the entity and selecting Properties.

Sometimes after modifying a dimension or dragging sketch geometry, the results may be far from what you were expecting. This would be a good time to use the Undo command. The Undo icon will undo the last command performed, just like AutoCAD's Undo command. AutoCAD's Undo function is one of the best around, and has gotten quite a few AutoCAD users out of many a sticky situation.

SolidWorks' undo function is not quite as extravagant as AutoCAD's, but it gets the job done. The Undo icon in Solid-Works is found just to the left of the Rebuild icon and looks like a counterclockwise sweeping arrow. This follows standard Windows icon conventions.

It is generally recommended to not rely too heavily on the Undo command. This will help to develop good skills. If the Undo command is used too often, there will come a time when it will not be there when needed. This is because SolidWorks will not allow an unlimited number of undo commands to be carried out. After a rebuild or after exiting a sketch (which actually forces a rebuild), the Undo command list is usually cleared and is not available.

This is not as bad as you might think. AutoCAD needs a powerful undo option because its editing capabilities are not as powerful as SolidWorks'. When it is possible, as in SolidWorks, to edit any aspect of a solid model at any time in the design process, an undo feature is less important.

Even with SolidWorks' power and flexibility, the fact remains that it is nice to be able to undo a command or two once in a while. If modifying a sketch gives you unexpected results, click on the Undo icon and try again. The drop-down list box attached to the Undo icon will display a list of commands that can be undone. Selecting a command from this list will undo that command, as well as every command above the selected command in the list.

↦ **NOTE:** *Use the Undo command if required, but be cautious. SolidWorks does not have a Redo command.*

Sketch and Relations Functions

The following sections describe various sketch and relations functions. These include modifying a sketch, adding relations, displaying and deleting relations, the Scan Equal function, the Constrain All function, and deriving a sketch.

Modify Sketch

The Modify Sketch function allows for translating, scaling, and rotating an existing sketch. Existing dimensional or geometric sketch constraints cannot be overridden. Therefore, fully defined sketch geometry is not a good candidate for the Modify function. If a sketch is anchored to the origin point, it cannot be translated because it is anchored. Use common sense before implementing the Modify command.

If sketch geometry is unconstrained, the sketch can be translated, scaled, or rotated. The sketch can be modified from inside or outside sketch mode. You need to be editing a sketch, wherein you would then click on the Modify icon (shown in the following illustration) or click on Tools/Sketch Tools/Modify. If you are not currently editing the sketch, you can alternatively select the sketch from FeatureManager.

The Sketch toolbar usually contains the Modify Sketch icon.

└─The Modify Sketch icon

The Modify command contains the Modify Sketch dialog box, shown in the following illustration. Be aware that your Sketch toolbar may not contain the Modify icon, in which case you could customize the Sketch toolbar and add the icon to it. As mentioned previously, use caution when customizing Solid-Works if you are still fairly new to the software.

Modify Sketch dialog box.

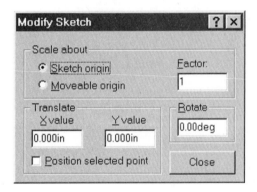

The following are some of the functions that can be performed with the Modify Sketch dialog box.

- To translate (move) a sketch, perform the following steps.

  1. Specify an X and/or Y value in the dialog box and press the Return key, or

  2. Hold the left mouse button down and drag the sketch by moving the mouse.

- To rotate a sketch, perform the following steps.

  1. Specify a rotation angle in the dialog box and press the Return key, or

2. Hold the right mouse button down and dynamically rotate the sketch by moving the mouse.

- To scale a sketch, perform the following steps.

    1. Type a value into the dialog box and press Return. Note that the sketch is scaled using the part's original origin point as a base point.

    2. If you want to use the moveable (black) origin, select "Moveable origin."

- To relocate the moveable origin with relation to the sketch geometry, perform the following steps.

    1. Place the cursor over the black square at the moveable origin point, where the X and Y axes meet.

    2. Drag the origin to the desired location.

- To mirror the sketch geometry, perform the following step.

    1. To mirror across the X or Y axis, right click over the black square on the tip of the X or Y axis on the moveable origin. You can also flip the sketch (mirror across the X *and* Y axes) by right clicking over the black square located on the moveable origin.

Exit out of the Modify Sketch dialog box when finished, or just press the Escape key. Do not forget to rebuild your part when finished.

Adding Relations

Geometric relationships are used to define a geometric association between sketch entities, as shown in the following illustration. These relationships can be as powerful as dimensions in defining design intent for a part. Properly determining and defining geometric relationships will produce a part that will more significantly incorporate the design intent you require.

*Six examples
of geometric constraints.*

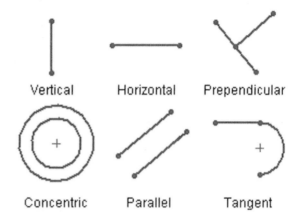

Vertical   Horizontal   Prependicular

Concentric   Parallel   Tangent

NOTE: *The terms constraint and relation mean the same thing. They are used interchangeably throughout this text. The icon used for adding constraints is Add Relations.*

As previously explained, AutoCAD does not use intelligent relationships between sketch entities. Take a look at another example and this fact should become quite clear. Assume you have a circle drawn in AutoCAD. The drawing requires that another circle be drawn concentric to the first, meaning they have the same centerpoint. This task is easy enough to accomplish by snapping the centerpoint of the second circle to the centerpoint of the first. Now the two circles are concentric.

Now say that the second circle's grips are enabled. If the center grip is selected, the circle can be moved to any position on the screen and the first circle remains where it was. Remember, this is a fundamental difference between AutoCAD and SolidWorks. If the second circle were moved in SolidWorks, the first circle would move with it to maintain the relationships you set.

Much like dimensions, conflicting constraints can be added that overdefine a sketch. It is up to you to incorporate the proper relationships to achieve the desired design intent. The dimensions or constraints already applied to the sketch will determine how the geometry will change when relationships are added. The following is a list of geometric relationships available in SolidWorks, and a description of each.

Horizontal	Constrains lines, centerlines, or two or more points horizontally.
Vertical	Constrains lines, centerlines, or two or more points vertically.
Collinear	Constrains two lines, line and face, planes, or axes so that they are aligned linearly.
Perpendicular	Constrains two lines, or a line and an edge, at 90 degrees.
Tangent	Constrains an arc, spline, or circle tangent to another entity.
Midpoint	Constrains a point to the center of a line, centerline, or edge.
Coincident	Constrains a point to a line, arc, circle, or nearly any other entity type.
Symmetrical	Constrains two entities to be symmetric about a centerline.
Coradial	Constrains two circles or arcs to have the same radius and centerpoint.
Parallel	Constrains two lines, or one line and an edge, to be parallel to each other.
Concentric	Constrains two circles and arcs to have the same centerpoint with relation to each other or a vertex point.
At intersection	Constrains a point to the intersection of two lines.
Equal length/radii	Constrains two lines, arcs, or circles to be of equal length or radius.
Fix	Constrains an entity to the current location. This acts as an anchor.
Pierce	Constrains an entity to pass through, or pierce, a point at the position the point resides on its sketch plane.
Merge points	Similar to Coincident, constrains two sketch endpoints to become coincident (merged).

Many geometric relationships are added automatically when you sketch, as discussed earlier in this chapter. This function can be disabled, but it is not recommended. If you want to experiment with this setting, click on Tools/Sketch Tools and select Automatic Relations from the pull-down menus.

If Automatic Relations is disabled, the relationships graphically displayed by the cursor while sketching will not add relationships. This is not normally a desired function because none of the entities created while sketching will contain any intelligence.

⇥ **NOTE:** *Do not forget to turn Automatic Relations back on if you turn this off for any reason.*

Like nearly any other function in SolidWorks, you can choose to select entities before or after opening the Add Relations dialog box, shown in the following illustration. If there is something else selected already, either click in an open area devoid of geometry or simply press the Escape key if you want to deselect items. To add geometric relationships to selected entities, perform the following steps.

1. Click on the Add Relations icon or select Add Relations from the Tools/Sketch Relations menu.

2. Select the entities to be geometrically constrained.

3. Select the appropriate geometric relationship from the Add Relations dialog box.

4. Click on OK.

Add Geometric Relations dialog box.

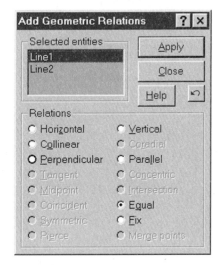

SolidWorks attempts to guess what relation you want to add. Sometimes the software guesses correctly, sometimes not. Simply select the appropriate relation if it is not already selected. In addition, if a relation already exists between the selected entities, that relation will appear as a dark circle in the Add Relations window. See *Perpendicular* in the previous illustration for an example of this effect.

More than one relationship may be defined for a sketch entity. The system displays only those relationships applicable to selected geometry. SolidWorks is intuitive enough to know what constraints are geometrically possible and can be added for selected entities. If the constraints you expected to see are not active, it means one or more entities were not selected correctly, or possibly that you have too many entities selected. You can see what entities have been selected by viewing the "Selected entities" list box area (salmon colored) in the Add Relations window (see the previous illustration).

Displaying and Deleting Relations

The Display/Delete Relations function displays geometric relationships already placed on a sketch entity. You can view each relation and delete unwanted relationships. A geometric relation is usually associated with another sketch entity, but not always. For instance, a line can be constrained to be horizontal by itself. If there are two or more entities involved and the relation is deleted from one of these entities, the relation is removed from all entities involved.

The entity in question can be selected prior to clicking on the Display/Delete Relations icon. This brings up a dialog box showing the relations for that entity. If no entity is selected first, the same dialog box appears, showing all relations to any entities in the sketch. Once again, whether you select objects before or after the Display/Delete Relations window is open is up to you. The Display/Delete Relations dialog box is shown in the following illustration.

*Display/Delete Relations
dialog box.*

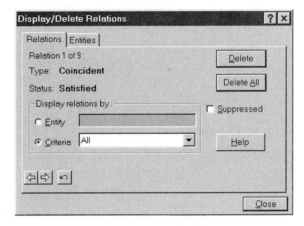

To view a feature's geometric relationships, perform the following steps.

1. Select the Display/Delete Relations icon, or select Display/ Delete Relations from the Tools/Sketch Relations menu, or right click over the entity in question and select Display/ Delete Relations.

2. Select an entity to view its relations. You can also click on the Criteria option and select All to view all relations in a sketch.

3. Click on the left or right blue arrows to cycle through each relationship for the selected entity or entities. The Type option will inform you of the relation type and Status will tell you the condition of the relation, such as satisfied or conflicting.

4. If needed, select Delete to remove the geometric relation shown, or select Delete All to remove all relations from the entity or entities.

5. Select Close to close the dialog box.

If you want to view (or search for) specific types of constraints within the sketch, continue with the following steps.

1. Select the Criteria option.

2. From the drop-down list, select the criteria that best suits your requirements. Options available are All, Overdefining/ Not Solved, and Dangling (External or Defined In Context).

3. Select Next or Previous to cycle through each relationship for the selected criteria.

4. Select Delete or Delete All as needed.

5. Select Close to close the dialog box.

↪ **NOTE:** *The term* dangling *refers to dimensions or constraints that no longer refer back to a valid entity. This usually occurs when features are deleted. If, for example, a dimension is referring to a particular feature that has been deleted, the extension line of the dimension is left dangling because it is referencing a missing object.*

Other options when using the Display/Delete Relations function are to suppress a particular relationship or to undo your last action. You will notice that there is an Undo button present on the Display/Delete Relations window near the bottom left corner. The Suppress option is very useful when trying to troubleshoot an overdefined sketch because you can temporarily suppress a relation to see how it will affect the sketch geometry.

Entities tab of the Display/ Delete Relations window.

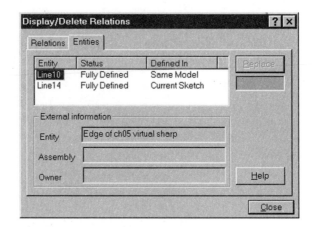

There is another aspect of the Display/Delete Relations window that should be mentioned, and that is the Entities tab shown in the previous illustration. The Entities tab will display more in-depth information on the currently specified relation. For example, if you were to view relation 3 of 8, you could then click on the Entities tab and see which entities were actually being constrained and if there were any external references to other geometry.

The numbers shown attached to entity names in the Entities tab are used by SolidWorks as internal identification numbers and really are not that significant to the user. What is beneficial, though, is the ability to replace one entity with a new one and essentially "swap" the entity being constrained to. This is very convenient when there are dangling or unsolved relationships in a sketch. To replace a bad relationship with another, perform the following steps while viewing the Entities tab of the Display/Delete Relations window.

1. From the list area, select an entity you want to replace (swap out). This would typically be a dangling relationship.

2. Select a replacement entity from your model. It should then appear in the Replace box.

3. Click on the Replace button.

If the Status of the entity you are replacing had been dangling, it should now read "Fully Defined." This means you have successfully swapped out the problem entity with a valid one. At this time, you could close out of the Display/Delete Relations window.

Scan Equal

The Scan Equal function can be a time-saving shortcut for adding relations. It reviews a sketch for entities of equal length or radii, and allows you to set geometric relationships automatically. You can set equal two or more lines of equal length, two or more arcs or circles of equal radius, and even a line with the same length as the radius of an arc or circle (which would be the "both" option).

Scan Equal can be used when there are a number of entities that should be grouped via the equal length/radii constraint. These entities need to be the same size for the function to recognize them as equal length or radius. This command is not used that often in practice, but has been included here for reference. It is sometimes convenient to use this command for imported geometry that needs to have constraints added. To scan a sketch for equal length or radii, perform the following steps.

1. Select the Scan Equal icon or select Scan Equal from the Tools/Relations menu. This opens the "Scan for equal radii and line lengths" dialog box, shown in the following illustration.

2. Select Set Equal to constrain the highlighted set of entities. If the dialog box is obscuring your view of the work area, drag it out of the way.

3. Select Find Next to move to the next set of entities and repeat the process.

4. Select Close to quit this command.

"Scan for equal radii and line lengths" dialog box.

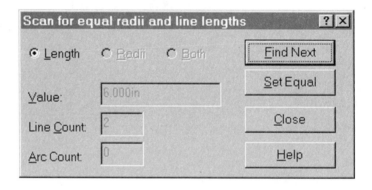

Constrain All

The Constrain All function is used to add geometric relationships to unconstrained imported DXF or DWG files. The entire sketch is reviewed, and applicable geometric constraints are added to the sketch, which you are informed of onscreen, as shown in the following illustration. Only basic constraints will be added, such as horizontal, vertical, perpendicular, parallel, and tangent. To add geometric constraints to an unconstrained sketch, perform the following steps.

1. Select Constrain All from the Tools/Relations menu. The system will display a dialog box showing how many constraints were added to the sketch.

2. Select OK to continue.

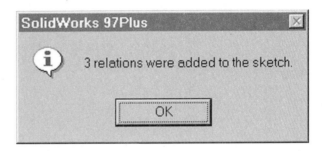

*SolidWorks informing
you how many constraints
(relations) were added.*

It should be noted that the sketch must not contain any constraints or dimensions whatsoever. This is why this function is best used after importing AutoCAD DXF or DWG files. It is always best to add constraints and dimensions to such files. AutoCAD dimensions will import, but they will not contain parametric information. It is actually best to turn off any extraneous layers, such as dimensions or text, before importing AutoCAD files into SolidWorks.

➭ **NOTE:** *Importing AutoCAD files is covered in Chapter 11.*

Derived Sketch

A derived sketch can be used to copy existing sketch geometry. This allows one sketch to drive a number of identical part features. The new sketch is dependent on the parent sketch. Any changes to the parent sketch are shown in the derived sketch. It should be noted that derived sketches do not contain any dimensions precisely for the reason that they are dependent on their parent sketch geometry. Therefore, no dimensions will automatically be created in the 2D layout, and reference dimensions will probably need to be added manually.

Sometimes the need for a specific design intent overrides the luxury of having SolidWorks automatically import dimensions into the 2D layout. This process is described here for reference. It should be noted that a derived sketch cannot be created until you get a little farther into creating feature geometry. You might find it convenient to come back to this section at a later time. To create a new derived sketch based on an existing sketch, perform the following steps.

1. Make sure you are not in an active sketch.

2. Select the parent sketch from FeatureManager using the left mouse button.

3. Hold down the Control key and select the planar face to which you want to apply the sketch.

4. Select Derived Sketch from the Insert menu.

5. Locate the new sketch as desired using sketch dimensions or constraints.

6. Build the feature, which is covered in Chapter 6.

An Interactive Sketching Exercise

This section deals with the selection of a sketch plane or face, as well as sketching entities, dimensioning to impart design intent, and modification of existing sketch entities. In the following exercise, a sketch is created for the features named Base Profile, Bearing Hole, and Adapter Mtg Bracket, shown in the following illustration. These features have rectangles surrounding their names so that you can identify them. The chapters that follow build solid features, create an assembly, and produce a drawing using this example.

Adapter bracket.

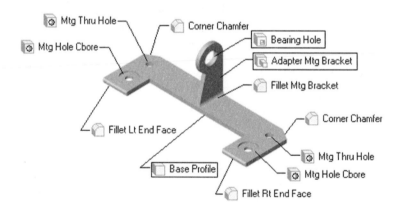

When starting a part, you should identify the features that constitute the part, such as the features named in the previous illustration. You should also consider the best method of locating and dimensioning the sketch geometry so that the dimensions will incorporate your design intent. Default planes can be used as a basis for creating sketch geometry. Identification of the adapter bracket's planes is shown in the following illustration.

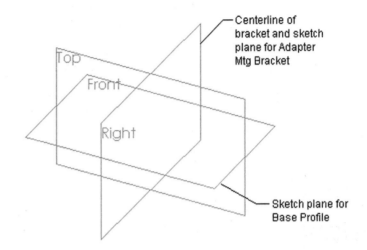

Identifying the best planes for the adapter bracket's initial features.

When selecting the base feature, review the part for a major shape that defines the part. For this exercise, the feature's base profile would be a good choice. This was created using one sketch, which keeps the sketch fairly simple and allows for easy modification.

The orientation of the base feature could be per the normal assembly position, a standard orientation based on drawing views, or according to a plane that makes viewing the part in an isometric view easier. This example uses the initial orientation of the bracket in the final assembly.

Do not worry too much about the part's final orientation in any future assemblies because individual parts can be moved, rotated, and mated as needed when the time comes. To start sketching, select the plane named Front and click on the Sketch icon. The Orientation function can be used to sketch in a normal (2D) view orientation. The sketch can be created in a 2D or 3D orientation.

When creating the sketch entities, do not be concerned with their precise sizes. Dimensions that will control the size and position of the sketch will be added later. The following sequence of illustrations shows the steps used to create the sketch. Note how the cursor changes when various types of information are sketched.

Sketch creation, step 1.

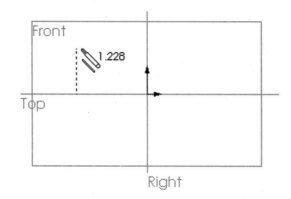

Sketch creation, step 2.

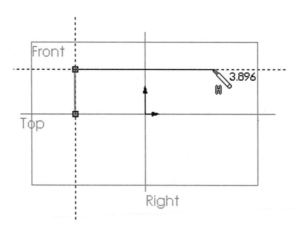

Sketch creation, step 3.

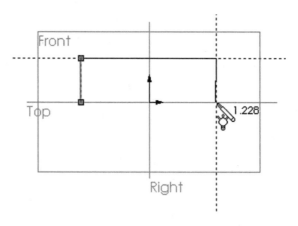

Sketch creation, step 4.

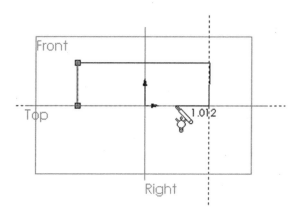

Sketch creation, step 5.

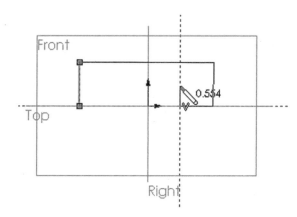

Sketch creation, step 6.

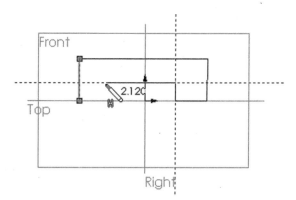

Sketch creation, step 7.

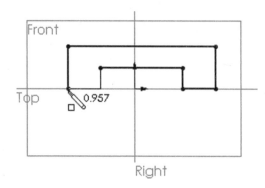

Now that all of the sketch entities have been created, as shown in the following illustration, the dimensions that control the size of the sketch can be added. Note the black sketch lines in the sketch geometry. This means the lines are constrained. This happened during sketching, and if you were paying attention to the cursor, you were already are aware of this. The black lines are constrained coincident with the origin point. Because the origin is anchored to one location, anything constrained to it will be further restricted in movement.

Completed sketch profile.

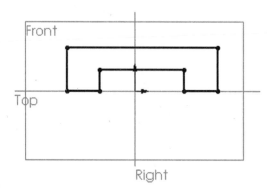

When adding dimensions, keep in mind which dimensions may need to be changed later. Placing dimensions is important because they drive the geometry of the sketch. This is not required in AutoCAD. If the SolidWorks user is concerned about the people in the manufacturing department having enough dimensions to build the part, consider this: When a sketch is fully defined, it needs no more dimensions to define the geometry within the sketch.

In addition, when a detailed drawing is created, all of the dimensions placed on the part will be brought into the drawing. Rest assured that there will be enough information on the drawing to complete the project. If not, you can add reference dimensions in the part or in the drawing.

The completed sketch has all sketch entities shown in black. This is another visual cue that all sketch entities are fully constrained. Fully constraining sketch geometry is not necessary, but is a good practice. Underconstrained geometry can sometimes behave erratically. The following illustration shows a fully defined sketch.

Fully defined sketch geometry.

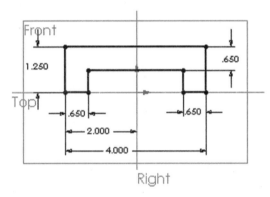

The next step would be to create the solid feature. This is explored in Chapter 6. The following illustration shows a base feature for a solid model.

Base feature for a solid model.

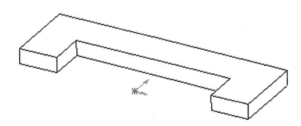

Summary

Sketching is the foundation of parametric, feature-based solid modeling. Building strong sketching skills is an important step for a new user and should not be taken lightly. The method used to create sketch entities, and to dimension and constrain geometry, will determine the design intent of the feature, as well as how easily the model can be modified.

Be aware of what the cursor and SolidWorks visual cues are telling you during the sketch process. Focus on how best to dimension and constrain sketch geometry to allow for easy editing and to capture the design intent of the feature.

6 Parts

Introduction

This chapter discusses how to create and modify solid features within SolidWorks. These features are described using graphical examples and step-by-step instructions. The 2D sketching techniques and tools learned in Chapter 5 will be applied in this chapter to create solid features that add or subtract material from a part. The order and method of creating these features determine the parametric characteristics of the part, and define the design intent imparted to the solid model.

There is a wide range of feature types explored in this chapter. Because SolidWorks is a design tool, this chapter will also explore the capabilities present in SolidWorks for making design changes to a feature's definition and dimensions.

Prerequisite

This chapter requires that you have a good working understanding of the basic sketch tools and techniques needed to create valid 2D sketch geometry. Defining geometry by adding dimensions and constraints should also be understood, along with the system feedback used in creating sketch geometry and the techniques used for understanding and correcting an over-defined sketch. How to correctly select specific entity types, such as an edge or face, is also necessary.

Content

The "Reference Geometry" section describes reference geometry (e.g., planes and axes) used within SolidWorks. Reference geometry used on a regular basis is covered early in the chapter, with some of the more obscure reference geometry commands covered later. The section "Types of Features" describes the various types of SolidWorks solid modeling features, both applied and sketched. The "Modifying a Feature" section describes how to modify an existing feature through various means.

Features can be changed, added, or removed—anywhere in the design process—after a part has been defined. The section "Advanced Part Topics" describes some of the more advanced topics related to part modeling. This section introduces these features and provides insight as to how they can be applied to a design.

The "Thin Features" and "Sheet Metal Parts" sections describe thin features and how SolidWorks can create a sheet metal part from a thin feature. SolidWorks contains functions specific to thin feature and sheet metal parts. The sheet metal functionality defines both the folded model and the flat pattern created by SolidWorks automatically from the thin feature part.

Objectives

After reading the "Reference Geometry" section, you should understand the uses of planes and axes and be able to identify the methods used to create these reference entities. With completion of the section "Types of Features," you should understand the various types of solid features and how they can be applied, and you should understand when different types of features should be created.

When you have completed the "Modifying a Feature" section, you should understand the various methods for redefining or editing existing features. You should also understand how and why features would be suppressed and how rollback can be used to organize features into functional groups.

Upon completion of the section on advanced part features, you should understand what design tables are, how to select a configuration created by a design table, what equations are and how they are defined, what a configuration is and how it can be used to define optional versions of a part or assembly, what a base part is used for, and how a new part is derived from an existing part.

When you have finished the "Thin Features" and "Sheet Metal Parts" sections, you should understand what a thin feature part is and how it is created, and the difference between a thin feature and a sheet metal part. You should also know the components of a sheet metal part and how they can be used to create, modify, and document sheet metal parts.

Part Planning

Because a solid model consists of many different features that have many complex dependencies, planning is required to ensure that a flexible part is created that incorporates all of the design intent. A parametric solid model requires more planning than a simple wireframe because of the many interdependent relationships that can be created.

The main elements of a part should be identified and prioritized. Specifically, a base feature and initial sketching plane should be determined prior to beginning a part. Functional features should be created in order of importance. Less important features should, when possible, be created near the end of the part. This helps minimize impact if these features are changed or deleted.

Similar design features should be functionally grouped. For example, if a set of drill holes are going to be placed around the perimeter of a part, and counterbores and chamfers added, it would make sense to add all of the holes first, then all of the counterbores, then the chamfers.

This is much more logical than adding a hole, then a counterbore, then a chamfer, then another hole, and so on, especially if all of the chamfers are the same size and can be added as part of the same feature definition. This makes the part more easily understood, mostly because like features are grouped and the part itself is created on the CAD system in a manner similar to its physical creation.

The Rollback function can be used to analyze a part, whether created by you or another individual. Rolling a part back accomplishes what many of us wish we could do when creating real prototype models, which is going back in time to add a feature you wish you had added earlier, or simply to see the process of how the features were created. When analyzing a part, the Rollback bar can step through the part one feature at a time, graphically displaying the part, feature addition by feature addition.

Because FeatureManager is a chronological list of features, Rollback can step back through time to a point before certain features were created. It would be great if we were all perfect planners, knowing exactly how a part should be built. However, we are not. With Rollback, if you forget to put in a feature, you can step back in time and add it at any step in the process. The model will update accordingly, as if the feature had been added at that point in time.

When creating a solid model, you need to determine how many features should be used to define a part, and in what order. Making a part with too many simple features can produce a part in which finding specific features later on becomes quite a chore. A part with a small number of complex features produces an inflexible model with sketches that are difficult to control. A good rule of thumb is to work between these two extremes. Group geometry functionally, keeping in mind the possibility of changes and modifications that may need to be made later.

General Feature Creation Order

As discussed in Chapter 5, the best plane should be chosen with which to create the first sketch. The first sketch should be a profile that will most accurately define the overall shape of the part. After this base feature is created, you create features that more significantly define the part's shape, such as bosses or holes. Adding features less likely to be modified early in the creation process will result in less likelihood of downstream effects causing unforeseen problems with other features.

Draft and shell features should be added at a fairly early stage because of their geometric nature. It is best to add fillets toward the end of the design process if this is considered feasible in the design of the part. Fillets add complexity to a part, making for a larger draw on system resources both computationally and graphically. In general, you should leave complex features such as threads or variable radius fillets for the end of the design process. If at all possible, try to maintain the following order when adding features once the base feature has been created.

- Create the main functional part features such as bosses and cuts. Create the most important features and those least likely to change.

- Insert any draft or shell features.

- Insert fillets and complex geometry last.

If the solid model will not allow you to follow the suggested creation order, do what you have to do. From a software standpoint, SolidWorks does not care about the order. The creation order is a suggestion only, not a requirement. When creating a solid model from an AutoCAD standpoint, the order in which geometry is added to the part is not as important an issue. Because there are no parametrics or external references to existing geometry, there are no internal conflicts that may arise within the part.

For example, if a rectangular base plate is created in AutoCAD, and a circular boss is then placed on top of that plate, it makes absolutely no difference if the base plate is removed or if the circular boss had been created before the base plate. In Solid-Works, the base plate could not be deleted because the boss is dependent on it. The order of creation of features is important for this and other reasons, which are explored in this chapter.

Types of Features

The first feature in a part is called its *base feature*. This is usually an extrusion but can be any sketched feature type. All subsequent sketched features that add material to a part are called *bosses*. If material is being taken away, the sketched feature is considered a *cut*. There are other feature types that do not require sketch geometry, and these features are known as *applied features* because they are directly applied to the solid model.

AutoCAD allows the creation of various types of primitive shapes, and offers various methods of creating extrusions and revolved features. In the SolidWorks sense of the word, these cannot be called features because the AutoCAD shapes do not have their own unique attributes the way a SolidWorks part feature does. That is, an AutoCAD solid does not contain information pertaining to dimensions, end condition specifications, external associations to existing geometry, and so on.

Simple base feature.

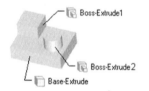

Boss features defined by extruding sketch geometry.

Boss

Cut

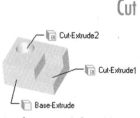

Cut features defined by extruding sketch geometry

The shape and orientation of the base feature (see illustration at left) should be chosen carefully because all other features are extensions of the base feature. The plane used to sketch the base feature's profile should correspond to the front, top, or right-hand face of the part.

This is not a requirement, only a good technique for new users who may find adjusting to the 3D qualities of solid modeling a bit confusing. If you do have a design in mind, decide what would be the best way to orient the design in space, then decide what the best plane for the base feature's profile would be and sketch on that plane.

Assume you wanted to create a simple part that has a small plate with various cuts and extrusions placed on it at various locations. In AutoCAD, the process of building a solid model would consist of creating solid shapes and then performing a number of Boolean operations. In SolidWorks, the same thing is accomplished by sketching the first profile and using that profile to create the feature. This process of creating a sketch and then defining the feature is performed repeatedly to build up the solid model. A simple rectangular profile could have been used to create the base feature shown in the previous illustration.

Instead of performing Boolean unions or subtractions, Solid-Works defines parametric bosses or cuts. These two categories of sketched features are described in the sections that follow.

A boss adds material to a part. A boss can be defined by using the Extrude, Sweep, Revolve, or Loft commands. The illustration at left shows three examples of simple boss extrusions, one of which is titled Base-Extrude because it is the base feature.

A cut subtracts material from a part. A cut can be defined by using the Extrude, Sweep, Revolve, or Loft commands. The illustration at left shows two examples of simple cut extrusions.

Bosses and cuts are almost complete opposites. Procedurally, these two commands are almost identical. Do not be misled by the terminology. Simply remember that a boss adds material and a cut takes it away.

Solid Features

Once the decision has been made whether to add or to remove material, it must then be decided what classification of feature should be used to complete the task. This probably sounds more difficult than it is. In most cases the choice is obvious once you know the four main feature classifications. The following is a summary of the information on feature types covered to this point.

- Features can be either bosses or cuts. Bosses add material and cuts remove material.

- Features can be sketched features or applied features. Sketched features require the user to create a sketch; applied features do not.

- Features are of four main types: extrusions, revolves, sweeps, and lofts.

Perhaps 90 percent of everything done in SolidWorks is based on these four feature types.

You also need to brush up on the five main steps involved in creating features. These steps will be repeated as each sketched feature is created and the solid model built.

1. Select a plane or planar face.

2. Enter sketch mode (by clicking on the Sketch icon).

3. Create the sketch.

4. Add dimensions and constraints.

5. Create the feature.

As discussed in Chapter 5, AutoCAD's function analogous to the sketch plane is the User Coordinate System (UCS). The UCS plane would be specified, and the geometry would be created, whether it be wireframe geometry or solid geometry. Accuracy is very important in AutoCAD because the ability to modify a dimension to reshape the part does not exist.

If working with a solid model, geometry must be added or subtracted to achieve the desired shape of the model. If a mistake is made, it can be very difficult to correct. Even if you were the most accurate and error-free AutoCAD user, design changes would still be a fact of life.

Modifying a solid model in AutoCAD would require major work. An example is a boss that needs to be resized and relocated. An additional solid would have to be added to the file and subsequently subtracted using a Boolean operation. An additional boss would then be created, and an additional Boolean union could then be performed. In SolidWorks, a dimension or two could be changed as needed.

For the descriptions of the four feature classifications that follow, it is assumed you have already completed the first four steps mentioned previously. In other words, the sketch has been created, is fully defined, and you are ready to create the feature.

Extrude

Extrude is probably the most commonly used of all features. This particular feature extrudes a profile a specified distance. Either a boss or cut can be extruded. The sketch plane should be either a plane or a planar face on which a 2D profile will be created. A closed profile, by default, will extrude as a solid feature. An open profile will always extrude as a thin feature. (Thin features are covered later in this chapter.)

Follow the general guidelines for sketching presented in Chapter 5. Summed up, these guidelines state that a sketch should be a closed, non-self-intersecting profile. There will be exceptions to this rule, which are discussed later, but in general the rule is a good one to follow. An extruded boss is shown in the following illustration.

Extruded boss.

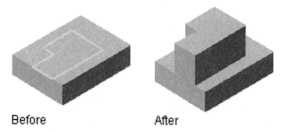

Before   After

To create an extruded boss feature from an existing sketch profile, perform the following steps.

1. Select Insert/Boss.

2. Select Extrude to open the Extrude Feature dialog box.

3. Select End Condition Type. Blind is the default selection, which is very common. Leave this set to Blind for now.

4. Specify a depth by entering a value in the Depth field, or use the up/down increment arrows.

5. Observe the preview on the screen and click on Reverse Direction if necessary.

6. Select OK to complete this function.

To create an extruded cut feature, perform the previous steps, but use Insert/*Cut*/Extrude instead.

End Condition Types

End conditions are nothing more than a way of determining how far a sketch will be extruded. The end condition types may vary slightly, depending on whether you are performing a cut or a boss, and depending on other variables (discussed later in this chapter). The following are the SolidWorks end condition types. The illustration that follows shows end type results.

Blind	Extrudes a profile a specified distance.
Through All	Extrudes a profile through the entire part. Most commonly used for cuts.
Up to Next	Extrudes up to the next surface encountered.
Up to Vertex	Extrudes up to a selected vertex point, which you must specify.
Up to Surface	Extrudes up to a surface you specify.
Offset from Surface	Extrudes a profile to an offset distance from a surface you specify.
Mid-Plane	Extrudes a part equally in both directions.

End conditions and what the results might look like.

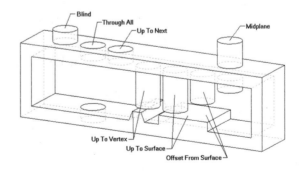

Other Extrude Attributes

There are other options that can be found in the Extrude dialog box, such as Reverse Direction, the function of which should be self-evident if you are looking at the preview onscreen. The following is a list of these options and the functions they perform.

Depth	Enter the extrusion depth or click on the up and down arrow buttons to increment or decrement the depth value. This is not always present, depending on the end condition type.
Reverse Direction	This is a toggle switch that reverses the extrusion direction.
Flip Side To Cut	This toggle switch is only available during a cut operation. It toggles the portion of the material to be cut away.
Selected Items	This box is only active if there are items to select, such as with the end condition types Up To Vertex or Up To Surface. You should select the appropriate entity type, which will be listed in this box.
Draft While Extruding	Check to add draft to the extruded feature.
Angle	Enter the draft angle or click on the up and down arrow buttons to increment or decrement the draft angle value.
Draft Outward	This is a toggle switch that controls whether a draft will be inward or outward.
Both Directions	Check to select whether the extrusion should be extruded in one or two directions. This is different from the Mid-Plane end condition type in that Mid-Plane allows for extruding in equal amounts in opposite directions, whereas Both Directions allows for specifying separate end condition types for either direction.
Settings For	When Both Directions is selected, the Settings For pull-down menu is used to define the properties for Direction 1 and Direction 2.

AutoCAD does have an Extrude command. If you are familiar with this command in AutoCAD, the equivalent SolidWorks functionality will seem similar. The Depth option in the Solid-Works End Condition dialog box is analogous to AutoCAD's "Height of extrusion" option while performing the Extrude command. Both programs have the ability to add draft during the extrusion process. There is no need to convert geometry into regions or anything of that sort in SolidWorks. The program is a solid modeler by nature; therefore, regions and wireframe geometry are not necessary.

Revolve

The Revolve function creates a boss or cut feature from a profile revolved about a centerline. A centerline entity must be used for creating this feature. A common mistake for a new SolidWorks user is to attempt to create a revolved feature with an axis. This will not work. In addition, a centerline must exist in the current sketch, not one reused from an existing sketch. A closed profile, by default, will extrude as a solid feature. An open profile will create a thin feature.

There are a few general rules regarding revolved features. First, there must be a centerline in the current sketch that will become the revolved feature. In addition, you will want to adhere to the following guidelines when creating a revolved feature. The illustration at left shows a revolved feature.

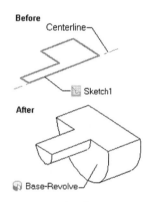

Boss revolved 180 degrees.

- The sketch should not touch the centerline at one isolated point. This results in invalid geometry. Think of an object having substance but no thickness. Such an object might exist in theory, but not in the real world, where mathematics define solid geometry. If the solid model requires mathematical definition with imaginary numbers, it cannot be created in SolidWorks.

- The sketch should not cross the centerline. This results in self-intersecting geometry.

Revolved features are previewed in the graphics area. If the Revolved Feature dialog box is obscuring your view, drag it out of the way. To create a revolved feature from an existing sketch profile, perform the following steps.

1. Select Insert/Boss.

2. Select Revolve to open the Revolve Feature dialog box.

3. Enter the angle of rotation.

4. Toggle Reverse to change the rotation direction if needed.

5. Select OK to complete the operation.

To create a revolved cut feature, perform the previous steps, but use Insert/*Cut*/Revolve instead.

End Condition Types

There are end condition types associated with revolved features, just as there are with extruded features, but not nearly as many. The following are revolved feature end condition types.

One Direction	Revolves a profile in one direction only. You must specify the angle.
Mid Plane	Revolves a profile in both directions at the same time. The angle you specify is the total angle revolved.
Two Directions	This allows you to specify angles for both clockwise and counterclockwise directions.

Other Revolve Attributes

The Revolved Feature dialog box contains an option to revolve as a thin feature. Thin features are discussed in a later section. For now, the following are the options you need to be aware of when dealing with the Revolved Feature dialog box.

Angle	Enter the angle to revolve the sketch profile.
Reverse	Click to reverse the revolve direction (clockwise or counterclockwise).
Revolve As	Click to select whether to revolve a solid or thin feature. This option is only accessible if you begin with a closed profile.

As with AutoCAD, SolidWorks requires a way of defining the axis of rotation when revolving geometry. AutoCAD requires two points to define the axis of rotation, whereas SolidWorks requires a centerline. You must also supply the rotation angle in each program. Keep in mind that SolidWorks does not accept negative dimensions. Instead, you must use the Reverse check box to alternate between clockwise and counterclockwise.

Sweep   A swept feature is created using a closed profile (sweep section) and a trajectory curve (sweep path). The profile (sweep section) is used to define the shape of the swept section, and the trajectory (sweep path) defines the direction and path of the sweep. Optionally, a guide curve can be used to control the sweep profile.

As with any other feature created with SolidWorks, there are a couple of simple rules that should be followed to ensure the feature is created without error. The Sweep function is considered a slightly more advanced feature of SolidWorks. There are a larger number of computations that must be performed, and sometimes the graphics are more difficult to display because of the shading requirements that need to be met. The guidelines to keep in mind while performing a sweep are as follows. The illustration that follows shows a simple swept feature.

- *The sweep trajectory start point must reside on the same plane as the sweep profile.* This condition does not always have to be established with a constraint. If the trajectory's start point just happens to be on the profile's plane, the sweep will work, as long as the other listed conditions are met. Sometimes, however, you may find it necessary to implement the Coincident constraint.

- Basically, you have two options. If you create the sweep path first, you must then position a plane at one end of the path with which to create your profile. Your other option is to create the profile first, in which case you must then create your path second, making sure to add a coincident relationship between the start of the sweep path and the profile's sketch plane. Either method is fine and the choice is totally up to you.

- *A sweep trajectory must not be self-intersecting.* This simply means that the trajectory should not come into contact with (cross over) itself.

- *A swept feature must not intersect itself.* This sounds similar to the previous guideline but is actually different. Imagine a helix. The helix would be spring-like in appearance and would not intersect itself. However, if a large enough circle were used as a profile (with a diameter greater than the pitch of the helix), the circle would intersect itself as it swept along the helix. This constitutes self-intersecting geometry, which is not permissible in SolidWorks.

➥ **NOTE:** *There is nothing wrong with features intersecting other features. It is when a single feature intersects itself that errors occur.*

The following is a step-by-step procedure for creating a swept feature, shown in the illustration at left. In this example, a boss will be created, but you could create a swept cut as well. Keep

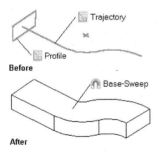

A simple swept feature.

in mind that two sketches must be created before a sweep can be performed.

1. Create the sketch that will define the profile. (A sweep profile must be closed. There is no option for creating a thin feature during a sweep operation.)

2. Create the sketch that will define the sweep trajectory. Usually this would be an open trajectory, but that is not necessary. For instance, a circle can be swept along another closed profile, such as a circle, to create a donut.

3. Exit from the final sketch.

4. Select Insert/Boss/Sweep to open the Sweep dialog box.

5. Click in the Sweep Section box and select the sweep profile from the graphics area or from FeatureManager.

6. Click in the Sweep Path box and select the sweep path from the graphics area or from FeatureManager.

7. Click on OK to complete the operation.

To create a swept cut feature, perform the previous steps, but use Insert/*Cut*/Sweep instead.

Sweep Guide Curves

Guide curves can be used to alter a sweep section. It is mandatory that the sweep section have a pierce relation added in order to attach the sweep section to the guide curve. If this were not the case, the sweep profile would have no association with the guide curve and would not know enough to "follow" the guide curve, thereby altering the shape of the profile.

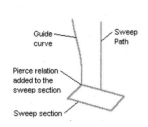

Elements of a swept feature using a guide curve.

A guide curve can be defined by a sketch or feature edge, or through the use of 2D or 3D curves defined by *x-y-z* coordinates. The sweep path and the guide curve do not need to be the same length. The length of the sweep is determined by the shorter of the two objects. More than one guide curve can be used to alter the shape of the sweep section. The previous illustration shows the elements of a guide curve sweep. To create a swept feature using a guide curve, perform the following steps.

1. Create a guide curve.

2. Create the sketch that will define the profile. It is required that the sweep profile be closed. Use the Pierce relation to pierce the sketch profile with the guide curve.

3. Create the sketch that will define the sweep trajectory. This profile is typically an open trajectory if the plan is to implement a guide curve.

4. Exit from the last sketch.

5. Select Insert/Boss/Sweep to open the Sweep dialog box.

6. Click in the Sweep Section box and select the sweep profile from the graphics area or from FeatureManager.

7. Click in the Sweep Path box and select the sweep path from the graphics area or from FeatureManager.

8. Select the Advanced tab.

9. Select the guide curve(s) from the graphics area or from FeatureManager to use for the sweep.

10. Click on OK to complete the operation.

Optional Guide Curve Creation

Creating a guide curve does not need to be a feat of technical expertise. A guide curve can be a simple sketch used to further define the shape of a profile as it sweeps along a trajectory. If needed, a guide curve can also be a set of coordinates that defines a curve through 3D space. This obviously results in a much more complex solid feature. To create such curves, refer to the section "Curves" later in this chapter.

Other Sweep Attributes

There are other options you may have noticed in the Sweep dialog box. Most of these options are not used on a regular basis. The Sweep command itself is not as commonly used as Extrude or Revolve. However, there are some features of Solid-Works that are just very nice to have around when needed. This section covers most of the additional options that can be implemented when performing a sweep.

There are two tabs contained in the Sweep dialog box: Sweep and Advanced. The Orientation/Twist Control selection box under the Sweep tab contains a number of options, and controls how the sweep profile moves along the path and guide curves. The options are as follows.

Follow Path	The profile will remain tangent to the sweep path.
Keep Normal Constant	The profile will remain parallel to the original profile sketch plane during the course of the sweep.
Follow Path and 1st Guide Curve	The original angle between the profile and path remains constant, and the twist is based on a vector between the path and the first guide curve (rarely used).
Follow 1st and 2nd Guide Curves	The original angle between the profile and path remains constant, and the twist is based on a vector between the first and second guide curves (rarely used).

There are additional optional settings in the Advanced tab. They are the following.

Up/Down	Click Up or Down to move the selected guide curve up or down the guide curve list box to change its order of priority. This may result in a somewhat different shape.
Show Intermediate Profiles	This option will calculate a number of intermediate profiles that represent cross sections of the solid geometry. It is essentially a preview of the swept feature. This gives you an idea of what the outcome of the sweep will look like without SolidWorks having to perform all of the calculations necessary to complete the entire operation.
Maintain Tangency	This option is only relevant if the entities in the sweep profile are tangent, meaning the individual line or arc segments in the profile. Maintaining tangency may reduce segmentation lines in the resultant swept surface.
Advanced Smoothing	If the sweep section has circular or elliptical arcs, checking this option will result in the final surface being somewhat smoother.
Start Tangency	Tangency condition settings only take effect if guide curves are used. None is the default setting. If Path Tangent is used, the sweep is tangent to the path (perpendicular to the profile) at the start or end of the path. If Direction Vector is used, the sweep is tangent to a selected curve. If All Faces is selected, the profile will be tangent to all faces of existing geometry (start of the sweep only).
End Tangency	See Start Tangency.

During a sweep, geometric conditions can sometimes cause a feature to either fail or not be produced. When defining the sweep section, sweep path, and guide curves, you may create geometry that cannot be geometrically solved. The other factor is control of tangency, orientation, and twist. These options can be changed to alter the sweep geometry created along the sweep and guide curves.

If the sweep section turns or twists too sharply, the geometry may not be valid. Usually, this is caused by self-intersecting geometry. Take a simple scenario, where a circle 10 millimeters in diameter is swept along a path. The center of the circle is positioned on the path. Now visualize the path having a bend in it that is less than 5 millimeters. In other words, the path contains a bend that is less than the circle's radius. During the sweep, the circle would begin to kink in on itself as it followed its tangent course along the path. Once again, you would have self-intersecting geometry, as shown in the illustration at left.

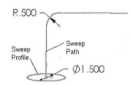

This geometry would result in a self-intersecting sweep.

The following are troubleshooting tips for dealing with failed sweeps.

- Review the sweep section, sweep path, and the optional guide curves to determine if the sketch geometry can physically produce a solid feature. See if it can be corrected by modifying the sketch geometry of the sweep section, sweep path, or guide curve.

- Ensure that the sweep path start point is on the same plane as the sweep section.

- Ensure that the correct sketches are listed in the Sweep Section and Sweep Path list boxes and that they are not reversed.

- If the sweep is a boss, make sure it touches existing geometry. If it is a cut, make sure it is not cutting empty space. Otherwise, a disjointed-feature error will be encountered.

Sweeping in SolidWorks is much more powerful than in AutoCAD. Sweeping a region in AutoCAD would normally be done with the Extrude command, specifying the Path option. A spline can be used as the path to create a variety of swept shapes. The profile is automatically aligned perpendicular with the path. These are not so much options as limitations.

There are no such limitations when creating a swept feature in SolidWorks. The Sweep command is a separate command and not just an extension of the Extrude function. A SolidWorks profile does not need to be perpendicular to the trajectory, guide curves can be used, and the profile can be swept tangent to the trajectory or with its orientation left constant throughout the sweep.

Loft   Loft is used to create a solid boss or cut from multiple closed profiles. Multiple guide curves can be optionally selected, as with the Sweep feature. A common "connection point" for each profile should be consistent to avoid twisting the solid.

The sketching planes for each profile do not have to be parallel, and each section does not need to have the same number of edges. For example, a loft can be created between a circle and a square, a series of ellipses, or between a triangle and a point, to name a few. Optionally, a guide curve can be used to define a loft direction. The guide curves can be used to alter the loft profile or as a guide for twist. A guide curve can be defined by a sketch, part edges, or construction curves. The illustration at left shows a lofted feature.

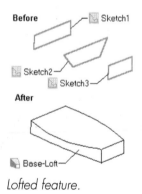

Lofted feature.

Creating a Lofted Feature

To create a feature defined by multiple closed profiles (a lofted feature), perform the following steps.

1. Create the sketch profiles. There must be at least two profiles, and they must be closed. Make sure you exit out of sketch mode after the last profile has been created.

2. Select Insert/Boss/Loft to open the Loft dialog box.

3. Select the sketch profiles in the order you want the loft to occur. The sketches can be selected from FeatureManager, but this is not recommended. If selected from FeatureManager, twisting may occur. If the sketch profiles are selected from the graphics area, you can select each profile near a common vertex and this will help SolidWorks determine the loft and how the profiles should be connected.

4. Select OK to complete the loft function.

To create a lofted cut feature, perform the previous steps, but use Insert/*Cut*/Loft instead.

Other Loft Attributes

Lofting is not very complicated from a user's standpoint. From a software standpoint, however, it is quite an achievement. There are options that allow you to define some very interesting shapes. The following is a list of these options and what they perform.

Profiles	Field displays the name of selected loft profiles. Pick inside this field to add or remove sketch profiles.
Guide Curves	Field displays the name of the selected guide curves. This is an optional field. Guide curves do not need to be used. Pick inside this field to add or remove guide curves.
Maintain Tangency	This option is only relevant if the entities in the sweep profile are tangent, meaning the individual line or arc segments in the profile. Maintaining tangency reduces segmentation lines in the resultant lofted surface.
Advanced Smoothing	Check if you want SolidWorks to approximate loft sections, thereby resulting in a smoother surface. This only works when there are arcs in the loft profiles.
Close Along Loft Direction	Check to connect the last profile to the first profile. This requires at least three profiles and the loft should be moving in a circular fashion, not linearly.
Up/Down	Click to change the order in which the profiles or guide curves are listed. This is usually determined during the selection process, but can be altered here.

Advanced Loft Attributes

The Advanced tab of the Loft dialog box allows the user to specify tangency conditions, much like when performing a sweep command. There is also the option of employing a centerline, which is very similar to a guide curve. Guide curves are used in helping to shape a profile as it is swept or lofted. Centerlines are more useful for simply pushing the loft along a certain direction. Centerlines do not have to be centerline entities. They can be lines or arcs, but are called centerlines because they often flow through the centers of similar segments in all loft profiles. The following list contains the options you will see in the Advanced tab of the Loft dialog box.

Center Line	Used to guide the loft direction as the loft blends between the profiles.
Show Sections	Only available if a centerline is used. Shows a preview of the loft cross sections. Use the slider bar to increase the number of sections shown in the preview.

Start Tangency	None is the default setting. If Normal To Profile is used, the loft is perpendicular to the profile at the start or end of the loft. If Direction Vector is used, the loft is tangent to a selected curve. If All Faces is selected, the loft profile will be tangent to all faces of existing geometry at the start or end of the loft.
End Tangency	See Start Tangency.

Following are some graphic examples to help you understand what the various advanced loft options can achieve. The following illustration shows a basic loft with the tangency conditions set to the default value None.

A basic loft operation between two sketch profiles.

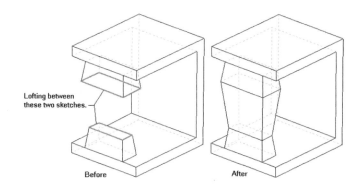

In the illustration at left, the tangency conditions have been set to Normal To Profile for both the start and end conditions of the lofted feature. If a more specific direction vector needs to be specified, the Direction Vector option can be used, which would allow the user to select a curve (typically a line or edge) to determine the tangency direction for the start or end profiles (not shown).

Both tangency conditions set to Normal To Profile.

In the following illustration, the tangency conditions have been set to All Faces for both the start and end tangency conditions. Notice how the surfaces of the lofted feature initially start off tangent to the faces surrounding the original sketch profiles.

Both tangency conditions set to All Faces.

Loft Centerlines

As mentioned in the previous section, the Centerline option can be used to help guide the direction the loft takes as it moves between profiles. Unlike the Guide Curve option, Centerlines do not require the Pierce constraint. For this reason, Centerlines are easier to use. In the illustrations that follow, you can see firsthand how centerlines can alter the shape of a lofted feature.

The first of the following illustrations shows what you might have before actually completing a loft. In the second of the following illustrations, the two semicircles are lofted together without employing the centerline option. The loft takes on a straight line from one profile to the other.

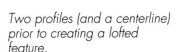

Two profiles (and a centerline) prior to creating a lofted feature.

The same two profiles after creating a lofted feature.

In the following illustration, the Centerline option is used, and the lofted feature takes on a much different shape. Notice how the loft seems to follow along the original curve.

The same two profiles as shown in the previous image, but additionally using the centerline.

Very interesting and complex shapes can be created using the loft process. However, it is not a foolproof procedure. The following are a few tips to help you when troubleshooting a failed loft.

- Review the loft profiles and the guide curves to determine if the geometry created is physically attainable. In other words, can it exist in the real world? It is possible to create geometry that is impossible to manufacture, but it is not possible to create geometry that cannot exist in the real world. Change the geometry to see if it can be corrected by modifying the sketch geometry of the loft profiles or the guide curves.
- Ensure that the profiles were selected in the correct order.
- Select profiles from the graphics area instead of FeatureManager.

- Ensure the loft profiles were selected using the same relative location on the profile sketches. The loft preview line should show you what will be the common vertex points between profiles.
- Loft profiles work best if all profiles have the same number of segments.

There is no counterpart in AutoCAD for a SolidWorks Loft feature. There is not even a valid workaround for creating similar solid shapes. About the best that can be done is to use some of AutoCAD's surfacing commands to try to achieve a shape similar to what might be created using the SolidWorks Loft command.

The following table describes the characteristics used to determine the solid feature used for a boss or cut. Use this list if you have a particular shape or profile in mind but are not certain how to best create the feature. Sometimes there is more than one choice for a particular shape or profile. This table is meant to get you pointed in the right direction, not as a rigid set of rules.

Options	Extrude	Revolve	Sweep	Loft
One profile	x	x		
One profile and one trajectory			x	
Multiple profiles				x
Cylindrical feature	x	x		
Spherical feature		x		
Toroidal features		x	x	x
Pyramidal or conical features				x
Project geometry along a curve			x	
Able to use optional guide curves			x	x
Drafted feature	x	x	x	x

Feature Names

Naming features makes a model easier to understand because the named features are then much easier to find in FeatureManager. Keep names short, and as meaningful as possible. Meaningful names help document the purpose and design intent of a feature. To rename a feature, perform the following steps.

1. Perform a slow double click on the feature name in Feature-Manager. This is identical to renaming a file in Windows Explorer.

2. Type in the name for the feature.

3. Press Enter (on your keyboard).

You may use spaces in feature names. Usually you can get away with using odd characters in names, but it is best to stick with alphanumeric characters, spaces, and underscores. Dimension names are typically D1, D2, and so on. Full dimension names include the name of the feature attached to them, almost like a surname. The following are the default system names for dimensions, features, and components.

- Dimension names are found in the Dimension Properties dialog box. Right click on the dimension to access its properties. The following are examples of the *full* dimension names.
    — D1@Sketch1 (if a sketch dimension)
    — D1@Feature_name (if a feature dimension)
    — If a dimension name belongs to an assembly, the name of the component the dimension belongs to must be attached to the dimension as well. An example of a full assembly component dimension name may look like the following.
    — D1@Boss-Extrude1@Part1
- Part feature names are found in FeatureManager. The following are examples of feature names. (This you should already know, and is included here for reference.) Slow double click on a feature name to rename it.
    — Base-Revolve
    — Cut-Extrude2
    — Shell1
- Assembly component names are found in the FeatureManager of Assemblies. The following are examples of component names. These would actually be the names of the part files, followed by the instance number. The instance number is appended to the name automatically by SolidWorks so that the software can tell components apart from one another.
    — Part1<1>
    — Widget<4>

By defining meaningful names, you can associate a feature's function, grouping, or purpose by reviewing the defined name and creation order. There is an option that prompts you for a feature name after it has been created. This option is selectable by checking the "Name feature on creation" field of the General tab in the Tools/Options menu. The following are examples of SolidWorks default names and what you might use for a new name.

- *Default name:* D2@Sketch2
  New name: Diameter@ScrewProfile
- *Default feature name:* Base-Extrude
  Renamed feature: Mounting Plate

The first example alludes to the fact that dimension names can be altered just as feature names can be. This is covered in the section "Design Tables" later in this chapter.

Because AutoCAD is not a feature-based program, there is no such thing as a feature, and therefore no need to name one. Likewise, there is no distinction between a feature and a sketch. An AutoCAD file is typically a wireframe, a 2D layout, or a solid model. Feature names are irrelevant.

Parent/Child Relationships

Parent/child relationships are created when a feature references an existing feature (e.g., sketch plane or feature edge), or when a new feature cannot exist without an existing feature. A *parent* is a feature other features have referenced or are dependent on. This reference could be defined by dimensioning to the edge or face, by using the Convert Entities or Offset Entities sketch tools, or by sketching the feature on the face of another feature.

A *child* is a feature that references another feature. Child features are dependent on parent features. If a parent feature is deleted, the children are not able to exist because they are dependent on the parent. The FeatureManager design tree is used to query existing parent/child relationships.

Unnecessary parent/child relationships should be kept to a minimum. When selecting edges or faces for dimensions or geometric references, you should be aware that a relation to the selected entity has been created. If the parent is removed, the child will no longer be able to exist.

Always select edges or faces in a view orientation that allows a clear view of what is being selected. Selecting an edge viewed by a normal projection may select the edge or face, or another feature directly behind the desired entity. Using an isometric orientation can produce better results, simply because it is easier for you to see what is going on.

Undesirable Relationships

If unwanted relationships do exist, the features can be redefined to eliminate the dimension or geometric reference that created the relationship. When investigating parent/child relationships, items to look for include a sketch plane, converted or offset edges, geometric relations, or dimensions. The easiest method for investigating parent/child relationships is to right click on the sketch or feature in FeatureManager and select Parent/Child. All parent/child relationships will be displayed. To display the parent/child relationships for a sketch or feature, perform the following steps.

1. Right click on the sketch or feature in FeatureManager.

2. Select Parent/Child.

3. Select OK to close the Parent/Child dialog box when you are finished.

To remove an unwanted parent/child relationship, you must remove any references to the parent geometry. This can sometimes be accomplished by editing the relations of a sketch (see Chapter 5). In other cases, it is impossible to remove the parent/child relationship. Take this simple example: if a hole is created through a block, the hole is a child of the block and the block is a parent to the hole. Without the block, there will be no hole. Removing references is not an option.

Parent/child relationships are a derivative of a feature-based program. It could be argued that AutoCAD has a form of parent/child relationships. For instance, if a hole were made through a block, the hole could not exist without the block. Therefore, the block must be a parent of the hole. This is a correct assumption in the AutoCAD environment, but the implications of this relationship are more substantial in SolidWorks.

Because features are created in a specific chronological order in SolidWorks, a parent feature exists at a place in time before the child. This in itself is not remarkable. However, the time a feature was created can in essence be altered so that it exists at an earlier stage in the design process. Features can be reordered to exist at different locations in time.

The parent/child relationship plays an important role in regard to this reordering capability. A child feature cannot be reordered to exist before its parent. AutoCAD does not have anything in its software code similar to this. In AutoCAD, Boolean operations can be performed at any time without regard to the order of events. As long as the desired shape is obtained, that is all that matters.

Reordering

Now that you have an understanding of parent/child relationships, take a closer look at reordering features. Reordering a feature is the act of moving a feature from one position in FeatureManager to another. A feature can be moved so that it occurs at an earlier or later position in time. What you cannot do is reorder a child feature so that it exists before its parent.

Before reordering a feature, sometimes it is best to check the parent/child relationships first (as described in the previous section). This will allow you to see how far up in FeatureManager you can move the feature in question. To reorder a feature, perform the following steps.

1. Place the cursor over the feature to be reordered in Feature-Manager.

2. Drag the feature to its new position. SolidWorks will automatically rebuild the part with the features reordered in the new sequence.

When dragging a feature to be reordered, pay close attention to the small white arrow that appears. The arrow will look as though it is pointing downward and to the left. What this is telling you is that the feature being reordered will be positioned below the feature on which it is dropped.

Other Feature Functions

The following sections describe other SolidWorks feature functions. These include holes, fillets/rounds, chamfers, shells, ribs, draft, and mirror.

Holes

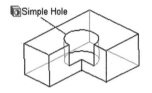

Simple hole.

The Simple hole function can be used to quickly insert a single hole on a model face. The Hole Wizard is used to define complex hole geometry (e.g., countersunk or counterbored) using a set of easy-to-use menu selections. In either case, the hole definition can be modified or redefined after insertion. The hole location can be precisely defined by dimensioning or constraining the sketch (typically the hole's centerpoint) either during or after insertion. Otherwise, the hole location is defined by the location of the "pick" when selecting the face on which to place the hole. A Simple hole is shown in the illustration at left.

Adding a Simple Hole

A simple hole can easily be added to a planar model face using the Simple hole function. The benefits of using Simple hole is that you do not have to worry about starting a sketch first, or even creating a sketch (albeit a simple circle). SolidWorks does this for you. To add a circular hole to a part, perform the following steps.

1. Pick on the face where the hole is to be created.

2. Select Insert/Features/Hole/Simple from the pull-down menus.

3. Specify End Condition Type, Depth (if required), and the Diameter.

4. Select OK to complete this function.

Addition of the Diameter field in the Hole Feature window.

You should have noticed when running through this routine that the dialog box for a Simple hole looks almost identical to that for extruded features. The main difference is the addition of the Diameter data field in the Hole Feature window (see the illustration at left).

It should be noted that after using this procedure to add a simple hole, the locating sketch must then be edited in order to accurately locate the center of the hole. You may decide it is easier simply to create a circle and use Cut/Extrude instead. This is totally a matter of personal preference and will not affect the outcome of your model in any way.

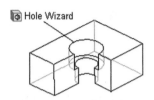
Hole Wizard

Counterbored hole using Hole Wizard.

Using the Hole Wizard

The following is a partial list of the types of holes that can be automatically created using Hole Wizard. The illustration at left show a counterbored hole created with Hole Wizard.

- Simple (this creates a simple straight-hole feature)
- Tapered
- Counterbored
- Countersunk
- Counterdrilled
- Simple drilled
- Tapered drilled
- Counterbored drilled
- Countersunk drilled
- Counterdrilled drilled

The following illustration shows a screen shot of the Hole Wizard using the countersunk drilled hole type option. In the case of a more complex hole type, Hole Wizard will save you steps.

Hole Wizard.

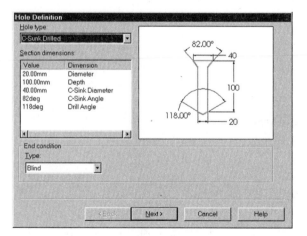

To create a hole using Hole Wizard, perform the following steps.

1. Pick on the face where the hole is to be created.

2. Select Insert/Features/Hole/Wizard from the pull-down menus.

3. Select the type of hole to be created from the Hole Type drop-down list.

4. Select the End Condition Type (again, from the drop-down list).

5. Double click on a dimension value in the Section Dimensions area to modify the necessary dimensions for the hole you are creating.

6. Select Next to continue. This will open up the Hole Placement window.

7. Locate the sketch for the hole using dimensions and/or constraints to position the locating point of the hole. Do this *before* you click on the Finish button.

8. Select Finish to complete the procedure and insert the hole.

There are two sketches created when using Hole Wizard. The first sketch is the locating sketch, which consists of nothing more than a point. This point can be dimensioned or constrained just like any other sketch geometry.

The second sketch is created by SolidWorks automatically using the parameters entered by the user. This sketch can be edited just like any other sketch in SolidWorks. The definition of the hole can also be edited if more significant changes need to be made. Use the right mouse button to edit the definition, and the original Hole Wizard dialog boxes will appear.

Other Hole Attributes

The various options that might appear in the Hole Wizard dialog boxes are totally dependent on what end condition type is selected. These options are identical to the End Condition options listed in the section covering extrusions earlier in this chapter.

The parameters that must be specified in the last dialog box are dependent on the hole type selected. For example, if a countersunk hole type is selected, a countersunk depth and diameter must be entered. These values are shown in a preview that tells you which dimensions are related to which parameters (see the previous illustration).

Again, AutoCAD has no such counterpart to the SolidWorks Hole Wizard, although there are third-party vender applications that will add this functionality. In order to reproduce the functionality Hole Wizard provides, the AutoCAD user would have to recreate the exact opposite of the hole to be created. It would probably be easiest to draw half the profile in AutoCAD, change the profile into a region, then revolve the region and perform a Boolean subtraction operation once the desired hole shape was moved to the appropriate location on the part. It is much easier, however, to plug the information into a dialog box such as those provided in SolidWorks' Hole Wizard and let SolidWorks do the work for you.

Fillet/Round

The Fillet/Round feature will create a fillet or round on specified part edges. SolidWorks does not care whether it is a fillet or a round, and there is no need to specify one or the other. If an interior edge is selected, a fillet will be applied, and if an exterior edge is selected, a round will be applied. There is absolutely no difference in the implementation of this command from a user standpoint. From this point on, the Fillet/Round command will simply be referred to as Fillet.

Using the Fillet Function

The creation order and grouping of fillets will affect how a part can be modified. This order and grouping will also determine tangency conditions for the fillet. Selecting multiple edges allows you to add fillets for all edges selected at one time. This works fine as long as all selected edges will have the same radius. Edges with fillets of differing radii must be added as independent features. Large radius fillets are usually added before smaller radius fillets.

Some fillets should be added to the model toward the end of the modeling process. Many fillets are inserted for cosmetic reasons and should not burden the model early in the design process. This will result in a performance gain. It may also limit unnecessary parent/child relationships and keep the model less complicated, which makes for easier editing.

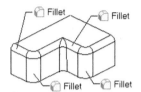

Filleted edges.

The creation order and grouping of fillet edges will determine the blending of fillets. Blending is the manner in which fillet intersections are formed. When inserting fillets, you should group edges that are similar in function and radius. Grouping too many fillets into the same feature can make the part difficult to modify. In contrast, if every edge were made into a separate feature, there would be too many features in the part. Try to find a reasonable middle ground. The illustration at left shows fillet features.

Fillet Feature dialog box.

When adding a fillet, it is possible to select either an edge or a face. If a face is selected, every edge on the face will be filleted. The edges or faces to fillet can be selected either before or after the Fillet command is entered. This is the case with nearly all SolidWorks commands. As a reminder, you need not hold down the Control key when selecting more than one entity if a dialog box is open. For this reason, it is easier to select entities after accessing the Fillet command. The Fillet Feature dialog box is shown in the following illustration. To insert a constant radius fillet, perform the following steps.

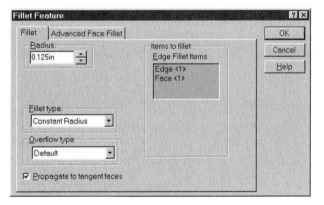

1. Select Fillet/Round from the Insert/Features menu.

2. Select the edges or faces to be filleted.

3. Enter the fillet radius.

4. Leave Propagate To Tangent Faces checked if the fillet should continue along every tangent face it encounters.

5. Select OK to accept the changes and close the dialog box.

Variable radius fillets can also be added to a part. The main difference in creating a variable radius fillet is how the radii are entered. To insert a variable radius fillet, perform the following steps.

1. Select Fillet/Round from the Insert/Features menu.

2. Select the edges to be filleted.

3. Select Variable Radius from the Fillet Type drop-down list.

4. Leave Propagate To Tangent Faces checked if the fillet should continue along every tangent face it encounters.

5. Select a vertex from the list box and enter the radius. This must be done for each vertex.

6. Select OK to continue.

Variable radius fillets have an additional option that allows for creating a smooth or straight transition. An example of each is shown in the following illustration. Descriptions for these options and others are given in the next section.

The difference between smooth and straight variable radius fillets.

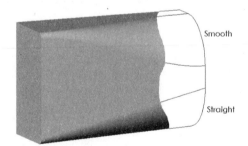

Other Fillet Attributes

Not all of the additional options available in the Fillet dialog box are straightforward and self-explanatory. The following is a list of these additional attributes, with explanations of each.

Propagate Along Tangent Edges	Check to continue the radius (fillet) along all tangent edges until a non-tangent corner is encountered. If this option is not selected, the fillet will continue in a linear fashion at each end of the edge selected until it runs out of material to fillet. Adjacent edges will be mitered as needed to blend with the filleted edge.
Smooth Transition	Available with the variable fillet type only. The tangent lines of the fillet will be parallel with the filleted edge at each end of the fillet.
Straight Transition	Available with the variable fillet type only. The radius varies linearly.

The following are Fillet Type attributes.

Fixed Radius	Fillet with one radial value.
Variable Radius	Fillet with multiple radial values. Radial values can be set for each vertex of the selected edges.
Face Blend	Extends two or more faces and fillets the intersection. This fillet type can swallow up entire faces without resulting in geometry errors.

The following are Overflow attribute types.

Default	The system selects which method to use, depending on the geometry selected.
Keep Edge	Blends the target surface smoothly, but the fillet surface may be broken. In other words, the fillet surface will change in order to accommodate the edge encountered by the fillet.
Keep Surface	Blends the fillet surface smoothly, but the target surface may be broken. The fillet's radius will remain unchanged, and the edge where the fillet is overlapping will change to accommodate the fillet.

Additionally, there is an Advanced Face Fillet tab that can be used when employing a Face Blend fillet. This advanced tab allows you to select Hold Lines that determine the radius of the fillet as faces are being filleted.

AutoCAD's Fillet command is similar to SolidWorks' in that edges are selected and a radius is supplied. AutoCAD's Chain option is similar to SolidWorks' Propagate To Tangent Faces option. When used, these options will continue the fillet along tangent edges. This is where the similarities end. SolidWorks allows you to select from a wide variety of options that control the attributes of the fillet, such as the overflow characteristics when a fillet runs off the face of a feature, thereby interfering with the fillet's face properties.

As previously mentioned, variable radius fillets and face blends that cannot be accomplished in AutoCAD can be created in SolidWorks. In addition, the Parasolids solid modeling kernel that SolidWorks uses is much more powerful than AutoCAD's. Try this simple experiment: Create a solid block in AutoCAD and fillet one edge at one-quarter inch. Now create another fillet using the Chain option and place it on the edges that wrap around one end of the first fillet. Give it a radius greater than one-quarter inch, and AutoCAD will return an error. SolidWorks will complete the Fillet without flinching, and blends the two fillets at the corners.

Chamfer

Chamfer will create a beveled edge on selected edges at a given distance and angle. Chamfers can be performed on multiple edges at one time, similar to the fillet function. The two parameters that must be entered are distance and angle or two distance values. If using the distance-angle method, the distance is the distance from the edge being chamfered, and the angle is the angle of the beveled edge as measured from the face where the distance was applied. If using the distance-distance option, the preview arrow indicates the direction in which the first distance measurement will be applied.

Using the Chamfer Function

Chamfer features do not slow down your computer's performance as fillet features do. This is largely because fillets require rounded edges with a large computational overhead, and chamfers do not. The illustration at left shows examples of chamfer features. To create a chamfer on selected edges, perform the following steps.

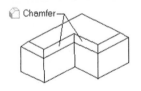

Chamfer features.

1. Select Chamfer from the Insert/Features menu.

2. Specify the chamfer type.

3. Select the edges or faces to be chamfered.

4. Enter the chamfer parameters, such as Distance and Angle values.

5. Checking Flip Direction will reverse the direction the distance is measured. In the case of the distance-distance option, the arrow indicates the first distance.

6. Select OK to continue.

Other Chamfer Attributes

There are a number of options available, depending on the type of chamfer added. As with fillets, it is easiest to select entities after the dialog box is open because the Control key does not have to be held down to select multiple entities. Likewise, selecting a face will chamfer all edges on that face. Chamfers will always propagate along tangent faces. Unlike the Fillet command, there is no switch for toggling face propagation on or off.

Flip Direction	Only available with the Angle-Distance option. Dictates the direction in which the distance measurement will be applied.
Equal Distance	Only available with the Distance-Distance option. Automatically makes the second distance equal to the first.
Vertex-Chamfer	The third type of chamfer listed in the Chamfer Type drop-down list. Allows for creating a chamfer on a vertex point and enables a third distance measurement, for which the user must supply a value.

AutoCAD release 14 does not have a vertex chamfer command. Instead, the user would have to modify the UCS and create a primitive solid that could be subtracted from the main body of the part. SolidWorks, on the other hand, provides you with a user-friendly dialog box in which to enter information.

Shell   The Shell feature creates a constant wall section on the inside or outside of a solid part. At least one face must be selected as the open face. The open face is the face removed during the Shell operation. Multiple shell thicknesses can be defined as an added option. In SolidWorks terminology, this is known as a *multi-thickness shell.*

Using the Shell Function

The most important decision that needs to be made with a shell is when to insert the shell feature. It is usually desirable to create a shell early in the design process. A shell essentially hollows out a solid model; therefore, it follows that if many features have been added, the shell operation will have to work overtime to shell out many features that do not necessarily need to be shelled.

SolidWorks is very intuitive when it comes to the shell process. For instance, if a rib has been added to a part and the rib has a thickness greater than half the shell wall thickness, the rib will be included with the shell. What happens if the shell wall thickness increases beyond half the rib's thickness?

One might think the shell operation would fail, but the software is intelligent enough to know it cannot shell the rib and ignores it. Nevertheless, play it safe and shell geometry as soon as you can. This will keep the part from becoming too complex before the shell feature is implemented. The illustration at left shows a shell feature.

A shelled part.

To create a shell, perform the following steps.

1. Select Shell from the Insert/Features menu.

2. Pick the faces to be removed during the shell process.

3. Enter the wall Thickness for the shell.

4. Select OK to continue.

Multi-thickness Shell

A multi-thickness shell creates a shelled part that has varying wall thickness. The wall thickness will not vary gradually. Rather, different faces can be chosen that will have specific wall thicknesses associated with them. To create a shell with multiple wall thicknesses, perform the following steps.

1. Select Shell from the Insert/Features menu.

2. Pick the faces to be removed during the shell process.

3. Enter the wall Thickness for the overall shell. This will be the default wall thickness.

4. Click in the Multi Thickness Faces box and select additional faces on the model that will require a different wall thickness.

5. Enter the wall thickness by highlighting the desired face in the Multi Thickness Faces box and specifying the thickness. Do this for each face listed.

6. Select OK to continue.

Other Shell Attributes

Another shell attribute is Shell Outward. This option specifies whether the wall thickness will remain on the inside of the part or be added to the outside. If wall thickness is added to the inside of the part, it is as if the part has been "hogged out" and a wall thickness left behind. If adding wall thickness to the outside of the part, it is as if the entire original model has been removed, and a wax coating of a desired thickness remains. The following guidelines might help if you are experiencing difficulties creating a shelled part.

- At least one face must be removed for a shell feature.
- Surfaces can be selected or deselected after the dialog box has been brought up.
- The face or faces for a multiple-thickness shell must be selected after the dialog box has been brought up.
- Click the cursor in the Multi Thickness Faces field prior to selecting faces that are not to have the default thickness.
- Prior to selecting OK, make sure there is a selected face in the Faces To Remove field.

AutoCAD does not have a shell command. It would be possible in AutoCAD to create a solid, copy it, rescale the copy, and then subtract it from the original, but this is a very crude workaround at best.

 Rib

The Rib tool uses an open sketch with which to create a thin-walled rib feature. The rib sketch does not need to be geometrically constrained to the walls of the part. The advantage to the Rib tool is that a very simple sketch can be used to produce this feature. SolidWorks does most of the work needed to create the rib.

Using the Rib Tool

The sketch geometry for creating a rib must be on a plane that intersects the solid geometry the rib will extend to. The rib sketch should be within the extents (boundary) of the current part geometry. In other words, the endpoints of the rib sketch should not protrude outside the existing part geometry. This is not an absolute requirement, but is good practice. The Rib command has limited extrusion end conditions compared with the

Extrude function, but allows for a bare minimum of input from the user in order to create the rib. The following illustration shows a rib feature and the sketch that was used to create it.

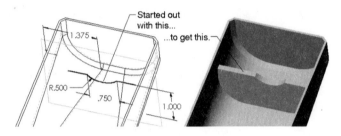

A rib feature.

There are only a couple of guidelines that must be followed in order to implement the rib feature. New users tend to find this function difficult, but it is easy once you have been through it a couple of times. The main point to remember is that the new rib will follow parallel to the sketch plane, and not extrude perpendicular to it. If you can remember this, half the battle is over. To insert a rib, perform the following steps.

1. Select a sketch plane that intersects the solid model. If necessary, first create a plane. This plane should extend in the same direction in which you want the rib to extrude (in other words, parallel to the rib).

2. Select the Sketch icon or select Sketch from the Insert menu.

3. Create a sketch that will define the height of the rib with respect to the face it should be extruded to. The sketch should fit within the boundaries of existing geometry. If sketching a simple line, its length is irrelevant.

4. Add dimensions or constraints as needed.

5. Select Rib from the Insert/Features menu.

6. Enter the Thickness for the rib and specify the direction for the wall thickness. Click on Next.

7. Enter the material direction by selecting Flip Side of Material and observing the preview arrow. Click on Next Reference to move the preview arrow from one sketch segment to the next. This is only relevant if the sketch contains greater than one segment and draft is being added.

8. Check Enable Draft if required and specify a value for the draft. If necessary, select Draft Outward as well.

9. Select Finish to complete the rib.

Other Rib Attributes

In its simplest form, the Rib tool is easy to implement. The previously listed steps take into consideration the possibility of draft and multi-segmented sketches. If you are creating a rib with a single line segment, the procedure is significantly simplified. All of the additional options encountered in the Rib dialog boxes are included in the list that follows, to eliminate any confusion.

Single Side	Check to create the rib on one side of the sketch plane only.
Reverse	Check to reverse the direction for a single-sided rib. The direction arrow indicates the direction in which the rib thickness will be applied.
Mid Plane	Check to create the rib split equally on both sides of the sketch plane. This is the most common setting.
Thickness	Enter the rib thickness by picking inside the field and entering the value, or pick the up/down arrows to increment/decrement the thickness value.

The second dialog box contains the following options.

Flip Side of Material	Check to reverse the extrusion direction for the rib. The preview arrow indicates the direction of extrusion. Obviously, this should be toward the part.
Next Reference	This option only applies to a rib sketch that has more than one segment and is having draft applied to it. Click Next Reference to move the preview arrow to the next sketch segment. The segment with the preview arrow is the segment that will measure the rib thickness specified in the previous dialog box (see Thickness).
Enable Draft	Check to create draft on the rib feature.
Angle	Enter the rib draft angle by picking inside the field and entering the value, or pick the up/down arrows to increment/decrement the draft value.
Draft Outward	Check this feature to draft outward. If drafting inward, keep in mind that drafting to the point of making the sides of the rib intersect each other will result in an error.

It is possible to create the same types of ribs in AutoCAD that SolidWorks creates using the Rib command. However, it is much easier in SolidWorks due to the high degree in which SolidWorks automates the routine. The dialog box is helpful and steps you through the process. When using the Rib tool, all you have to do is draw a line, add a dimension, and plug in a few parameters when asked. SolidWorks does the rest, going as far as including draft if needed.

Draft   Draft is used to create a tapered surface on selected faces from a neutral draft plane or parting line. Draft is used to add an angle to a part face. Draft is typically used by designers of injection-molded parts and castings to allow the tool to release from the part (or the part to release from the mold).

Neutral plane draft is defined by selecting a plane or face to be used as the neutral axis of the draft feature. Selecting a curve to define the axis of rotation for the draft face creates parting line draft. The Split Line function can be used to create a curve that can be used with the Parting Line function to create nonplanar parting lines. The following illustrations show a neutral plane draft and a parting line draft.

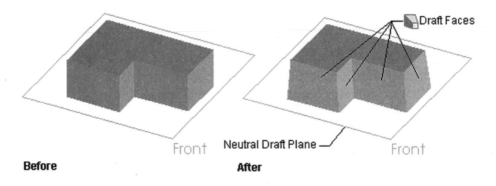

Neutral plane draft.

Parting line draft.

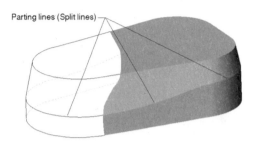

The terms *parting line* and *split line* deserve a little elaboration. The actual name of the command used to create a parting line is called Split Line. This is due to what is technically taking place. SolidWorks is physically taking a face and splitting it into two faces. Hence the name Split Line. However, the edges that are created are often used as parting lines.

Both the Neutral Plane and Parting Line methods of creating draft are included here. First, to create a drafted face using a neutral draft plane, perform the following steps.

1. Select Draft from the Insert/Features menu.

2. Select Neutral Plane as the Type of Draft.

3. Click in the Faces to Draft list box and select the faces to be drafted.

4. Enter the Draft Angle.

5. Click in the Neutral Plane list box and select a neutral plane that will define where the draft will begin. This neutral plane will also dictate the direction the mold will be pulled away from the part.

6. The Reverse Direction field can be checked to reverse the draft direction. As stated previously, the preview arrow points in the direction the mold will be pulled away from the part.

7. Select OK to create the draft.

The Split Line commands will be discussed later in this chapter. The method for creating draft using a Split Line (parting line) has been included here for reference. To create a drafted face using the parting line option, perform the following steps.

1. Select Draft from the Insert/Features menu.

2. Select Parting Line as the Type of Draft.

3. Click in the Parting Line list box and select the parting lines that will be used for drafting.

4. Enter the Draft Angle.

5. Click in the Direction of Pull list box and select an edge that will indicate the direction the mold will be pulled away from the part. You may optionally select a planar face whose perpendicular vector (or *normal*) will define the direction vector.

6. The Reverse Direction field can be checked to reverse the draft direction. The large preview arrow points in the direction in which the mold will be removed from the part. The smaller preview arrows point to the individual faces that will be drafted.

7. Select OK to create the draft.

The previous description shows how draft can be added as a separate feature. However, you have seen other ways in which draft can be applied. One method is to incorporate draft directly into sketch geometry. This can be done by using angled lines and then extruding the sketch. Draft can also be applied during the extrusion operation itself. (See the section on extruding in this chapter.)

Draft is not an option in AutoCAD, at least not as a command. There are valid workarounds for this function, and draft can be added when extruding a profile, but there really is no comparison. AutoCAD has no way of creating draft from a parting line that may be a curved parting line running around the outside of a curved part. Trying to create something such as this is a scary prospect in AutoCAD, but quite easily performed in Solid-Works.

Mirror   In Chapter 5 you learned that sketch geometry can be mirrored in SolidWorks. Feature geometry can also be mirrored. Additionally, there are options for mirroring an entire part to create a symmetrical part or to create a completely new mirror-image part. These functions are discussed in the following section.

Mirroring Features

The Mirror Feature function is used to create a copy of a feature mirrored about a plane or planar face. The new feature is a child of the parent feature. Any changes to the parent feature will be reflected in the mirrored feature upon rebuild. Mirror Feature can be used to mirror a single part feature or groups of features that have already been mirrored or patterned. When mirroring sketch geometry, a centerline suffices, but a plane or planar face must be used to create a 3D mirror of solid feature geometry. The following illustration shows a mirrored feature.

Mirrored feature.

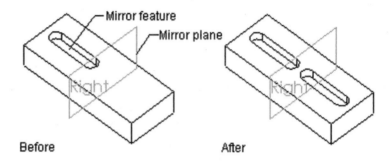

To mirror individual part features, perform the following steps.

1. Select Mirror Feature from the Insert/Pattern/Mirror menu.

2. Click in the Mirror Plane field and select a plane or a planar face from existing feature geometry.

3. Click in the Items to Copy field and select the features to be mirrored. More than one feature can be mirrored at once. A preview will be displayed.

4. Select OK to continue.

Geometry Pattern Option

There is only one additional option available when mirroring features and that is the Geometry Pattern option. Typically, when mirroring or patterning features the end condition of the features being copied is taken into consideration. This is not the case if the Geometry Pattern option is checked. All that is copied is an exact duplicate of the feature's geometry, but no end condition data is taken into account. This would make a drastic difference if, for example, you were patterning a feature whose end condition was Up To Surface.

If this concept is still difficult to conceive, picture this: when performing a geometry pattern, the copies will look exactly like the original feature. This is much more obvious when performing a pattern, rather than mirroring features. (Linear and circular patterns are discussed later in this chapter.)

If an AutoCAD model has not yet been turned into a complete solid, it is possible to mirror individual components in the model. However, because an AutoCAD solid does not consist of features, it is impossible to mirror specific portions of an AutoCAD solid. SolidWorks allows mirroring features at any time. Like AutoCAD, a plane is needed that can be used to mirror across. AutoCAD does not require an actual plane, just the representation of one, such as three points, which would define a plane. SolidWorks actually requires a plane entity or a planar face on a part.

Mirror All

Mirror All creates a mirrored feature that is a reversed copy of the active part and is attached to the original part about the mirror face. The new feature is a child of the parent feature. Any changes to the parent feature will express themselves in the mirrored feature upon rebuild. This function is used to define half a symmetrical part and produce the other half by mirroring the geometry. This can reduce the time required to model, change, or add features to symmetrical part models. A part face must be selected to define the mirror plane. The following illustration shows the Mirror All function in action.

A part created with Mirror All.

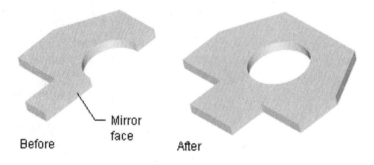

Before   Mirror face   After

The steps needed to perform this operation are quite simple and are as follows.

1. Select the planar face on the part to be mirrored.

2. Select Insert/Pattern/Mirror/Mirror All.

This command is most similar to AutoCAD's 3D Mirror command in that it will allow you to mirror a part about a plane to create a new part. In SolidWorks, the mirrored part must be attached to the original. This is not necessary in AutoCAD, and there is no relationship to the original, as in SolidWorks. If an entire part needs to be mirrored in order to create a new "left hand" part, SolidWorks' Mirror Part command, described in the material that follows, is necessary.

Mirror Part

Mirror Part is used to create a new derived part that is a reversed copy of the active part. This function requires a plane or planar face in order to create the new part. A new part is created, and the Save or Save As command can then be used to redefine the name and file location. Any modifications to the original part will also update the mirrored part because the newly created mirrored part is a child of (dependent on) the original parent part. As features are added to the mirrored copy, the original part is not updated. The following illustration shows a mirrored part.

A mirrored part.

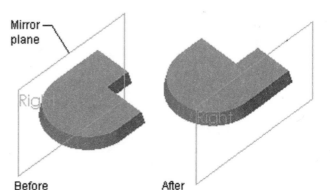

To create a mirror copy of a part, perform the following steps.

1. Select a face or plane that defines the mirror plane for the new part.

2. Select Mirror Part from the Insert menu. The new part will be created and become the active part. The existing part will not be altered in any way.

3. Select Save from the File menu to rename the derived part.

Be aware that the new part is dependent on the original. This means that moving, renaming, or deleting the original will make it more difficult or impossible to open the mirrored part. Solid-Works gives you an option to Browse for a parent file in such cases. However, if the original has been deleted, you will be out of luck, and the mirrored file might as well be deleted from the hard drive.

Pattern Features

A linear pattern creates linear copies of a feature by adding parameters that control the number, direction, and distance between the first feature (parent) and the copies (children). Linear patterns can be made in one or two directions at a time. A circular pattern creates copies of a feature by adding parametric parameters that control the number, direction, and angle between the first feature (parent) and the copies (children).

All features created by a linear or circular pattern are dependent on the parent geometry. Any changes to the parent geometry are reflected in the pattern features. Individual pattern members may also be deleted after the pattern has been inserted. This is discussed in the section "Pattern Instance Deletion."

Patterning features is more efficient than either creating a large number of features or creating one large feature through the use of a very large and often complex sketch. It is also quicker and easier to change a parameter and rebuild the model than to change a large number of individual features and dimensions. The patterned features are easier to modify simply by redefining the original feature. Pattern features also produce fewer features than creating each identical feature separately. The following illustration shows a linear pattern.

Linear pattern.

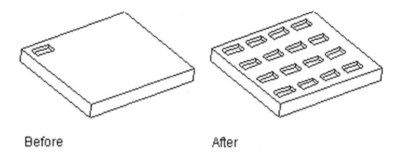

Before   After

To create a linear pattern, perform the following steps.

1. Select Linear Pattern from the Insert/Pattern/Mirror menu.

2. Select the features to be patterned. This is most easily accomplished by selecting the features in FeatureManager rather than in the sketch area, for the sole reason that items are named out for you, which means that selection is not impeded by view orientation or other entities on the part.

3. Select an edge or linear dimension that defines the vector for the First Direction. When this is done, the Edge/Dim Selected check box will contain a check mark. Reverse Direction can be checked to reverse the pattern creation direction. A preview arrow displays the pattern creation direction.

4. Enter the Spacing. This is the distance from the start of one instance to the start of the next.

5. Enter the Total Instances.

6. If a two-directional pattern (typically rectangular) is required, select Direction 2 and repeat steps 3 through 5.

7. Select OK to continue.

Circular patterns can be created in a similar fashion. However, one difference with this function has to do with the items that can be selected to determine the pattern direction. Because this is a circular pattern, a linear dimension cannot be used. Instead, an angular dimension, edge, or axis must be selected. For the sake of continuity, the Edge/Dim Selected check box retains the same name. A circular pattern is shown in the following illustration.

Circular pattern.

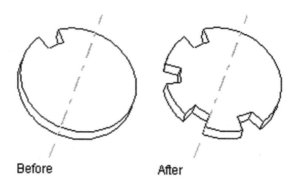

Before   After

To create a circular pattern, perform the following steps.

1. Select Circular Pattern from the Insert/Pattern/Mirror menu.

2. Select the features to be patterned. Again, for best results use FeatureManager to select features.

3. Select an axis, edge, or angular dimension that defines the center of the pattern rotation. When this is done, the Edge/Dim Selected check box will contain a check mark. Reverse Direction can be checked to reverse the pattern creation direction to either clockwise or counterclockwise.

4. Enter the Spacing. This is the angle between instances.

5. Enter the Total Instances.

6. Select OK to continue.

7. It is fairly difficult to mess up a pattern because SolidWorks shows you a preview. You can see what the outcome will look like before the pattern is created. What the preview may not show you, however, is the correct end condition of the features being previewed. The preview will usually appear as though you are creating a geometry pattern, whether or not the Geometry Pattern option is checked. To read more about the Geometry Pattern option, see the previous section on mirroring geometry and the section that follows.

SolidWorks' Pattern Attributes

The following list includes the pattern attributes previously discussed, as well as some that have not.

First Direction	The direction to which the values shown will be applied (linear only). Select Second Direction to define a linear pattern in two directions.
Reverse Direction	Pick to reverse the direction of the pattern.
Spacing	The spacing between pattern features.
Total Instances	Total number of features in the pattern, including the parent feature.
Items to Copy	Displays the numbers of features that will be patterned.
Vary Sketch	This is a somewhat obscure option. Sketch geometry for the feature being patterned must be constrained to existing edge geometry, typically by using the Offset Entities sketch tool. As the feature is patterned, its dimensions change to maintain the relationship with existing edge geometry. This option works with linear patterns only.
Equal Spacing	When checked, the Spacing option changes to Total Angle and the user can specify the total angle encompassed by the pattern. The spacing, in this case, would be dependent on the Total Instances.
Geometry Pattern	When checked, copies just the geometrical shape of the original features. End condition data is not taken into account.
Instances Deleted	Displays the deleted instances in the feature pattern. The feature number is displayed by row and column if linear, or by number if circular.

SolidWorks is very similar to AutoCAD when it comes to creating patterns. AutoCAD names this function differently, but it accomplishes nearly the same thing. AutoCAD's Array command is analogous to SolidWorks' Pattern. Features are the only entities that can be patterned in SolidWorks, however, and not sketch geometry. AutoCAD users might initially make the mistake of trying to pattern sketch geometry in SolidWorks. Make sure to turn any sketch geometry into a feature before patterning. After that, almost every other option is the same, such as number of instances to pattern, distance between instances, and so on.

AutoCAD's resultant pattern will be a set of individual entities with no connection to one another. SolidWorks will create parent/child relationships; therefore, any change to the original will result in the patterned features updating automatically. This is actually very similar to creating a block pattern in AutoCAD, modifying the block, and then redefining the block to have it update all of the occurrences of the block in the drawing.

Pattern Instance Deletion

Portions of a defined pattern may be omitted by deleting a pattern instance. After the pattern (linear or circular) has been inserted, pattern features can be deleted or added back to the pattern. The pattern feature to be deleted must be selected within the graphics window by selecting a pattern feature face. This is because individual pattern instances are not listed in FeatureManager and simply cannot be selected any other way. The following illustration shows a pattern deletion.

What a pattern might look like after having instances deleted.

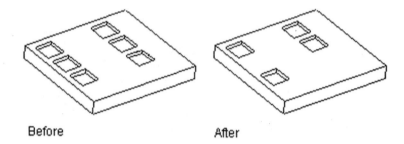

Before   After

To delete a selected pattern instance, perform the following steps.

1. Select a face on the patterned features to be removed.

2. Press the Delete key or select Delete from the Edit menu.

3. Select Delete Pattern Instances from the Options field (this is the default option). The following illustration shows the dialog box displayed when deleting a pattern instance.

4. Select OK to delete the selected instances.

Pattern Deletion dialog box.

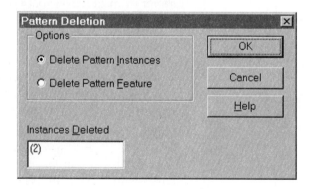

A very nice option available to anyone deleting pattern instances is the ability to bring the instances back again. To retrieve a deleted pattern instance, perform the following steps.

1. Right click on the pattern feature in FeatureManager and select Edit Definition.

2. Select the desired instance from the Instances Deleted field.

3. Press the Delete key or select Delete from the Edit menu.

4. Select OK to continue.

AutoCAD's Undo command is the closest function available that approximates SolidWorks' ability to retrieve deleted pattern instances. Where AutoCAD's undo information is stored in memory, SolidWorks' deleted pattern instance information is stored within the file. What this means to the user is that if you end your AutoCAD session you can say goodbye to your undo capabilities. SolidWorks' deleted pattern instances, on the other hand, can be retrieved anytime whatsoever. In summation, the following are points to keep in mind regarding linear and circular patterning of features.

- Any changes to the original feature are reflected in the patterned features.

- The number of features selected for the pattern appears in the Items to Copy field.

- Portions of the pattern may be deleted by selecting a face (rather than an edge) of an instance.

- The Instances Deleted field displays the deleted instance by number for circular patterns, and by row and column for linear patterns.

Reference Geometry

Reference geometry is used to define special features known as reference entities. These features are (for the most part) simple in nature and can be used for a variety of purposes. Reference geometry can be used to define and document important part and assembly faces, create curve or surface type entities, and many other very useful functions, some of which will be discussed in this section. All nonsolid geometry is considered reference geometry and can be found in the Insert/Reference

Geometry menu. The following are two important uses of reference geometry.

- When a part face is not available to act as the sketch plane, a plane can be created for the 2D sketch and the feature can be created from the new location.

- When a sketch face will be used by more than one feature, it is helpful to create a common reference plane because this minimizes parent/child relationships. All sketches will then reference the plane, as opposed to the face of an existing feature. Creating a common reference plane also allows you to name the plane, which helps document the design intent of the part.

Planes

Planes are reference entities used to define construction surfaces for sketching and model creation. Planes can also be used to define important part functions or mating surfaces. The default plane names can be changed under the Tools/Options menu by selecting the Reference Geometry tab. The names of the default planes can be defined so that all new parts (or assemblies) will have the plane names that correspond to the system views SolidWorks creates automatically.

The names should be changed to Front, Top, and Right (in that order). To change the names of the planes in the current part (not future parts), the slow double click method of renaming items in FeatureManager must be used. The default planes with typical names are shown in the following illustration.

Default planes.

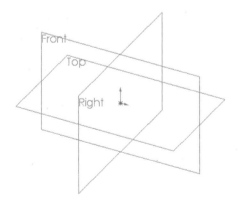

User-defined planes should be renamed to help document the intent of the plane. This visually helps anyone viewing the document to understand the use of the plane. For example, if a plane were defined to create a sketch for a mounting bracket feature, the name of the plane might be Mounting Bracket Plane.

The visual size of a plane's border can be changed by selecting the plane, then dragging the handles to resize the plane. It should be noted that the visual representation of a plane's border makes no difference as to where sketch geometry can be created. All planes theoretically extend infinitely in all directions.

To turn on or off the ability to view planes globally, select the Planes option in the View menu. To view planes or hide planes independently, right click on the plane and select Show or Hide.

Planes can be created through the use of existing geometry. For example, the laws of basic geometry state that any three points can be used to define a plane. This is one of seven creation methods available for defining a plane. Different geometry is required, depending on what method is used to define a plane. The following series of illustrations shows seven planes created seven different ways, each using its own geometry.

Plane offset from an existing plane:

An offset plane.

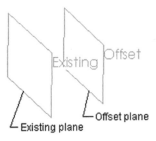

Plane inserted at an angle from an existing plane passing through an edge:

Plane created using the At Angle option.

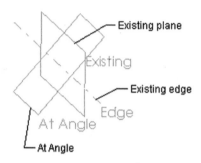

Plane using three sketched points or vertices:

Plane created using the 3 Points option.

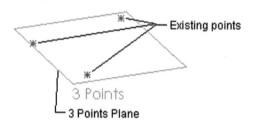

Plane parallel to an existing plane and passing through a vertex:

Plane created using the Parallel Plane @ Point option.

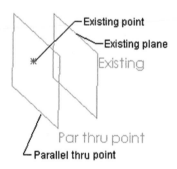

Plane passing through an edge and a sketched point or vertex:

Plane created using the Line & Point option.

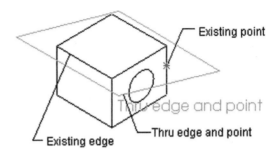

Plane perpendicular to a line (known as a curve) passing through a sketched point or vertex:

Plane created using the ⊥ Curve option.

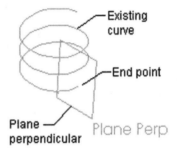

Plane tangent to a cylindrical or conical face at the position where a reference plane passes through the cylindrical or conical face:

Plane created using the On Surface option.

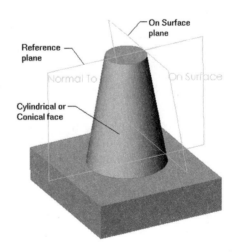

Resizing a plane has already been discussed, but for reference, the steps for resizing a plane follow. Remember that a plane's size is for visual aid only and that the plane actually extends infinitely in all directions. To resize or move a plane, perform the following steps.

1. Select the plane with the left mouse button.

2. Select one of the plane's resize handles with the left mouse button and drag to resize the plane. These handles appear at each corner of the plane boundary and at the center of each boundary segment. Select the plane boundary to move the plane instead of resizing.

AutoCAD does not make use of planes. The closest analogy would be modifying the UCS to create entities on a different plane. However, as an entity in the drawing database, a plane entity does not exist in AutoCAD.

Axes

An axis is used for dimensional reference and circular pattern creation. All cylindrical and conical features and surface sections have an axis. These axes are called *temporary axes*. Axes that are created by the user are referred to *as reference axes* (sometimes as *user-defined* axes). To display temporary axes, check Temporary Axes in the View menu. To display reference axes, check Axes in the View menu. At left is an image of an axis created with one of the six axis creation methods available in SolidWorks.

Axis

Axis created in SolidWorks.

An axis should be renamed to help document its intent or use. For example, if an axis were created to denote the *x*-axis centerline of a part, the axis might be named Centerline X. The extents (size) of an axis can be changed by selecting the axis and dragging one of its handles to change the axis length.

As previously mentioned, there are various ways of defining an axis, just as there are ways of defining a plane. The method used depends on the reference geometry available for creating the axis. The following are the various SolidWorks methods of creating an axis, with a graphical representation of the entities used to create each axis.

Axis created using a temporary axis:

Axis created using the One Line/Edge/Axis option.

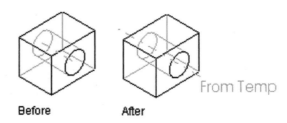

Axis at the intersection of two planes:

Axis created using the Two Planes option.

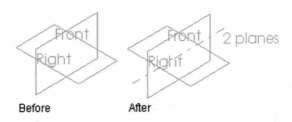

Axis using two points:

Axis created using the Two Points/Vertices option.

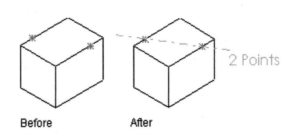

Axis using a cylindrical surface:

Creating an axis using the Cylindrical/Conical Surface option requires selecting one cylindrical or conical surface.

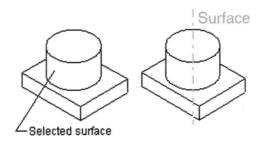

Axis using a point and a surface:

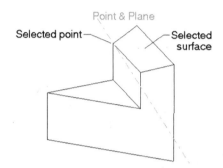

Axis perpendicular to a surface through a point using the Point & Surface option.

The following are other facts to keep in mind when creating or using an axis.

- Axes can be resized and renamed. Temporary axes cannot.

- Axes can be shown or hidden just like planes. Temporary axes cannot.

- Reference and temporary axes may be used for adding reference dimensions, relationships, plane creation, and assembly constraints.

- Axes are displayed with a "+" symbol when viewed normal to an existing axis (i.e., head-on).

Coordinate Systems

There are occasions when it is desirable to reference another set of coordinates other than SolidWorks' default *x-y-z* axes and coordinate system. For example, when checking the material properties of a part, SolidWorks uses the part's origin point to determine properties such as the moments of inertia and center of mass. By defining a different coordinate system, you can tell SolidWorks to reference an *x-y-z* reference point other than the default world coordinates.

To create a user-defined coordinate system, you must specify a vertex point as the new *x-y-z* (0,0,0) reference point and edges that will represent the *x*, *y*, or *z* axes. To create a user-defined coordinate system, perform the following steps. The illustration that follows shows the Coordinate System dialog box.

Coordinate System dialog box.

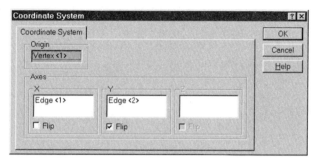

1. Select Reference Geometry/Coordinate System from the Insert pull-down menu.

2. Select a vertex point to establish the new 0,0,0 reference point.

3. If desired, select an edge to define the *x* axis. You may also choose to define the *y* or *z* axis in the same way. Use the Flip options for any of the axes in order to change their directions if required.

4. Select OK to create the new coordinate system.

Once the new coordinate system is defined, it will appear in FeatureManager. Make sure you have enabled the display of Coordinate Systems globally in the View menu, or you will not be able to see your new coordinate system.

Surfaces

Surfaces can be defined or imported to define reference surfaces for solid feature creation. For example, surfaces can be used as a depth reference for an extruded or cut feature, thickened to create solid geometry, and used to cut through existing geometry. Surfaces and solid features both use 2D sketches to define their geometry. The surfacing functions are located in the Insert/Reference Geometry menu.

The four basic surface creation functions (extruded, revolved, swept, and lofted surfaces) use the same creation procedures as their solid feature counterparts. Take, for example, the Extruded Surface function. If a sketch is created and this operation is begun, the same dialog box used for creating an extruded solid feature is opened. However, you still need to determine an end condition and depth, and whether or not draft should be applied. Refer to the "Types of Features" section in this chapter for a further explanation of the various dialog boxes encountered when creating extruded, revolved, swept, or lofted features.

For some time now, AutoCAD has offered the ability to create a variety of surface types. Some very complex and interesting surfaces can be created. SolidWorks also has the ability to create a variety of surfaces, but what can actually be accomplished with those surfaces is somewhat limited. This holds true for AutoCAD as well. There is a distinct difference in the nature of the geometry, and in the nature of software programs, when discussing solid modelers versus surface modelers. Normally, the two do not mix.

Combination surface/solid modeling packages are usually the domain of high-end software. AutoCAD has an add-on package called AutoSurf that allows for the creation of surfaces. There are third-party add-on surface creation packages for SolidWorks as well that increase the functionality of the program. Basic surfacing, however, is already available in SolidWorks. The series of illustrations at left and following shows examples of what the various SolidWorks surface types might look like.

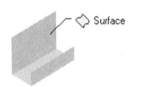

Extruded surface.

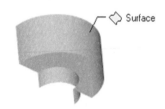

Revolved surface.

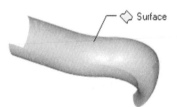

Swept surface.

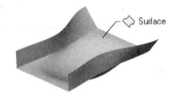

Lofted surface.

Planar Surfaces

A Planar surface is one of the easiest of the surface types to create in SolidWorks. Its name implies the surface type, and this type of surface is very simple by nature. It is possible to use any existing feature edges to define the new planar surface. The only requirement is that the edges all reside on the same plane. To create a planar surface, perform the following steps.

1. Select Reference Geometry/Planar Surface from the Insert menu.

2. Select the edges you want to use to define the planar surface. The edges must reside on the same geometrical plane.

3. Select OK to complete the operation.

Offset Surface

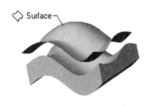

Offset surface.

An offset surface does not have a solid counterpart; therefore, the creation of such a surface will be explained here. Offset surfaces require an existing surface or solid face from which to offset. In the illustration at left, the top face of the solid part was offset a distance above the part to create the surface. To create an offset surface, perform the following steps.

1. Select the face or surface to be offset. If adjacent faces are selected, they will remain adjacent after the offset.

2. Select Insert/Reference Geometry/Offset Surface.

3. Specify an offset Distance.

4. Use the Reverse toggle if necessary to reverse the offset direction.

5. Click on OK when done.

Imported Surface

Imported surfaces are usually imported as IGES files; however, ACIS (.SAT) and Virtual Reality Modeling Language (VRML =*.WRL) files can also be imported using this particular function. Files imported in this fashion are left as surface files that can be used to edit existing solid geometry or to create new solid geometry. Generally speaking, single surface files are brought in using this method. Multiple surface IGES files, when imported via the IGES File Open command, are knitted by SolidWorks to form a solid.

↝ **NOTE:** *This solid creation for file import purposes is covered in more detail in Chapter 11.*

Radiate Surface

One of the more powerful and most valuable of the surfacing commands available in SolidWorks is the Radiate Surface command. With Radiate Surface, it becomes possible to radiate edges of a model outward to form a new surface. This new surface can prove invaluable to mold-makers, because it can be used to accurately cut a mold at the desired parting line. The radiated surface need not be created from simple straight lines. Curved parting lines of any sort can define a radiated surface, an example of which is shown in the following illustration.

Radiated surface from a nonlinear parting line.

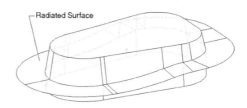

Very often the Radiate Surface command is used in conjunction with the Knit Surface command, described in the next section. It is also typical to create a radiated surface from a parting line that was defined using the Split Line function, described later in this chapter. You may want to skim over these other sections to get a feel for these other commands and how they relate to each other. To employ the Radiate Surface command, perform the following steps.

1. Select Reference Geometry/Radiate Surface from the Insert pull-down menu.

2. Select a Reference Plane. The radiated surface will radiate parallel to the Reference Plane. The large preview arrow will point perpendicularly to the radiate direction.

3. Click in the Edges to Radiate list box and select the edges that will be radiated. You will see a small preview arrow for each edge selected. These smaller preview arrows will point in the radiate direction.

4. Specify a Radiate Distance.

5. Select OK to complete the operation.

There are two options associated with the Radiate Surface command. These are described in the following table.

Propagate Along Tangent Faces	Checking this option allows for selecting one edge and having SolidWorks select all other edges that are tangent to the first. This greatly simplifies the selection process.
Reverse Direction	Reverses the direction the selected edges are radiated.

Knit Surface

The Knit Surface command is another commonly used surface command that makes mold making a much less tedious process. Knit Surface does exactly what its name implies: it knits multiple surfaces into one larger surface. This "knitted" surface can then be used to extrude up to when creating a mold, or it can be used to cut a mold.

To knit surfaces, you must select them from the work area or from FeatureManager. When creating a surface from many smaller surfaces for use with mold making, it is beneficial to use the Seed Face option. When using the Seed Face option, all adjacent faces used to form the knit surface are selected automatically. The only requirement prior to using a seed face is that the Radiate Surface command has been performed previously.

In the following material you will take a look at two methods of creating a knitted surface. The first method involves the basic Knit Surface command, and the second method shows the special characteristics the Knit Surface command exhibits when employing a radiated surface. To knit surfaces, perform the following steps.

1. Select Reference Geometry/Knit Surface from the Insert pull-down menu.

2. Select the faces and surfaces to be knitted. They do not have to be on the same plane, but they must touch.

3. Click on OK to complete the process.

At this point you will notice a new feature in FeatureManager called Surface-Knit. This surface feature will consist of all selected faces and surfaces. To create a knitted surface using the Seed Face option, perform the following steps.

1. Create a Radiate Surface (see steps in the previous section).

2. Select Reference Geometry/Knit Surface from the Insert pull-down menu.

3. Select the radiated surface from FeatureManager or the work area as the surface to be knitted. The Seed Face list area will become active.

4. Click in the Seed Face list area and select one face on the model. All faces adjacent to this face will be selected, and this selection process will propagate all the way to the radiated surface. Automatically selected faces will not appear in the list box.

5. Click on OK to complete the process.

The benefit of using a seed face is that you do not have to select many faces to be knitted. SolidWorks takes care of that task for you. AutoCAD R14 really has nothing similar to this command. To achieve surfacing capabilities similar to SolidWorks, an AutoCAD user would have to use an add-in program, such as AutoSurf. The Mechanical Desktop software created by AutoDesk also has surfacing capabilities built into it.

Projected Curves

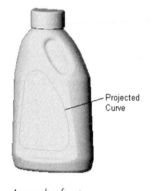

Projected Curve

A result of using Projected Curve.

The Projected Curves command projects a sketched curve onto a model face. This curve can be used to define sweep trajectories and to project sketches onto a nonplanar face, to name two possibilities. In the illustration at left, Projected Curve has been used to project a sketch onto the nonplanar face of a container. To project a curve onto a face, perform the following steps.

1. Create a sketch. The profile can be open or closed, as required.

2. Exit the sketch.

3. Select the sketch and the face to be projected to. Remember to hold down the Control key when selecting more than one entity.

4. Select Projected Curve from the Insert/Reference Geometry menu.

It is also possible to create a 3D spline by projecting two orthogonal 2D curves into one 3D curve. The resultant curve will have the shape of each of the separate curves when viewed perpendicular to their respective sketch planes.

Helical Curve

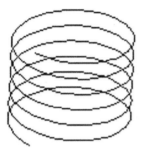

Helix curve.

The Helix function creates a reference curve that can be used to define a sweep trajectory for thread profiles and springs. A helix can be used as a trajectory for a cut or boss sweep. A circle is used to define the diameter of the helix. The helix can then be defined by specifying the pitch and revolution, height and revolution, or height and pitch. It should also be pointed out that a spiral can be created from within the Helix dialog box.

The sweep sketch profile must start at the endpoint of the sweep trajectory (the helix) or be pierced by the helix when performing the actual sweep once the helix has been created. If a plane does not exist at the start point of the helix, one must be created before sketching the sweep profile. Use the Perpendicular Curve option to create a plane for a helical sweep, as described earlier in this chapter. A helix curve is shown in the illustration at left. To create a helix, perform the following steps.

1. Sketch a circle that defines the diameter of the helix.

2. Select Helix from the Insert/Reference Geometry menu.

3. Specify the parameters for defining the helix (see the following section).

4. Select OK to create the helix.

Helix Properties

Before creating a helix, it would benefit you to know a little bit more about the parameters involved. The following list of helix properties available in SolidWorks should help you on your way when creating a helix.

Defined By	Defines which combination of pitch, height, and revolution will be used. The three alternatives are Pitch & Revolution, Height & Revolution, and Height & Pitch. You may also define a spiral.
Pitch	Pitch, with reference to threads, is the distance from thread peak to thread peak. The distance on the helix is measured along the height of the helix from revolution to revolution (coil to coil).
Height	Total height of the helix.

Revolution	Number of helix revolutions.
Taper Helix	Check to apply a taper angle to the helix.
Angle	Taper angle. Enter a new value in the field or use the up/down arrows to increment/decrement the value.
Taper Outward	Check to taper the helix outward rather than inward.
Starting Angle	Angle to start the helix measured with respect to the circle used to define the helix diameter. Enter a new value in the field or use the up/down arrows to increment/decrement the value.
Reverse Direction	Check to reverse the helix creation direction.
Clockwise	Check to create the helix in a clockwise direction.
Counterclockwise	Check to create the helix in a counterclockwise direction.

There are usually some very good AutoLisp routines that can be found to create geometry such as spirals (see the following section) and helix shapes. However, there is no counterpart command to create a helix from within the AutoCAD program itself. In addition, generally speaking, most Lisp routines require command line input and do not make very good use of dialog boxes, such as the dialog box for creating a helix or spiral in SolidWorks. Once again, these curves are parametric and can be easily modified in SolidWorks.

Spirals

Spirals are created from within the Helix dialog box. This option is found in the Defined By drop-down list box. If Spiral is selected, only Pitch and Revolution need to be specified. CW or CCW can still be specified. Clicking on Reverse will create the spiral inside the circle instead of outside. Starting Angle has the same effect it does when creating a helix.

Split Lines

Split Line is used to create a nonplanar curve for the Draft feature, to create variable radius fillets on edges without vertex points, and anywhere it is necessary to break a single surface into two or more surfaces. The Parting Line option in the Draft command will use the curve created by the Split Line function to create draft. The split line can be defined on multiple part faces at once. Many designs require the parting line of the injection-molded or casting tool to jog the mating surfaces between the ejector and stationary side of the tool. The illustration at left

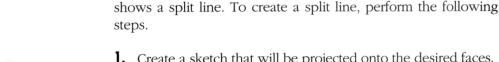

shows a split line. To create a split line, perform the following steps.

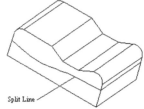

A split line that could be used for adding draft.

1. Create a sketch that will be projected onto the desired faces. The sketch should extend beyond any edges to be projected to, or at least be coincident with those edges.

2. Select Split Line from the Insert/Reference Geometry menu.

3. Select the Projection split line type.

4. Select Next to continue.

5. Click in the Sketch to Project list box and select the sketch (this is only required if you were not in an active sketch previous to entering the Split Line command).

6. Click in the Faces to Split list box and select the faces to be split by the sketch.

7. Click on Single Direction if projecting in only one direction.

8. Ensure the projection direction preview arrow is pointing toward the part.

9. Select Finish to complete the process.

You may have noticed when walking through the steps for creating a split line that there was another type of split line in the first dialog box, named Silhouette. The Silhouette method allows for creating a split line on a cylindrically shaped part without the need to first sketch something. When using the Silhouette option, select a cylindrical face you want to split, and select another plane or face that will represent the direction of pull. The direction of pull will be perpendicular to the plane or face selected.

Make sure you are not in a sketch when trying to use the Silhouette option. You can predict where the split lines will be because they are the same as the silhouette edges of the cylindrical face being split with respect to the direction of pull. In other words, if you view the part using the direction of pull plane as the Normal To plane (see "Introduction to View Orientation" in Chapter 3), the silhouette edges of the part will be the split lines.

Curves

There are two methods of creating curves through points in space in order to achieve a 2D or 3D curve: Curve Through Reference Points and Curve Through Free Points. Curve Through Reference Points is used to create a curve based on existing points in a part. Curve Through Free Points is used to create a curve from a set of data points you specify, or from a set obtained through a file. These points can be read from an external text file with a TXT or SLDCRV extension. The following illustration shows a curve created through reference points.

Curve created through reference points.

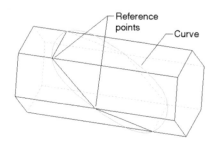

Make sure you are not in an active sketch before attempting to insert a curve using either method described in the following material. Otherwise, the functions will be grayed out. To create a curve through a set of existing points, perform the following steps.

1. Select Curve Through Reference Points from the Insert/Reference Geometry menu.

2. Select the points that define the curve. These can be vertex points or sketch points.

3. Select OK to continue.

To create a curve from an external point file, perform the following steps. An external point file list is shown in the illustration that follows. It is the dialog box you will see when initiating this command. Make sure to examine the preview displayed on your screen as points are entered in the dialog box.

1. Select Curve Through Free Points from the Insert/Reference Geometry menu.

2. To begin entering points, double click anywhere in a blank line (below the XYZ headings) and begin entering the *x-y-z* coordinates for the curve.

3. If importing coordinates from a text file, click on the Browse button. The coordinates will be imported once the file is selected.

4. Highlight a row and click on Insert if another row of *x-y-z* coordinates needs to be inserted into the existing set. The new row will appear above the highlighted row.

5. If coordinates need to be deleted, highlight the row and press the Delete key.

6. Select OK to continue.

Creating a curve "on the fly" by entering external reference points.

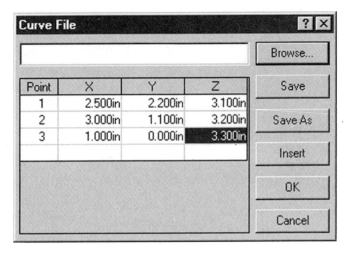

This is an area in which AutoCAD outperforms SolidWorks. Creating a 3D spline entity in AutoCAD is as easy as entering the Spline command and punching in coordinates onscreen. Grips can be used to modify the spline. SolidWorks does offer a way of creating 3D splines, and it is similar to AutoCAD's method, but the process requires a few more steps. There is no way of "dragging" the spline geometry in SolidWorks, although the definition of the spline can be modified and coordinates changed on the fly.

Modifying a Part

This section describes the functionality available within Solid-Works to change the definition or references for an existing feature. One of SolidWorks' strengths is the ability it gives you to change or modify a feature once it has been created. The ability to easily redefine a part or assembly can mean the difference between being productive and having to redo existing work.

This area is where the main differences between AutoCAD and SolidWorks really start to become apparent. Because of Solid-Works' parametric nature, any dimension (for that matter, every aspect of a part definition) can be accessed at any time and modified to reshape a part. The ease and flexibility this provides is essential to a designer. AutoCAD does not have anything equivalent to this functionality, but in all fairness, AutoDesk's Mechanical Desktop software does have parametric capabilities.

Edit Definition and Other Edit Functions

The properties and attributes used to define a feature (i.e., end condition, depth, values, creation direction, and so on) can be changed after the original feature has been created. The same dialog box that came up during the creation of the feature is used for Edit Definition. This is a significant advantage because there is no new dialog box or set of commands that must be learned strictly for editing a feature's definition. To change a feature's definition, perform the following steps.

1. Right click the feature in the FeatureManager design tree.

2. Select Edit Definition.

3. Redefine the feature's original properties as desired.

4. Click on OK when finished.

It should be noted that you can also right click on the feature, or a face belonging to a feature in the work area, in order to bring up the Edit Definition menu item. This is an alternative to right clicking on items in FeatureManager, but right clicking in FeatureManager usually works better for the simple reason that every feature is spelled out for you and you know exactly what you are right clicking on. Right clicking on faces in the work area to edit a definition sometimes accesses the definition of an unexpected feature. This is because the face you right clicked on might belong to a different feature than the one you thought it did.

Edit Sketch

A sketch can be redefined after creation. The reason for redefining a sketch instead of modifying dimensions would be to add or subtract sketch entities or to make a number of "what if" changes without rebuilding the rest of the part. To change a sketch used to define a feature, perform the following steps.

1. Expand the feature tree if necessary in order to see the sketch below the feature in question. Do this by clicking on the "+" sign in front of the feature name in the FeatureManager design tree.

2. Right click on the sketch in the FeatureManager design tree.

3. Select Edit Sketch.

4. Redefine the sketch as needed.

5. Rebuild the part when done to see the changes. This can be done by clicking on the Rebuild icon or just exiting the sketch.

Edit Sketch Plane

The sketch plane used to create a 2D sketch can be changed after the feature has been defined. This allows the sketch feature to be changed to another face or plane, thereby changing its orientation. To change the sketch plane for a sketch, perform the following steps.

1. Expand the feature tree if necessary in order to see the sketch below the feature in question. Do this by clicking on the "+" sign in front of the feature name in the FeatureManager design tree.

2. Right click on the sketch in the FeatureManager design tree.

3. Select Edit Sketch Plane.

4. Select the new sketch plane.

5. Select Apply to continue.

List External References

When a feature is created referencing another assembly component, this is referred to as an *in-context feature*. SolidWorks identifies these references within the FeatureManager design tree and allows you to recall the referenced components to change the in-context features. To list external references, perform the following steps.

1. Right click on the sketch, feature, or part in the Feature-Manager design tree.

2. Select List External Refs. A dialog box will appear listing all applicable external references.

3. Select OK to close the dialog box.

Dynamic Editing

One of SolidWorks' strong points has always been its editing capabilities. When it comes to editing features, you have a number of options. You can always use the procedure outlined in the previous section for editing the definition of a feature. Or you could always just double click on a feature to access its dimensions. Another option is to use SolidWorks' dynamic editing capabilities.

To dynamically edit a feature or sketch, you must use the Move/Size Features icon on the Features toolbar. This is a toggle switch; therefore, once you click on the Move/Size Features icon it stays on until you click it again. To dynamically edit a feature or sketch, perform the following steps.

1. Select the Move/Size Features icon.

2. Select a feature to be edited. This can be done from Feature-Manager or in the graphics area. The Move/Size handles will appear, shown in the following illustration.

3. Drag one of the handles to edit the desired feature.

4. Select the Move/Size Features icon once again to turn it off when finished.

The "handles" that appear when using Move/Size Features.

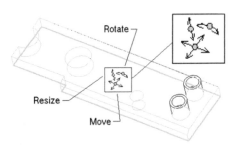

There are a maximum of three handles that will appear when using Move/Size Features, but you may not see all three. This really depends on the underlying sketch and type of feature. For example, if a cut has a Through All end condition, the resize handle will not appear. If the end condition had been Blind, on the other hand, the resize handle would be available. By dragging one of the handles, it is possible to change the definition of the feature, such as the extrusion depth. It is also possible to change, for instance, the rotation of a sketch.

Certain operations, such as dynamically rotating a sketch, may require that constraints on the original sketch be deleted. Solid-Works will warn you if an operation is being performed that will affect the underlying constraints or dimensions, in which case you will be able to cancel the operation, delete the constraints, or leave the constraints dangling. The term *dangling* is used to describe a dimension or constraint that cannot be solved and is therefore just hanging out in space without really accomplishing anything (to put it simply).

Whatever method you use to edit a sketch or feature depends on user preference. No particular method is any better than another for any specific reason. It just depends on what you find easiest to use.

Undo

Undo reverses the effects of recent actions. This function is available in all modes (i.e., sketch, part, assembly, and drawing). It may be accessed by pressing the Control + Z keys or by clicking on the Undo icon. Undo also allows for returning sketch geometry to its original state after dragging the sketch. This allows you to investigate many "what if" scenarios without committing to any of these modifications.

AutoCAD's Undo command is quite a bit more powerful than SolidWorks'. The knowledge that the Undo function is there sometimes gives the user more confidence than they might otherwise have. If you are an AutoCAD user accustomed to using the Undo command frequently, you might want to try to break out of that habit.

SolidWorks has an Undo function, but not everything can be undone. It should be pointed out, however, that because of the powerful editing characteristics of SolidWorks, an Undo command is not as important as it is in AutoCAD, where editing is severely limited with regard to solids. Also bear in mind that SolidWorks does not have a Redo command. If a feature is created that is not quite what you expected, it is much easier to edit the feature than it is to delete it and recreate it in Solid-Works. If the feature is deleted, you must then recreate it if you decide that deleting was not the smartest decision. The Undo list box is shown in the following illustration.

⚬ **NOTE:** *Once a rebuild is performed, the Undo list box is wiped clean.*

Undo drop-down list box.

When multiple commands are selected from the pull-down list, all functions from the selected command and above are undone. Single or multiple commands can be undone, depending on the icon selected. Care should be taken when selecting Undo, as the command cannot be reversed (such as with AutoCAD's Redo function). To undo recent changes, perform the following.

1. Select the Undo icon or select Undo from the Edit menu.

To undo recent changes by selecting from an item in a list, perform the following steps.

1. Open the Undo list box (see the previous illustration).

2. Select an item on the list, which will undo the selected action and all actions above the selected item.

Rebuild

Rebuild is the function used to update dimensional changes and to rebuild a part anytime modifications have been made. You can select when to rebuild a part. A number of changes can be made without rebuilding the part. When all changes are complete, the part can be rebuilt to change the geometry based on new dimensions.

The Rebuild function is available within the Modify dialog box. For instance, when a sketch dimension is double clicked to modify the dimension, the Modify dialog box appears. The dimension value can be changed, and by selecting the Rebuild button, the changes are displayed in the part and the Modify box is left open.

The Modify dialog box is also used in sketch mode. The Modify box appears when you double click on a sketch dimension to change the value. Changes to sketch dimensions within sketch mode are automatically shown unless Automatic Solve is unchecked in the Tools/Sketch Tools menu (it is recommended you leave Automatic Solve checked). The difference between rebuilding each change is that a number of "what ifs" can be reviewed without rebuilding the rest of the part geometry. To rebuild any model parameter changes, simply click on the Rebuild icon or select Rebuild from the Edit menu.

Rollback

Rollback is used to roll a model back to an earlier creation state, prior to the position of the rollback bar, shown in the illustration that follows. The rollback bar is a FeatureManager design tree indicator of the current display state of the part or assembly. Parts and assemblies are shown in a sequential or history-based order in the FeatureManager design tree.

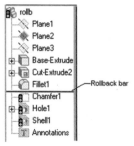

Rollback bar.

The rollback bar allows you to go back to an earlier creation state in FeatureManager's chronological history of events, or to step through the feature list to analyze a part to see how it was built. Any new features or assembly components will be inserted above the position of the rollback bar in the FeatureManager design tree.

Keeping features logically grouped can make interpreting and modifying a part or assembly easier. When modifying or adding features to a previously defined functional area, use Rollback to position the rollback bar in the desired area. New features for the functional group are added in the same location in the FeatureManager design tree.

The rollback function can also be used to edit an area of a part. Instead of modifying a feature and rebuilding the entire model, only the changed portion is rebuilt. This can speed up the modification of parts and assemblies because only a small number of features are rebuilt. This technique can be used very effectively on large parts and assemblies. This allows for many "what if" scenarios to be tried without rebuilding many unrelated features.

The rollback bar can be positioned at a desired location backward or forward. The bar can be repositioned at any time to easily view the model in many different states. By setting the rollback bar to the beginning and stepping through each feature, you can review how the part or assembly was created. There are two methods of rolling back a model. The first method is as follows.

1. Select the feature to define how far the model will be reverted (rolled back).

2. Select Rollback from the Edit menu.

When the mouse cursor is placed over the rollback bar, it changes into a small hand. To roll back a model using the second (and preferred) method, perform the following.

1. Select the rollback bar with left mouse button and drag it to the desired location in FeatureManager.

To cancel rollback mode (roll the part forward), perform the following.

1. Select the rollback bar with the left mouse button and drag the rollback bar to the bottom of FeatureManager, after the word *Annotations*.

Rollback is similar to a smart AutoCAD Undo command. It is as if a part can be undone without actually really undoing anything. Nothing is deleted or undone, but nearly the same thing is accomplished, as if you were going back in time, to a point before the rolled-back features were created.

Suppressing Features

Suppressing features temporarily blanks the display of selected features and removes them from memory. However, the features have not been deleted and can be redisplayed (unsuppressed). Suppressing features can be used to simplify a model for easier creation, to temporarily avoid unwanted parent/child relationships, or to minimize part complexity for part editing and complex assembly management.

When a feature is suppressed, the feature's children are automatically suppressed. The children can automatically be unsuppressed when the parent is unsuppressed by using the Unsuppress With Dependents command. Otherwise, the children remain suppressed. To suppress features, perform the following steps.

1. Select the features to be suppressed from FeatureManager. Hold the Control key down to select multiple features.

2. Select the Suppress icon or select Suppress from the Edit menu.

Another alternative is to use the suppress or unsuppress icons, typically found on the Features toolbar. You may have to customize your Features toolbar in order to see these icons. The suppress and unsuppress icons are shown in the illustration at left.

Icons used for suppressing and unsuppressing features.

☞ ***NOTE:*** *See Chapter 12 for customizing procedures.*

It is also possible to suppress or unsuppress a feature by accessing its properties. To do this, right click on the feature in FeatureManager and select Properties. You will find a check box option for changing the suppression state of the feature, though using the Edit menu or toolbar icons is easier. To unsuppress features, perform the following steps.

1. Select the features to be unsuppressed from FeatureManager. Hold the Control key down to select multiple features.

2. Select the Unsuppress icon or select Unsuppress from the Edit menu.

To unsuppress features with dependent children, perform the following steps.

1. Select features from FeatureManager to be unsuppressed. Hold the Control key down to select multiple features.

2. Select the Unsuppress With Dependents icon or select Unsuppress With Dependents from the Edit menu.

The following are guidelines for suppressing and unsuppressing features.

•   Children are suppressed when the parent feature is suppressed.

•   A suppressed feature will be shown in gray in FeatureManager.

•   The Control key can be used to select multiple features for suppression or unsuppression.

•   The Shift key can be used to select a range of features in the FeatureManager design tree. Select the first feature, hold down the Shift key, and select the last feature in the range. This also works for selecting parts in the assembly FeatureManager or for selecting files in Windows Explorer.

•   Features can also be suppressed or unsuppressed from the Property dialog box. Right click on the feature and select Properties.

•   Children are unsuppressed when the parent is unsuppressed only when the Unsuppress With Dependents function is used. Otherwise, the children have to be unsuppressed independently.

•   Parents are unsuppressed when a child is unsuppressed.

Advanced Part Features

Some of the topics covered in this section are not commonly used in all disciplines. Your particular company may find certain of these features very valuable. You might want to skim over this section to see if any of the topics pertain to your applications.

Scaling Parts

Occasionally you may find it necessary to scale an entire part model. This is often required when creating molds for parts that will have a shrink factor. It is possible to scale a part up or down, depending on your requirements. The dialog box used to scale a part is very straightforward. It is not possible to scale individual features, only the entire part. If you did need to resize an individual feature, you would simply change its dimensions. To scale a part, perform the following steps.

1. Select Insert/Features/Scale from the pull-down menus or click on the Scale icon.

2. Specify the Scaling Type, either About Origin or About Centroid.

3. Specify the Scaling Factor.

4. Click on OK to complete the scaling operation.

The Scale dialog box is shown in the following illustration. As you can see, there is not a great deal to it. When scaling a part, no dimensions are changed. As a matter of fact, all of the dimensions associated with the part will remain exactly as they were before the Scale command was initiated. This can sometimes be an issue, because if a drawing layout is made of a scaled part, the dimension extension lines will appear to be dimensioning to geometry that existed at a location before the scale was performed. This gives the effect of dimensions that are disassociated with the part geometry.

Scale dialog box.

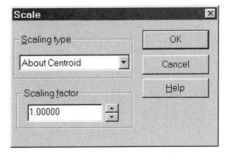

SolidWorks' Scale command is similar to AutoCAD's scale command. The commands perform the same function, but where AutoCAD can selectively scale specific portions of a model, Solid-

Works cannot. Once again, though, this is not an issue because you have parametric control of your SolidWorks model.

Dome

The Dome command is not one of the most commonly used commands, but it can come in very handy when needed. Applying a dome feature allows you to place a dome of varying height upon a surface, and to use the dome shape to either add material or cut material away from the model. The Dome dialog box, shown in the following illustration, is very simplistic in nature, and a sketch is not required when creating a dome. For that matter, to use the Dome command, you must not be in an active sketch.

Dome dialog box.

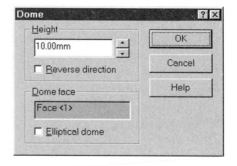

To use the Dome command, simply select a planar face you want to apply a dome to. If the face is circular or elliptical, you will see an additional option for an Elliptical Dome. Checking Elliptical Dome creates a dome feature with an axial cross section of half an ellipse. You will not see this option if a noncircular or nonelliptical face is selected. To create a dome feature, perform the following steps.

1. Select Features/Dome from the Insert menu or click on the Dome icon.

2. Select the face you want to apply a dome to. If you want to create a dome-type depression, check the Reverse Direction option. This will automatically remove material from the part.

3. Specify the height of the dome.

4. Click on OK to complete the dome operation.

There is no analogous command in AutoCAD to SolidWorks' Dome command. You would have to perform a number of workarounds in order to create the same type of feature created by SolidWorks' Dome function, possibly by revolving a portion of an ellipse and performing a Boolean operation.

Shape

The Shape function is even less likely to be used than the Dome command. It is also much more involved than the Dome command. There are many tweaks and adjustments that can be made in order to define the shape you are looking for. Essentially, the Shape command works as if you were to take a face of your model, rubberize it, and then apply pressure to the other side. You can control characteristics such as the amount of stretch the rubber has, the pressure, tangency constraints, and other characteristics. These options are explained briefly in this section. To shape a face using this "rubberize" technique, perform the Shape operation through the following steps.

1. Select Features/Shape from the Insert menu.

2. Select the face to shape. This does not have to be a planar face, but planar faces usually have the best results.

3. Click on the Controls tab and adjust the shape characteristics as desired. The preview will update to show you what the shaped face will look like.

4. Click on OK when you are satisfied with the resultant preview in order to create the Shape feature.

Shape features are impossible to dimension and must be documented by other means. How you document the shape of the feature is up to you or your company's standards and is beyond the intent of this book. The various options present in the Shape dialog box and available in the Control tab are shown in the following table.

Maintain Boundary Tangents	When checked, the shape feature maintains a tangency condition to the surrounding boundary faces.
Constrain To	Allows selecting points or curves that will define the shape more accurately. This is not a requirement, just an option.
Preview	Will update the preview of the shape when clicked.

The options present in the Control tab of the Shape dialog box are defined as follows.

Pressure	Inflates or deflates the surface to be shaped.
Curve Influence	If entities have been specified in the Constrain To list area, Curve Influence controls how much influence those entities have on the surface to be shaped.
Stretch	Controls how much stretch the surface has.
Bend	Defines how much the surface can bend. You might think of this control as the stiffness placed on the surface to be shaped.
Resolution	Controls the number of resolution grid lines that are used to define the shaped surface. Higher resolutions will require more processing time but will make for a smoother surface.

AutoCAD has no function that even comes close to SolidWorks' Shape command. It could be argued that you are not often going to need a feature that looks like somebody blowing a bubble. This is a valid point, but you never know when the power of SolidWorks' advanced functionality might prove very useful.

Annotations

Model annotations are used to define engineering characteristics and parameters directly on a part. The important features and characteristics can be defined and used later in the design process (e.g., drafting or manufacturing).

Model annotations can be used to add specific design intent during the modeling phase of the design. Often, this information is added near the end of the design phase. This allows the person who created the design to define important characteristics at the time the part and assembly are created. This information can also be used downstream to add important comments and design intent annotations without the need for detailed drawings.

•❖ **NOTE:** *For a complete explanation of how to create an annotation, see the "Drawing Symbols" section in Chapter 8.*

The methods used to define the annotation are the same whether they are created within a drawing, part, or assembly. The following illustration shows an example of model annotations.

Model annotations.

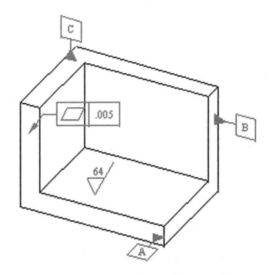

The following are types of model annotations.

- Cosmetic threads
- Notes
- Datum target symbols
- Geometric tolerance
- Surface finish symbols
- Feature dimensions
- Reference dimensions
- Weld symbols
- Datum feature symbols

There is an easy way in which you can turn model annotations on and off. Obviously, you would have had to have already added the annotations before you could display them. The steps that follow relate to turning annotations on or off and have been provided here for reference. To display model annotations, perform the following steps.

1. Right click on Annotations in FeatureManager.

2. Check Display Annotations.

To display only certain model annotations, perform the following steps.

1. Right click on Annotations in FeatureManager.

2. Select Details to open the Annotation Properties dialog box (see the following illustration).

3. Uncheck Display All Types.

4. Check the desired annotations to be displayed.

5. Select OK to continue.

Annotation Properties dialog box.

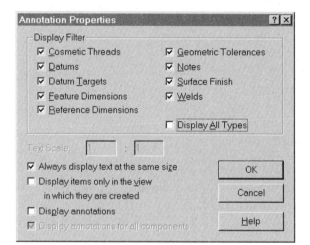

To insert a model annotation, perform the following steps.

1. Select the desired face or edge. This is where the leader arrow will be attached.

2. Select the desired annotation from the options in the Insert/ Annotations menu.

3. Enter the annotation object information, if applicable.

To edit a model annotation, perform the following steps.

1. Double click on the annotation.

2. Edit the object.

3. Select OK to continue.

As mentioned previously, different annotation types and how they are added are discussed in detail in Chapter 8. If necessary, you can flip forward to that chapter for instructions on adding individual annotations.

Any
questions
call me

773 481-8016

773 719-6967
(cell)

Tables

The Design Table function is used to insert an embedded Microsoft Excel spreadsheet to drive dimension parameters for feature suppression. Microsoft Excel 97 (or newer) is required to use this function, and no other spreadsheet type is currently supported, although any OLE-compliant software can have files embedded into a SolidWorks document.

Only one design table can be inserted into a part. The menu structure when inserting or editing design tables is the Microsoft Excel menu. In other words, when you insert a design table, you will see the Excel toolbars and menus, even though you are technically in the SolidWorks program. When design table insertion or editing is complete, the menu structure reverts to the SolidWorks menus. The following illustration shows a design table.

Design table.

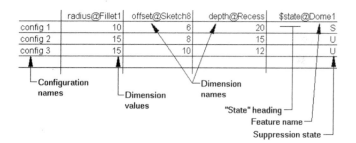

Function of Design Tables

Design tables can be used to change feature sizes or to suppress unwanted features. Design tables create part configurations and are used to create multiple models from one part file. A configuration can be used to define a subset, feature, or assembly component to create simplified parts and assemblies, exploded assemblies, or optional versions of a design. To recall a design table configuration, select the Configurations tab at the bottom of the FeatureManager design tree and select the desired configuration name. ConfigurationManager is shown in the following illustration.

ConfigurationManager.

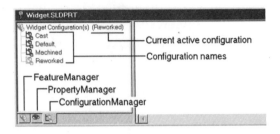

After insertion, the design table is embedded into the Solid-Works document and no longer references the original spreadsheet. Therefore, any changes to the original worksheet will not affect the part unless it is reinserted into the part. Design tables are embedded objects, not linked. A linked document differs from an embedded document by the fact that a linked document still references the values in the original document. Insert Object can be used to insert a linked OLE object into Solid-Works.

The value defined by the cross reference between row 1 (dimension or feature names) and column 1 determines design table configurations. The value in cell A1 is left blank. Any cell without a corresponding row and column value is ignored. Design table values and configurations should be documented so that anyone can view the design table and determine the intent for the configurations and any dimensional references. The best way to do this is to change sketch and dimension names prior to creating the design table, and create notes in the spreadsheet that explain what is being accomplished in the table.

The first row of the spreadsheet defines the names of the configurations (e.g., Machined, Cast, or Reworked). The second and subsequent rows control dimension values or feature display (e.g., D3@Sketch2 and Hole1). The cross-reference cell value can determine the size or visibility of a feature. Features can have their suppression or visibility controlled through the use of specific headings. The following table shows the various headings, the correct syntax for usage, and what each heading can accomplish.

Heading	Syntax	Definition
$STATE	$state@feature_name	Controls whether a feature is su̶ unsuppressed. SolidWorks recogn̶ "U" to mean unsuppressed and "S̶ suppressed.
$COMMENT	$comment	For use at the beginning of a column o̶ adding notes to a design table.
$USER_NOTES	$user_notes	Has the same function as the $comment ̶
dimension name	dimension_name@ sketch_name or dimension_name@ feature_name	The dimension name is the full name of the̶ you want to drive in the design table. Typica̶ would be D1@Sketch1 or D2@Boss-Extrude1.̶ Dimension names, however, are usually change̶ to creating a design table in order to better docu̶ the design table.

If you want only one configuration, a design table can still ̶ used to drive dimension parameters. The configuration name̶ Default is included with every part and can be redefined using design tables; however, it is best to create a new configuration in the spreadsheet by using a name other than Default. This is just a failsafe, so that the default configuration can be fallen back on if needed.

The design table should be properly annotated so that the design intent and dimension references can be clearly understood by others. However, this is not mandatory. If you do decide to add notes to your design table, make sure you add the proper headings. Otherwise, your design table may fail.

Dimension names inserted into a spreadsheet must be in the form of full dimension names. The full dimension name for a sample base feature, for instance, might be D1@Base-Extrude, and not just D1. Dimension names are case sensitive; therefore, the name must be entered exactly as shown. Dimension names can be found by selecting the dimension and pressing the right mouse button and selecting Properties.

The dimension name appears in the Full Name field. Dimension names can be copied and pasted into the design table using the right mouse button. By defining meaningful dimension, feature, and component names, you can associate the dimension's

...ign table with the feature or function it is associ-
...owing are some examples.

    ...2@Sketch2
        ...Diameter@ScrewProfile
            ...ude
        ...ounting Plate Profile

    ...a table, perform the following steps.

    ... Microsoft Excel spreadsheet that has configuration
    ...es starting in row 2, column 1, and extending down col-
    ...nn 1. Do not leave empty cells in column 1.

2. The dimension names or feature names start in row 1, column 2, and continue across row 1. It is best to copy and
   paste these dimension names into the spreadsheet to help
   eliminate typographical errors. Do not leave empty cells in
   row 1.

3. Enter dimension values for the respective revisions.

4. Save the design table when done. It can be helpful to save
   the design table with the same name as the part it will be
   embedded in. It might also be helpful to put the spreadsheet in the same directory.

5. Back in SolidWorks, select Design Table from the Insert
   menu.

6. Select the Excel Spreadsheet to be inserted. SolidWorks will
   inform you either of errors or that the table insertion was
   successful. If successful, SolidWorks will display the configurations created.

7. Edit the spreadsheet, if necessary. Click outside the spreadsheet in the sketch area when finished.

An alternative to using the previous steps would be to add a
new design table directly from within SolidWorks. It is best to
show the dimensions before using this particular procedure.
For parts with many features, this procedure may be impractical
due to the large amount of dimensions. To show the dimensions of the part, right click on Annotations in FeatureManager
and select Show Feature Dimensions. Once this is done, create
a new design table by performing the following steps.

1. Select Insert/New Design Table.

2. Select the first cell in which to place the first dimension. This would typically be cell B2.

3. Double click the dimension to place in cell B2. The next cell will become active, and you will be able to repeat the procedure.

4. Add the desired configuration names in column A, starting with cell A2.

5. Enter the dimensional values for the configurations.

6. Click once outside the spreadsheet in the graphics area to close the spreadsheet and return to SolidWorks.

This last method (just described) is a somewhat automated version of creating a design table. Use whatever method is easier for you. Once you have succeeded in creating a design table, you will find it necessary to alternate between those configurations. To activate a design table configuration, you would perform the following steps. The ConfigurationManager tab is shown in the following illustration.

1. Select the ConfigurationManager tab from the bottom of FeatureManager.

2. Double click on the desired configuration or right click the configuration name and select Show.

ConfigurationManager tab.

AutoCAD does not have the ability to import a design table in the same fashion as SolidWorks. It is possible to make database connections that achieve similar results, but it is not something to be attempted by the inexperienced.

Edit Design Table

Existing design tables can be added to, changed, or deleted. Design tables are embedded Microsoft Excel spreadsheets that contain information used to drive dimensional parameters for SolidWorks parts, as previously described. If configurations are added to a design table, the configurations are automatically added to the part by SolidWorks. Dimension and feature names can be added or deleted, and dimension values in the spreadsheet can be modified. To edit a design table, perform the following steps.

1. Select Design Table from the Edit menu.

2. Add, delete, or change values in the Microsoft Excel design table as needed.

3. To finish, click anywhere outside the design table.

To delete a design table, perform the following steps.

1. Select Delete Design Table from the Edit menu.

2. Select Yes to delete the design table. Select No to skip the selected item. Select Cancel to abort.

It should be mentioned that deleting a design table does not necessarily delete the configurations. The configuration names themselves will still be shown in ConfigurationManager. They will not be accessible, though, because all configuration data has been removed. To delete the configuration names, you must select one or more configurations from the list and press the Delete key on your keyboard.

Equations

Equations are used to define mathematical relationships between model dimensions. For example, an equation can be defined to set a part feature equal to half the total length, height, or width of the part. Another example might be adding an equation to adjust the distance between items in a pattern should the number of instances increase or decrease. The possibilities are quite numerous. The following illustration shows an example of an equation.

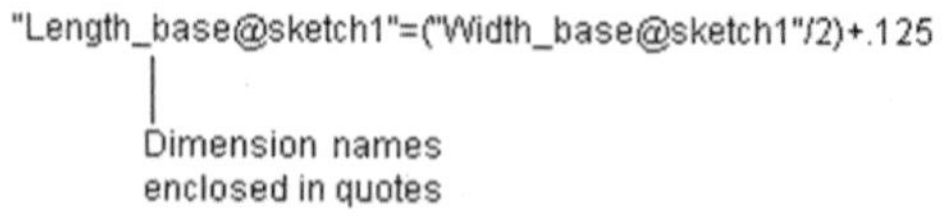

"Length_base@sketch1"=("Width_base@sketch1"/2)+.125

Dimension names
enclosed in quotes

An equation.

Dimensions and features should be renamed to help convey and document the design intent of the part. Without these names, it can be difficult to understand the intent of the design. Meaningful names can help a user understand the function or purpose of an equation when editing the equations. Without logically named features and dimensions, the user has to second guess design intent. By defining meaningful names, the user can associate dimension names found in the equations with what is being defined by an equation.

If any dimensions or feature names are going to be changed, this should be done before adding equations. Otherwise, the equations may be referencing dimensions or feature names that no longer exist (actually, they still exist, they have just been renamed).

Equations are not case sensitive, but equations can be finicky nonetheless. Dimension names in an equation need to match the dimension names on the part character for character. One mistyped character will cause the equation to fail. You must also use caution not to place spaces where none should be.

One other steadfast rule when creating an equation involves the location of the dimension names with respect to the equal sign. Any dimension name on the left of the equation is a driven dimension. Any dimension on the right is a driving dimension. Once a dimension is made driven by adding it to an equation, it cannot be altered by double clicking on the dimension in the sketch area.

Because of AutoCAD's lack of parametric technology, creating equations is simply out of the question. Intelligent relationships between geometry in AutoCAD cannot be obtained without the use of an add-on software package. To add an equation in SolidWorks, perform the following steps.

1. Select Equations from the Tools menu.

2. Select Add from the Equation Editor list box.

3. Double click on (or sketch in FeatureManager) the feature whose dimension you want to access.

4. Select the dimension. It will be placed in the New Equation dialog box.

5. Complete the equation by entering values, mathematical operators, or dimensions. It is recommended that only the mouse be used for this task.

6. Select OK to continue.

7. Click on the Rebuild icon or select Rebuild from the Edit menu.

To edit an existing equation, perform the following steps.

1. Select Equations from the Tools menu.

2. Select Edit All to edit the equations listed.

3. Edit the desired equation using standard word processing techniques (i.e., backspace, Delete key, and so on).

4. Select OK to exit out of all dialog boxes.

Editing equations opens up those equations to typographical errors. Use care when editing equations. Sometimes it is actually easier to delete an equation and just add a new one in its place. To delete an existing equation, perform the following steps.

1. Select Equations from the Tools menu.

2. Select the equation to be deleted.

3. Click on the Delete button.

4. Select OK to continue.

In summation, remember the following guidelines when working with equations.

- Dimensions driven by equation values are not editable by double clicking on the dimension.

- To determine the name of a dimension, right click on the dimension and select Properties. The name of the dimension can be found in the Full Name field. This typically is not needed when adding equations, because selecting a dimension enters its name in the Add Equation dialog box automatically.

- Equations can be deleted by selecting the equation and pressing the Delete key.

- Dimensions on the left side of the equation are driven. Driving dimensions are on the right-hand side.

Configurations

Configurations are used to simplify complex parts and assemblies into different named states or optional features or parts. A configuration allows you to temporarily suppress features or components that may not be required, or to show parts with various arrangements of features. Parts and assemblies will rebuild more quickly when configurations are used to suppress features that might require a large amount of computation.

Configurations are often used to suppress features or parts not currently necessary to the design process. A simplified part configuration can be used in assembly configurations to address large assembly management. Configurations might also be used to simplify a part for use with finite element analysis. This section covers part configurations only.

When a part or assembly is saved, the current configuration name becomes the default configuration. When the document is opened, this configuration is selected by default. Working-state configurations can be defined to minimize the number of components active when working on a single part. For example, you can use configurations to define a Structure Only assembly. This allows you to recall an assembly and review or modify the structure without rebuilding the complete assembly. The following illustration shows where configurations are found.

Where to find configurations.

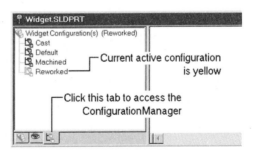

To toggle between FeatureManager and ConfigurationManager modes, select the desired icon at the bottom of FeatureManager, as described in the section "Design Tables" and shown in the previous illustration. The active configuration is shown to the right of the main (top) part/assembly ConfigurationManager icon. The configuration name is shown in parentheses to the right of the part or assembly name.

The reasons for creating a configuration are numerous, but the steps for creating a configuration are short. To create a new configuration, perform the following steps.

1. Select the ConfigurationManager icon at the bottom of FeatureManager.

2. Right click on the part name at the top of ConfigurationManager and select Add Configuration.

3. Go back to FeatureManager and suppress or unsuppress features or components as desired. (See "Suppressing Features," earlier in this chapter.)

4. When adding a configuration, you may decide to type in some comments regarding the configuration being created. This is certainly not mandatory, but feel free to add comments if desired.

It is possible to specify a particular configuration to be used while opening an existing part or assembly. The configuration must have previously been defined during an earlier SolidWorks session. To open a part or assembly using a defined configuration, perform the following steps.

1. Select the Open icon or select Open from the File menu.

2. Check the Configure option at the bottom of the menu.

3. Select the file to be opened and click on Open.

4. Select the configuration state to be used.

5. Select OK to continue.

To activate a configuration within a part or assembly, perform the following steps.

1. Select the ConfigurationManager icon at the bottom of FeatureManager.

2. Right click on the desired configuration and select Show Configuration. Optionally, double click on the configuration to be shown.

To delete a part or assembly configuration, perform the following steps. Be aware that the current configuration cannot be deleted.

1. Select the ConfigurationManager icon at the bottom of FeatureManager.

2. Select the desired configuration and press the Delete key.

Creating a new configuration and editing an existing configuration are essentially the same. The only difference is instead of adding a new configuration, you simply activate an existing configuration. Only the current configuration can be edited. To edit an existing part or assembly configuration, perform the following steps.

1. Select the ConfigurationManager icon at the bottom of FeatureManager.

2. Activate the desired configuration as previously described.

3. Suppress or unsuppress features or parts as needed.

In summation, the following list represents the important topics discussed in the "Configuration" section.

- The active configuration cannot be deleted. Make another configuration active and then delete the desired configuration.

- The Comment field is a good place to briefly describe the reason and function of the configuration. This field is found in the Add Configuration dialog box when adding a configuration.

- To make a configuration active, you can double click on the configuration name with the left mouse button.

- The active configuration name is displayed in parentheses following the part or assembly name in the FeatureManager design tree.

- Configurations can also be used in drawing views. A single drawing can display multiple configurations in the various views.

➥ **NOTE:** *Drawing views are discussed in Chapter 8.*

Base Parts The Base Part function uses an existing part file as the base feature for a new part. The original part becomes a single base feature in the new part. Any changes to the parent part will be reflected in the base part. A casting is a good example of a model for which a base part such as this would be used. The parent part is the raw casting. The new part would have the machined surfaces and tapped and drilled holes.

Another good example of when the Base Part function would be convenient is in the case of mold making. Once a cavity has been created in a mold die, the mold must be cut in half. The mold die can be brought into a new part as a Base Part and have the lower portion of the die removed in order to create the top half of the mold. This can be repeated for the bottom half of the mold. This is necessary because SolidWorks does not let you cut the mold die and keep both halves within the same part file.

Modifying a Base Part

Changes can be made to the parent part by right clicking on the Base Part feature in the FeatureManager design tree and selecting Edit In Context. This will open the parent part in a new window for editing. Any changes to the parent part will be reflected in the child part. The child part, in this case, would be the part in which the Base Part function was originally employed. The following illustration shows a base part.

Making use of a base part.

New part with added features

Original base part

An AutoCAD user might draw a comparison between a Solid-Works base part and an external reference (Xref). An Xref can be modified and then updated in the drawing file that incorporates the Xref so that the drawing will accept the new changes made to the Xref.

Differences Between Base Part, Cavity, and Derived Component Part

The Base Part function is used when a single part can be used to create a new part, as with a casting and the machined finished part. The only references that are developed are between the base part and the new part. This is a unidirectional associativity only. Changes made to the base part affect the new part, but not vice versa.

Cavity is used to create a cut in an assembly component while applying a shrinkage value. This requires assembly references to the cavity block and the design component. This function is typically used for creating a die (as one example). One or more components are "subtracted" from the die to create a cavity. In this case, if the design components are altered, the assembly also updates, thereby affecting the cavity feature within the die.

Derived Component Part is used when a part needs to be created from an assembly component part. To continue the example from the previous paragraph, the die could be derived from the assembly to create a new part. This new part could then be cut to create the top half of the mold. If repeated, you could then create the bottom half of the mold and then all that would be left would be to place the finishing touches on both halves of the mold.

Derived Component Part actually performs the same function as the Base Part command. With Base Part, you must begin a new part and then insert the Base Part. Using Derived Component Part, the part is selected first and then SolidWorks creates a new part using the selected part as the base part. It is really nothing more than a shorter version of the Base Part command. The associations are the same, meaning that if the originally selected part is altered, the new part that it was derived from will also be affected, but not vice versa.

⟶ **NOTE:** *The Cavity command and Derived Component Parts are discussed in greater detail in Chapter 7. If you do not understand those commands completely at this point, you can cross those bridges later when those topics come up in the next chapter.*

Implementing the Base Part Command

To use a previously defined part as the base feature in a new part, perform the following steps.

1. Start a new part by clicking on the New icon or selecting New from the File menu. There cannot be any solid features in the new part. Only nonsolid features (i.e., planes and axes) can exist in the part.

2. Select Base Part from the Insert menu.

3. Select the part file name to be used as the base part.

4. Select Open to continue.

Once the base part is brought into the new part file, it is listed in FeatureManager as the base feature. Because this base feature is totally dependent on the base part, it has no dimensions. Therefore, it cannot be modified parametrically from within the new file, only from the original file used to create the base part. However, new features can be added in the same fashion as any other part. Any modifications made to the new part are not reflected back to the parent base part.

Thin Features

Thin features are used to define constant wall thickness parts. This type of feature can be an extruded or revolved feature type. Thin features are created by default from open profiles, but can also be created from closed profiles. To create a thin feature for a closed profile, select the feature type as you normally would to create a feature (i.e., Extrude), then select Thin Feature from the box labeled Extrude As at the bottom of the Feature Definition dialog box.

Thin feature parts are often used to create sheet metal parts, although this is not a requirement. Adding the Bends feature to any part will define that part as a sheet metal part, which is covered in greater detail at the end of this section.

Thin-feature parts can be created in AutoCAD, but it is not an automated process as it is in SolidWorks. There is not really a counterpart for a thin-feature part in AutoCAD.

In the examples that follow, it is assumed that you have already created a sketch. To create a thin-walled extruded feature, perform the following steps. The illustration at left shows an example of a thin feature.

Simple thin-feature part.

1. Select Insert/Boss/Extrude and specify parameters as described in the "Extrude" section of this chapter.

2. In the Extrude As drop-down box, select Thin Feature. This is automatic if extruding an open profile.

3. Select the Thin Feature tab to define material thickness properties.

4. Select the thickness type from the Type field: One-Direction, Mid Plane, or Two Direction (see the section "Wall Thickness Properties").

5. Enter the Wall Thickness.

6. The Reverse option can be used to change the direction of the wall thickness. This option is only applicable for the One Direction type.

7. Select OK to continue.

To create a revolved thin feature, perform the following steps.

1. Select Insert/Boss/Revolve and specify parameters as described in the "Revolve" section of this chapter.

2. In the Revolve As drop-down box, select Thin Feature. This is automatic if revolving an open profile.

3. Select the Thin Feature tab to define material thickness properties.

4. Select the thickness type from the Type field: One-Direction, Mid Plane, or Two Direction (see the section "Wall Thickness Properties").

5. Enter the Wall Thickness.

6. The Reverse option can be used to change the direction of the wall thickness. This option is only applicable for the One Direction type.

7. Select OK to continue.

Thin-feature Properties

There are a number of options that will be encountered while creating a thin feature. Some of the options that follow have already been discussed, but are added here for convenience. These options are found under the Thin Feature tab.

One Direction	Add wall thickness a specified distance in one direction only.
Mid Plane	Add wall thickness an equal depth in both directions.
Two Direction	Add wall thickness of different thicknesses in opposite directions.

Reverse	Check to reverse the direction in which the wall thickness is added. Option available only for the One Direction type.
Wall Thickness	Enter the wall thickness or pick the up/down arrow buttons to increment/ decrement the wall thickness value. Option available only for One Direction and Mid Plane types.
Direction 1 Thickness	Enter the wall thickness for Direction 1. Option available only for the Two Direction type.
Direction 2 Thickness	Enter the wall thickness for Direction 2. Option available only for the Two Direction type.
Auto Fillet	Check to add a fillet to each sharp corner automatically. When the box is checked, a fillet radius value can be entered in the Radius field. The Fillet Radius field is only visible when the Auto Fillet box is checked. This option is only available for open profiles.
Cap Ends	Checking this option will place a cap on either end of the thin feature, creating a hollow part. When checked, the Cap Thickness box appears and a value must be added. This option is only available when creating a thin feature as a base feature using a closed profile.

Sheet Metal Parts

Thin features are commonly used to create sheet metal parts. The Bends function is used to add sheet metal features and attributes, but more importantly, to define the thin-feature part as a sheet metal part as far as SolidWorks is concerned. Certain types of sheet metal features (e.g., a cut across a bend) need to be added to a model in its unfolded state. Otherwise, the cut would not be true to shape or size when unfolded. The process used to unfold a sheet metal part is explored in this section.

The Bends function defines the thin feature as a sheet metal part and allows for inserting sheet metal parameters such as bend radius, bend allowance, bend order, and bend angle. The bend radii can be added to the thin-feature sketch or with the Bends function. The Bends function allows each individual bend to be edited when bends are added using the correct procedure. There is also the capability to add bend relief automatically.

Configurations can be used to show folded and unfolded parts. This is useful when creating detail drawings for sheet metal parts. This allows for a drawing to show the folded and unfolded views of the sheet metal part on the same sheet. The following illustrations show a sheet metal part and the sheet metal features in FeatureManager.

Sheet metal part.

Sheet metal features
in FeatureManager.

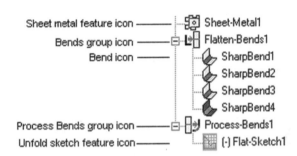

One advantage to creating a part using sheet metal features is that a flat pattern is produced automatically from the sheet metal part (flat patterns can be used to produce the part). Additionally, defining a part as sheet metal allows for creating features in the flat state (e.g., a corner relief along a bend) that could not be properly defined with features created in the bent state. Dimensions can be accurately obtained in the flattened state that automatically take into account any stretching that takes place when adding bends.

When creating a thin feature from an open profile, the radii can be added using Auto Fillet or by adding the fillets in the sketch. The latter is not recommended when creating a sheet metal part. If it is known that the part will become a sheet metal part, the Bends feature will take care of any bends needed and will offer more flexibility than the Auto Fillet option. Fillets added in a sketch are acceptable because they are usually incorporated into the sheet metal definition when bends are added. To create a sheet metal part, perform the following steps.

1. Create a thin feature using an open profile.

2. Select Bends from the Insert/Features menu. This opens the Flatten Bends dialog box.

3. Select a face that will remain stationary when the part is unfolded. Select a flat edge if unrolling a part.

4. Specify the Default Bend Radius.

5. Specify the Bend Allowance. You should leave the K-factor at .5 if the bend allowance is not known.

6. Check Use Auto Relief if desired and specify the Offset Ratio.

7. If Use Auto Relief is checked, select Rectangular or Tear from the list to specify what type of relief cuts to create.

8. Click on OK to accept the data.

There are many third-party programs for AutoCAD that will allow you to create sheet metal parts. Some of these programs are very good, but AutoCAD itself does not contain sheet metal functionality.

Modifying Bends

There are certain features that need to be defined in the flat pattern state. A sheet metal part is manufactured from a flat piece of metal. The flat part is then bent to form the completed part. SolidWorks allows you to add features in the formed or flattened states. To modify a bend, perform the following steps.

1. Right click on the desired Sharp Bend below the Flatten Bends feature.

2. Select Edit Definition.

3. Make modifications as needed to the Radius or Bend Allowance. Bend Order can be modified, but does not serve any higher purpose other than for documentation.

4. Select OK to continue.

To add a cut or tab to the flat pattern, perform the following steps.

1. Roll back prior to the Process Bends feature.

2. Add cut features or tabs to the flat state as desired. Actually, any feature can be added at this point, but remember that this is supposed to be a sheet metal part.

3. Move the rollback bar to the bottom of the FeatureManager feature list to reprocess the bend attributes and view the part in its folded state.

New thin features such as tabs can be added to an existing sheet metal part and have the bends added automatically. This is accomplished through the use of the rollback bar once again. To add new features to the sheet metal part and have the bends incorporated automatically, perform the following steps.

1. Roll back prior to the Sheet Metal feature.

2. Add features to the part as desired. When extruding a new feature, a Link To Thickness option will be present. Checking this option will link the thickness of the new feature to the thickness of the sheet metal part. If the sheet metal part thickness is changed, all linked features will change.

3. Move the rollback bar forward to reprocess bend attributes. Additional bends will be added to any new sharp corners.

Configurations can be used to show sheet metal parts in both the formed and flat patterns. The drawing can display both configurations. To prepare the flat pattern for use in a drawing, perform the following steps.

1. Select the ConfigurationManager icon.

2. Add a new configuration by right clicking on the part name in ConfigurationManager. A suggestion for the configuration name might be *Flat*.

3. In FeatureManager, suppress the Process Bends feature.

4. Select OK to continue.

Other Sheet Metal Attributes

The following list is a summary of the options encountered while adding bends and defining a sheet metal part.

Default Bend Radius	Specifies the default bend radius.
Thickness	Shows the part thickness. This is determined when you select a face to hold flat when defining a part as a sheet metal part. It is the thickness of the material measured perpendicular to the selected face. The Thickness cannot be edited from this dialog box.
Fixed Edge or Face	Select a face to hold flat or a linear edge to hold immobile during the process of inserting bends and defining the model as a sheet metal part.
Use Bend Table	Select this option to use a bend table from the *sldworks\LANG\English* directory (English is defined by the installed language type). The default file is named *SAMPLE.btl*. This ASCII text file contains various bend allowances based on the material thickness, bend radius, and bend angle. This file name can be copied to define various material types or manufacturing processes.
Use K-Factor	Select this option to define the bend allowance based on a ratio between the neutral plane and material thickness. Pick inside the field to change the K-factor value.
Use Bend Allowance	Select this option to define the bend allowance based on adding a set value to each bend. Pick inside the field to change the bend allowance value.
Use Auto Relief	Check to add bend relief to areas where material deformation may occur. If Use Auto Relief is checked, specify whether relief is to be rectangular or a "tear" relief through the use of the drop-down list.
Relief Ratio	This value determines how much bend relief will be applied. The default setting is .5. What this means is that the relief cut will be half (.5) of the material thickness.

K-factor (K) is equal to the distance from the inside face to the neutral sheet (t) divided by the total sheet thickness (T). The actual equation would be represented by $K=t/T$. More on K-factors can be found using the SolidWorks on-line help. Click on the Help button when inside the Bends feature (Sheet Metal) dialog box to access this information.

It is possible to unroll sheet metal parts, instead of simply adding bends. In the case of unrolling parts, the user must select an edge as opposed to a face to hold flat during the sheet metal definition process. Take for example a situation where the model consists of a cylinder or cone that almost forms a cylindrical (or conical) part, but the ends of the material do not quite meet. In SolidWorks, it would be possible to select an edge of an end of the "rolled" part you want to unroll. This process is analogous to grabbing the end of a roll of paper towels and letting the towels unroll across the floor.

There are additional toolbar icons (such as Flattened and No Bends) that can be added to the Features toolbar using the Commands tab in the Tools/Customize menu. Select the Features toolbar and drag the icons to the desired position within the Features toolbar. Do not attempt this if you do not feel comfortable doing so. Customizing is usually best left to users more familiar with the program.

Summary

SolidWorks has a number of features that make producing and redefining solid parts simple. SolidWorks offers you an easy-to-use interface for creating solid modeling features. The selection of the feature type, location, and creation order of features helps define the design intent of the part and determines the ease with which the design can be modified.

Reference geometry is used to create simple geometric references that can be used to provide a sketch plane or axis, or to define parametric reference locations for features. With seven different methods of creating reference planes and five ways in which to create an axis, you should be able to create nearly any reference geometry imaginable with which to begin a sketch and create a feature.

Design tables and equations can be used to add intelligence to part features. Multiple versions can be produced and maintained from one part through the use of design tables and configurations. Equations can be used to define relationships between dimensions. Configurations are a powerful, flexible tool that can be used to define simplified versions of a part or assembly, define optional or alternate versions of a design, and define exploded versions of an assembly.

Modifying features can be accomplished quickly and easily. Right clicking on a sketch or feature allows for editing sketch geometry, sketch planes, and definitions, and for checking parent/child relationships. Double clicking on a sketch or feature accesses its dimensions. Very rarely does anything ever need to get deleted with the powerful editing tools available. "Starting over" is a phrase that can be eliminated from your vocabulary with regard to creating parts in SolidWorks.

SolidWorks allows for the creation of solid geometry, but also has built-in functions for creating thin-feature parts and surfaces. Once a thin-feature part is created, it can be turned into a sheet metal part if desired. The various features of a sheet metal part (i.e., Bends, Flat Pattern, and so on) are identified by function within the FeatureManager design tree and can be used to make changes to the sheet metal part.

There are many similarities between AutoCAD and SolidWorks, but as you can see, there are many differences in the way the user interfaces with the software and its ease of use. The largest differences exist in SolidWorks' editing capabilities. When any aspect of the part can be edited at any time—whether it be a sketch, feature definition, dimension, or any other SolidWorks component—the term *design software* begins to have new meaning.

# 7 Assemblies

Introduction

This chapter discusses how to create assemblies and use these assemblies to create and control parts within SolidWorks. An assembly is a collection of parts and subassemblies that defines the relationships between assembly components. Subassemblies can be inserted into an assembly. Any assembly inserted into another assembly is considered a subassembly.

An assembly can be used to define or check part dimensions, features, movement between components, assembly interference, and clearances. Components are positioned within an assembly by inserting mate features. A mate feature describes a geometric constraint placed between the selected assembly components.

SolidWorks defines an assembly as a separate document type. The pull-down menu headings will not change when working in an assembly document, but the content of those pull-down menus will change. The logical placement of menu options and the adherence to the look and feel of part menus makes learning and using the SolidWorks assembly menus much easier.

Prerequisite

You should at this time have completed chapters 3 through 6 at a minimum. Of course, it would be better to have read the entire book up to this point prior to tackling assemblies. The concepts that should be firmly understood are how to select objects and correctly interpret system feedback by the use of the cursor, color schemes, and interacting with SolidWorks' FeatureManager. The implementation of constraints should be well understood, because this functionality will be directly applied to assemblies, but in a 3D perspective. In addition, how to create the basic feature types discussed in the previous chapter should be understood.

Content

The "Assembly Modeling Methodology" section describes several approaches that can be taken when creating an assembly. This section also discusses what an in-context assembly feature is and how it can be applied. The section on creating an assembly describes how to build an assembly and the procedures used for this process. This section discusses the two methods used to add an assembly component (Insert Component and New Part) and the methods used to precisely assemble (mate) these components.

The "Modifying Assemblies" section describes the methods that can be used to fix, move, and rotate assembly components. SolidWorks allows for underconstrained assembly components to move or rotate independently. This can be used to simulate assembly motion when the appropriate assembly mates are added. The ability to define and change assembly component properties is discussed, along with how to edit a part in context with the assembly (or open the part individually by selecting an assembly component). The "Advanced Assembly Features" section describes some of the advanced topics concerning assemblies. The "Example Assembly" section shows how a simple assembly can be created.

Objectives

When you have completed the "Assembly Modeling Methodology" section, you should understand the difference between top-down and bottom-up assembly modeling, be able to identify an assembly layout and the benefits that can be derived from using this type of layout, and what an in-context assembly feature is and how it is created. With completion of the section on creating an assembly, you should be able to insert parts into an assembly, create a new assembly component within an assembly, and mate an assembly component.

Upon completion of the "Modifying Assemblies" section, you should be able to move or rotate assembly components and view or change existing assembly component properties (i.e., file name, configuration, and so on). You will be able to edit an assembly component within the context of the assembly or open a part document within its own window from the assembly FeatureManager.

With completion of the "Advanced Assembly Features" section, you should be able to define a component pattern to create an array (linear or circular) of an assembly component using an existing part pattern or by defining a new pattern. You will also understand how an assembly feature can be defined to create a cut through assembly components for visualization or manufacturing purposes, check an assembly for interference between assembly components, understand the uses for the Insert Cavity and Derived Component part functions, explode an existing assembly to create a view of the assembly in the exploded state, and understand the uses for the Join function.

When you have finished the "Example Assembly" section, you should be able to insert an assembly component, mate two assembly components, edit an existing assembly component within the context of the assembly, and explode the assembly.

Assembly File Management

Assemblies consist of other SolidWorks parts and assemblies known as subassemblies. When creating new parts within the context of an existing assembly, external references to these documents are created by SolidWorks. Sometimes managing the large number of parts and subassemblies can be a daunting task.

The Copy Files option in the Find References dialog box (under the Files menu) can be used to create a copy of the assembly and all related components. All files in the assembly can then be moved to another location, such as over a network, and the externally referenced files will no longer be referenced by the assembly. Instead, the copied parts will be referenced, thereby eliminating what can sometimes be a huge amount of network traffic. This not only speeds up the overall network but makes working on the SolidWorks assembly that much more productive.

Component properties can be used to redefine file locations or replace assembly components. Parts currently within an assembly can be redefined by "pointing" SolidWorks to a new location for a replacement component. If the new component is similar in size and shape to the original part, the existing mate relationships will be maintained.

FeatureManager component (part) names can be different from the component file name when "Update component names when documents are replaced" is unchecked under the External References tab in the Tools/Options menu. However, if that particular part is replaced, the new part name will not update in FeatureManager. If this is unknown to another user, he or she might look for the wrong component.

The actual name of the part can be determined easily enough, simply by right clicking on the part, if there is any question as to the part's real name. SolidWorks gives you the opportunity to open individual parts within the assembly, and lists the actual part name in the right mouse button menu. Likewise, the path to the original part file can be discovered by right clicking on the part in FeatureManager and selecting Component Properties.

This entire dilemma of needing to keep track of files used within an assembly and which files are externally referenced are problems not normally encountered in most AutoCAD drawings. Questions such as "What assemblies will be affected if I move this file?" are typically not relevant in AutoCAD unless you have external references to other files. With SolidWorks, these external references happen automatically with *every* part file in an assembly.

With AutoCAD, it is possible to create an assembly of externally referenced drawings (parts). This simulates SolidWorks natural mode of operation. If the externally referenced parts are changed, they can be updated within the AutoCAD assembly drawing. The overall level of associativity, however, is much less dramatic than it is with SolidWorks.

Unlike SolidWorks, if part geometry is referenced, AutoCAD does not care what happens to the referenced geometry unless that geometry is an Xref. The new part maintains no connection to the referenced geometry. Changing an AutoCAD part"s dimensions will not automatically affect every assembly the part is in, unless said part is inserted as an Xref into every assembly. In addition, Xrefs cannot be edited within the AutoCAD file unless the association to the original externally referenced drawing is broken. If these differences between AutoCAD and SolidWorks were summed up, it could be stated that AutoCAD files are not bidirectionally associative. SolidWorks files, on the other hand, are all associative to one another. This includes Part, Drawing, and Assembly files.

Assembly Modeling Methodology

An assembly can be used as a virtual prototype for a product. The assembly can be designed to mimic the location of components to determine fit and function, allow for the analysis of assembly motion for assembly components, and be used as a visualization tool for the assembled and exploded states of the assembly.

There are basically two methods that can be used to create an assembly. The top-down assembly method starts with the assembly, with components then built within it. The bottom-up method builds the components first and then assembles the components into the assembly. Skeletons and assembly layouts use simple geometry (planes, axes, and assembly sketches) that can be used to drive the size, location, and orientation of assembly components.

An effective approach to assemblies is a combination of the two approaches. The reference geometry or skeleton can be defined within the assembly. This allows the designer to identify the important features and functions within the assembly. By focusing on the design intent early in the process, you can produce robust, functional assemblies. You use top-down assembly design when in-context assembly references will be used to interrelate assembly components, or when using assembly skeletons or layouts. You use bottom-up assembly design when the parts used to create the assembly already exist, or when using customer- or supplier-defined geometry.

Skeletons and Layouts

Reference feature skeletons and assembly layouts are techniques used to define common references for part features and assembly components that use simple reference geometry and assembly layout sketches. A skeleton uses simple geometry (i.e., planes, axes, and assembly sketches) to drive the location or placement of features and assembly components. An assembly layout uses an assembly sketch to drive part geometry.

The goal of a skeleton or an assembly layout is to make simple, intelligent information available with little overhead, using an underlying sketch to control the movement of parts. The following illustration shows a robot assembly for which planes and (more importantly) an assembly sketch were used to drive the positioning of the components within the assembly. The underlying sketch has dimensions attached to it, just like any sketch would. These dimensions can be altered, and because the components in the assembly have relationships to the sketch, the components will move to accommodate these relationships.

Robot arm assembly.

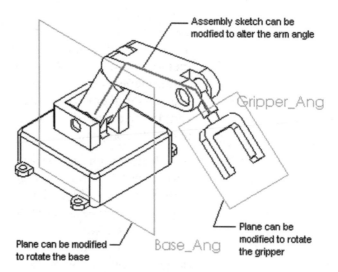

Assembly sketch can be modfied to alter the arm angle

Gripper_Ang

Plane can be modified to rotate the gripper

Plane can be modified to rotate the base

Base_Ang

This use of skeletons and assembly layouts to define the important mating features for an assembly design can be very useful. This helps the engineer or designer better plan how a design will function, and to describe movement to control the design intent and convey ideas to others. The simple sketches can also be renamed to help describe the purpose for the skeleton sketch and as a visual aid for documenting the purpose and function of a skeleton sketch.

Assembly Layouts

An assembly layout is an assembly sketch used to drive the location of assembly components. An assembly sketch can drive components created in or added to an assembly. An in-context assembly feature is a part feature created within an assembly. Part features that reference other parts or assembly geometry are directly related to (associated with) the features they reference.

Assembly layouts use the assembly sketches to control motion or location of assembly components. An example would be a linkage that follows a profile created within an assembly sketch. All components related via geometric relations (i.e., parallel, perpendicular, collinear, and so on) or dimensions move when the assembly sketch profile changes.

Using layout sketches reduces the amount of parent/child relationships. The components are related by one sketch in the assembly. Minimizing unnecessary parent/child relationships makes for assemblies (and parts) that are easier to manipulate and redefine.

The sketch can be fully or partially constrained, depending on how you want to modify the assembly once completed. You can control the level to which the sketch is geometrically or dimensionally constrained. Adding dimensions to the sketch allows you to precisely modify the geometry, whereas unconstrained or underconstrained geometry allows for a more free-form modification of the sketch geometry. Dragging the object handles of a sketch entity can be used to modify the sketch, and therefore the assembly. The following illustration shows an assembly layout sketch.

Assembly layout sketch.

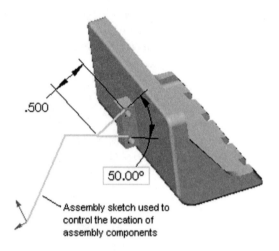

.500

50.00°

Assembly sketch used to
control the location of
assembly components

Assembly skeletons and assembly layouts are really the same
thing. These methods are ways of aiding in the assembly creation process and allowing for greater flexibility to the assembly
when finished. When assembly components are based on a
common object, such as an assembly layout sketch, you have
greater control over the assembly because many components in
the assembly are based on that common object.

These building tools are options, not requirements. They can be
useful when creating an assembly that will be used to show
movement. Many assemblies will never need the aid of these
tools and can be built piece by piece using mating constraints
and still show everything you are trying to show, such as movement or design configurations. It is suggested that you continue
with the rest of this chapter to get the full picture of assemblies
before attempting to implement an assembly layout sketch for
your particular assembly.

You will not find many similarities between AutoCAD and
SolidWorks when it comes to creating assemblies. One of the
largest portions of this chapter, and of creating assemblies, is
the application of mating relationships. These are intelligent
relationships between geometry, much like was done with constraints on 2D geometry within sketch mode. Because AutoCAD
does not make use of constraints, or in this case mating relationships, this chapter may present a new way of thinking.

Creating an Assembly

When creating an assembly, a plan should be developed to determine the method (i.e., top down or bottom up) used to create the assembly and the order of creation. The answer to the question of whether or not the assembly is going to be of the top-down or bottom-up variety is based on one condition. Are the parts built yet? If the answer is yes, or if the answer is that the parts will be built before the assembly is put together, you will be creating a bottom-up assembly. The bottom-up assembly is the easiest to create and requires two main steps, which follow.

• Bringing the components into the assembly

• Adding mates between components to move them into place

It is almost like putting together the pieces of a puzzle. The only tricky part for new users is figuring out what mates to use (covered later in the chapter). If the SolidWorks assembly is created or ordered in the same manner as the actual real-world assembly order, and you are already aware of how the assembly components should be placed together, putting the assembly together in SolidWorks should be fairly straightforward. When you plan an assembly, you should determine the following.

• What parts will go into the assembly?

• What is the assembly order for the product? The assembly should be defined from the first base component to the final assembly component, usually in the order it will be built in the real world.

• What is the function of each assembly component? Will an assembly skeleton or layout be needed to control these functions and define movement?

• What are the known additional features the assembly will need after it is built? These would be features that get machined into the assembly once it is put together.

• Will parts brought into the assembly be used in other assemblies? If so, will the original parts be modified later? It may be best to copy components to your project directory if the original may be used in other assemblies or modified by other users.

Another type of assembly is the top-down assembly. This assembly is created when the design of certain components is not going to be known until portions of the assembly are built.

The reason may be because measurements will not be known, or because certain factors in the shape of the part are difficult to determine and require existing geometry from components already placed in the assembly.

It should be noted that entire assemblies are usually not built this way. It is possible to build an entire assembly around a skeleton or assembly layout, as discussed previously, but this is not a typical real-world scenario. What is more common is when a few parts need to be built within the context of the assembly, usually because they must reference existing geometry. (Creating in-context parts is discussed later in this chapter.)

Starting a New Assembly

The steps for actually creating a new assembly are the same for creating a new part, with one minor difference. You would not need to be a rocket scientist to figure this out on your own, but the steps are listed here nevertheless for the sake of thoroughness. To create a new assembly, perform the following steps.

1. Select the New icon or click on New from the File pull-down menu.

2. Highlight the word *Assembly* and click on OK.

You will not notice much difference between the assembly window and a part window. There may be an Assembly toolbar that appears, and you will notice the MateGroup item in FeatureManager. In addition, the icon associated with the name of the assembly at the top of FeatureManager is slightly different than that for a part. Getting components into the assembly and mating them are the next objectives. These topics are covered throughout the rest of this chapter.

Assembly Creation Work List

Before actually running through the SolidWorks steps for inserting components into an assembly, you should review the factors to be considered before starting an assembly. The following is a step approach to creating an assembly.

1. Review the design concepts and determine the functional aspects of the assembly.

2. Determine the assembly method (top down or bottom up).

3. Create an assembly layout if desired or necessary for the assembly.

4. Determine what should be the first assembly component. This component is usually the main component in the assembly and the one that will remain stationary.

5. Insert components into the assembly.

6. Mate the components in the assembly.

7. Check for movement. This allows you to determine if the correct mates have been added.

8. Continue adding components, mating as necessary, and checking movement to determine interaction between components.

9. If at any time during the assembly process errors are encountered, such as overdefined components, search for the cause of the problem before continuing. Problems that are ignored will always get bigger!

Insert Component

Assemblies consist of parts and subassemblies. The assembly contains information on order and how the components are assembled. Parts or subassemblies can be inserted into an assembly by selecting the document window or by dragging the part or assembly icon from the Windows 98 or Windows NT Explorer into the assembly document window. The new component can be positioned precisely after insertion.

Assembly components can be left fully or partially unconstrained. These constraints can be added, edited, or removed at any time. Underconstrained components can be used to simulate assembly motion. The following illustration shows one method of inserting a component.

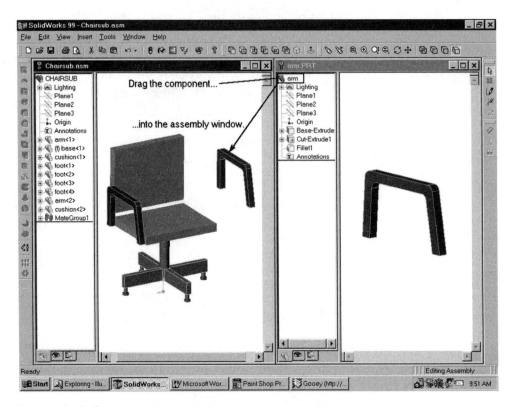

One method of component insertion.

There are four ways of bringing a component into an assembly, which are presented as steps in the material that follows. It is assumed you have already started a new assembly and no components have yet been inserted. As a reminder, be aware that the very first component brought into the assembly is usually the main component for the assembly.

By default, the first component will be fixed in space. In other words, its location will be locked and the component will not move relative to other components, or relative to the assembly planes or coordinate system. This can be altered at a later time. To unfix or "float" a component, right click on the component in FeatureManager and select Float. This is a toggle; therefore, right clicking on the same component a second time will allow you to Fix the component. The four methods of bringing a component into an assembly follow.

Method 1: Inserting a component when the component file is closed. To do this, perform the following steps.

1. Select Insert/Component/From File from the pull-down menu.

2. Select the type of file to be opened. Either Parts (*.sldprt) or Assemblies (*.sldasm) can be chosen.

3. Select the file to be opened.

4. Select Open to complete this function.

5. Select the location for the part/subassembly in the graphics window by picking in a blank area of the screen.

Method 2: Inserting a component when the component file is opened. To do this, perform the following steps.

1. Select the component name from the FeatureManager design tree of the part or assembly and drag the icon into the graphics window.

2. Release the mouse button.

Method 3: Inserting a component from the Windows File Explorer. To do this, perform the following steps.

1. Select the component file name from Windows Explorer and drag the icon into the graphics window.

2. Release the mouse button.

Method 4: Inserting a component when the component file is open and simultaneously utilizing Smart Mates (further covered in the section that follows). To do this, perform the following steps.

1. Select from the component the vertex, edge, or face you want to mate to the assembly.

2. Drag the component into the assembly by the selected vertex, edge, or face and drop the component onto the respective vertex, edge, or face of the assembly.

3. Use the Tab key on your keyboard to alter the alignment condition of the component you are bringing into the assembly. You must press the Tab key prior to letting go of the left mouse button.

Once the part has been brought into the assembly, it can be repositioned or mated (unless using Smart Mates). Just remember that the first component in an assembly is fixed by default. Other components will move in relation to the first component. One part can be brought into an assembly as many times as needed. If multiple occurrences of a component must be brought into the assembly, it is possible to drag an existing component from FeatureManager to the graphics window while holding the Control key down in order to create another instance of that component.

The different methods of inserting components into the assembly have their pros and cons. If you decide to drag components into the assembly via the Windows Explorer (method 3), you will find that you can insert multiple components at the same time. Merely select more than one file from Windows Explorer and drag any one of them to the assembly window.

If you insert a component using method 1, you have the opportunity to place the component directly on the assembly's origin point. This is particularly convenient when inserting the very first assembly component. It often will benefit you farther down the line if you position the first assembly component on the assembly origin.

Method 4 requires that you already have the component you are inserting opened in its own window. However, the extra mouse clicks involved in opening the component will very well pay off in the end. This is because as many as two mates can be added automatically just by dragging the component into the assembly. Read on for more on using of Smart Mates.

Utilizing Smart Mates

Smart Mates are a way of simultaneously inserting a component (part or subassembly) into an assembly and adding mates. As many as two mates can be added simply by dragging in the new component by the correct entity type and dropping it into the assembly window on the respective entity of the assembly. For example, if you drag a block by its edge into an assembly and drop it onto an edge of another block edge in that assembly, the two edges will be coincident. This would create a hinge-type mate between the two edges.

The Tab key on the keyboard plays an important role when using Smart Mates. When placing a component in the assembly, pressing the Tab key will alternate between alignment conditions. The component will appear to flip back and forth between alignment conditions.

There are a total of five different Smart Mates that can be created when dragging a part into an assembly. The trick to this functionality is to drag the proper entity and drop it onto the corresponding proper entity in the assembly. In the table that follows, the mate type is listed along with the entity that must be dragged (on the component being inserted) and the entity you must drop the component on (in the assembly).

Mate Description	*Drag this entity...*	*...onto this entity.*
Coincident between points	Vertex point	Vertex point
Coincident between edges	Edge	Edge
Coincident between planes or planar faces	Planar face or plane	Planar face or plane
Concentric	Cylindrical face	Cylindrical face
Concentric and coincident	Circular edge	Circular edge

It is possible to drag a vertex point or an edge onto a plane or planar face of a component in an assembly. This adds a coincident mate between the vertex point or edge and the plane or planar face. Typically this sort of mating relationship is not one that would be needed by most users, but it can be done.

It should also be noted that Smart Mates can be implemented even after a component has been brought into the assembly. In other words, it is possible to bring a component into an assembly using any method you desire, then mate the newly inserted component through the use of Smart Mates. Use the following steps to implement Smart Mates (it is assumed the component has already been inserted into the assembly).

1. Click on the Smart Mates icon on the assembly toolbar.

2. Double click on the entity of the component you want to mate to another component. This would typically be a surface.

3. Drag the component over to the component to be mated to. The component you are moving will appear transparent if shaded. Use the Tab key to reverse the alignment of the part being mated once it is in position. You must do this before letting go of the left mouse button.

4. Let go of the left mouse button to establish the mate.

Components can be completely mated in this fashion. Through a process of double clicks and dragging, you can add a series of mates quite easily and build your entire assembly! The down side is that you are limited to the five types of mating conditions as described in the chart shown in the previous section. Nonetheless, those five mates are very commonly used, and may very well be enough to get the majority of the mates in place.

Adding New Parts

When top-down design needs to be implemented, Insert/Component/New is the option to use. This function will create a new part in the active assembly. Insert/Component/New allows you to reference assembly components for geometrical references and constraints. The term *in-context editing* refers to constructing part geometry based on assembly geometry, literally building up the component *in the context* of the assembly. The new part can then be defined using existing assembly geometry references.

The advantages to creating and editing parts while in assembly mode is that the interdependent features between parts can be defined in such a way that when one of these feature changes in size or location, the other related features automatically update. Any references to external component geometry will create an in-context feature in the FeatureManager design tree. The in-context feature is a child of the parent feature. If the parent feature changes, the in-context feature will change. If the parent feature is removed, the in-context feature will fail to rebuild. When this happens, the part must be edited to remove any references made to external (parent) geometry.

In summation, an in-context part is advantageous because it will automatically update when referenced geometry updates. If an assembly is altered, a part created in the context of that assembly will change size or shape to accommodate the geometry referenced in the assembly. On the other hand, if geometry being referenced is somehow deleted, whether accidentally or on purpose, an in-context part referencing that geometry will display errors.

An Introduction to Error Recognition

An advantage SolidWorks enjoys during problem solving (i.e., references cannot be found, or become invalid due to the shape or size of related geometry) is that a display message will warn you of the failed reference and allow you to correct the problem. SolidWorks will also explain what happened to cause the failure.

Any problem area within the FeatureManager design tree will have an exclamation point (!) attached to it. Arrows pointing to the problem area also exist to guide you to the error. These locating arrows are especially helpful when an error is many levels deep within an assembly (e.g., a sketch for a feature in a part within a subassembly). To view the cause of the problem, right click on the problem area in FeatureManager and select What's Wrong. An explanation of the problem will be displayed.

InPlace Mates

An InPlace mate is added when inserting a component using the Insert/Component/New function. This special mate is added by SolidWorks automatically and the user has no choice in the matter. You could delete the mate afterward, but this is not recommended and will be explained shortly.

When an in-context part (a new part) is added to an existing assembly, it is almost always because you want to base that part on existing geometry that resides in the assembly. New features will be created that reference existing geometry. For this reason, the new part must remain stationary so as not to destroy the features that reference other geometry. Think what would happen if you molded wax over a mechanical part. Once the wax dried, you could not move the wax over the part without breaking it (unless possibly you were able to lift the wax straight up off the part). The point is, the wax is "externally referencing" the part. The wax cannot be moved without changing its shape. Likewise, SolidWorks adds an InPlace mate to keep new parts stationary.

In-context parts cannot be moved due to the InPlace mate, but they can be exploded for assembly-exploded views (covered later in this chapter). Do not associate SolidWorks' Explode function with AutoCAD's Explode command because they have nothing whatsoever in common. (As a reminder, in-context parts are created when you employ the Insert/Component/New function.) An InPlace mating constraint should never be removed unless you plan on removing all of the external references present in the part's feature geometry. This is typically not done, because it contradicts the whole reason for wanting to create an in-context part in the first place. The following illustration shows a part added using the New function.

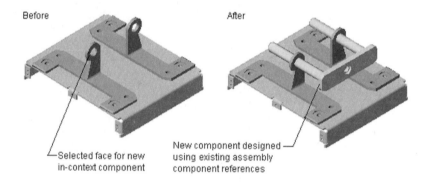

New part designed around existing geometry.

When a new part is created in an assembly, SolidWorks first asks you to give the part a name. This is because it is impossible to edit a part unless a part exists. Therefore, it must be named and saved. To create a new part in an active assembly, perform the following steps.

1. Select New from the Insert/Component menu.

2. Enter the name and file location in the Save As dialog box.

3. Select OK to continue.

4. Select a plane or planar face on which to begin sketching and to locate the new part in the assembly. SolidWorks will place you in sketch mode.

5. Create, dimension, and constrain the sketch.

6. Create the first base feature for the new part.

7. Create more sketch and feature geometry as needed to complete the part.

8. When finished creating the part, right click anywhere in the document window or on the assembly name at the top of FeatureManager and select Edit Assembly to return to editing the assembly.

As soon as the new part is inserted into the assembly, the assembly components will appear gray to let you know that you are no longer editing the assembly. Even though the assembly is still visible, it is the new part that is being edited. As new features are added to the part, they will appear salmon colored. This is SolidWorks' color scheme while editing a part in the context of an assembly. As soon as the right mouse button is clicked in the graphics window and Edit Assembly is selected, the components will regain their natural colors.

Part Color

When working with assemblies, it is a good idea to change the colors of different parts. Otherwise, differentiating between the components may become a difficult task, not to mention a strain on the eyes. AutoCAD, obviously, already has such functionality, typically controlled via layers.

If you are working in an assembly and want to change the color of an individual part, the procedure is somewhat different than if the part were opened in its own window. First, look at how to change a parts color (without the part being in an assembly). To modify a part's color in SolidWorks, perform the following steps.

1. Select Tools/Options.

2. Select the Color tab.

3. Select Shading and click on the Edit button.

4. Select a color, then click on OK to exit out of all dialog boxes.

There is a more straightforward method of changing the color of objects added to SolidWorks 99. This method involves using the Edit Color icon (which looks like a color palette) and is located on the Standard toolbar. The Edit Color icon can be used to change the color of a feature, selected faces, or a component in an assembly. You will probably want to use this method of color editing as it is easiest. The steps for using the Edit Color icon follow, but alternative methods will be spelled out in the rest of this section for reference purposes. To edit the color of a face, feature, or component, perform the following steps.

1. Select the Edit Color icon. This will open the Edit Color dialog box (or color palette as it is sometimes known).

2. Select the object you want to change the color of.

3. Select a color.

4. Choose whether to apply the color to a face, feature, or component.

5. Click on Apply.

6. Repeat steps 2 through 5 on any other desired objects.

7. Click on OK when finished.

The Edit Color dialog box is intuitive. It knows what is being edited, and if the user is in an assembly or part. For example, if an assembly is being edited and the face of a component is selected to have its color changed, the only option available that a color can be applied to is Component. If a part is being edited within an assembly and a face is selected, all three options of Face, Feature, and Component will become available. In short, the Edit Color dialog box does the job of all the other color editing options listed throughout this section. Some other alternatives follow, however, and you can use whatever method you feel most comfortable with.

If you are working in an assembly, here is another method of changing the color of a component. The procedure is somewhat different depending on whether or not you want the color change of the part to revert to the original part or only affect the part in the assembly. If you want the color change to affect the original part, you must be editing the part when the color change is made. To do this, perform the following steps.

1. Right click on the part in FeatureManager.

2. Select Edit Part. The part will temporarily turn orange.

3. Select Tools/Options.

4. Select the Color tab.

5. Select Shading and click on the Edit button.

6. Select a color, then click on OK to exit out of all dialog boxes.

7. Right click in the graphics window and select Edit Assembly. The parts will return to their original colors.

If you know you are creating components that will eventually wind up in an assembly, you should just go ahead and give parts different colors as you create them. This will save you from having to do it later when building the assembly.

The previous method will change the color of the part in the part file itself. If you want to change the color of the part in the current assembly only, perform the following steps.

1. Right click on the part in FeatureManager.

2. Select Component Properties.

3. Click on the Color button.

4. Click on the Change Color button and specify the desired color.

5. Click on OK to exit out of all dialog boxes.

This method will only affect the part in the current assembly and will *not* revert back to the original component. It is also the preferred method if you are working with parts that may be in other assemblies. After all, other people may like the current color scheme they are using and do not want the colors of their assembly components changed.

Assembly Mates

Assembly constraints are used to define the alignment characteristics of assembly components. Assembly constraints (known as mates in SolidWorks) allow you to define the assembly method for each component. The constraint chosen should simulate the actual assembly methodology. Using assembly mates that mimic the actual function of the component within the assembly allows you to mate assembly components in such a way that they can be manipulated for viewing their ranges of motion. Mating to planes instead of part faces has some advantages, including the following.

- Part features can be suppressed, or simplified configurations can be used for large assembly management without omitting features used to mate the assembly.

- Planes are less subject to change than part faces. Planes are not as dependent on existing geometry.

- Planes can be used to mate spherical or cylindrical parts that would typically have no planar faces that could otherwise be mated to.

Using a Distance mate allows you to move the component by changing a parametric value. This can be used to simulate component movement. This technique should not be used to explode a component. The Explode function should be used to define exploded assemblies, which will be discussed later.

There are a number of mating types, and they all have their purpose. Before adding mates, ask yourself how you want the part to move with respect to the part it is being mated to. If it were a pin, should it slide in and out of the hole? If it were a propeller, should it be allowed to spin on the shaft? If it is a hinge, should you allow for flexibility along a common axis? Perhaps the part should be firmly mated to another component without any freedom of movement. This greatly reduces the options and makes the task an easy one. The following illustration shows what will become a fully mated bracket on top of a box.

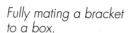

Fully mating a bracket to a box.

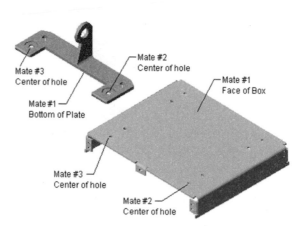

There are six ways a part can move. Different combinations of mates restrict certain movements, resulting in a limited degree of freedom of movement. The following are the six degrees of freedom of movement.

1. Translation along the x axis
2. Translation along the y axis
3. Translation along the z axis
4. Rotation about the x axis
5. Rotation about the y axis
6. Rotation about the z axis

What mates should be added to an assembly is dependent on how a given part should move in the assembly. The FeatureManager design tree object MateGroup is used to organize the component mates for an assembly. This MateGroup object can be thought of as a "folder" for keeping all of the mates organized. A component mate listed in the MateGroup describes the type of mate and the features mated. The following illustration shows the attributes of items listed in the MateGroup.

Mate attributes
of a mate feature.

Reference 1 and reference 2 in the previous illustration refer to the selected faces or entities the mate was placed between. To add mating relationships, perform the following steps.

1. Select the Mate icon (shown in the following illustration) or select Mate from the Insert menu.
2. Select the entities to apply the mate between. These are usually planes or planar faces, but can be vertex points or edges in certain conditions.
3. Add the desired mating constraint. Only one mate can be added at one time.
4. Select Apply to continue.

Mate icon on the Assembly toolbar.

Mates are usually added between planes or faces on components within an assembly. Mates can also be added between edges or vertices, but not as commonly. New users should attempt to add mates between faces if possible, as they are more easily understood for those having problems grasping the meaning of the mating relationships. Read on to get a better understanding of the various mates and what they do.

Mating Constraint Types

In Chapter 5, you learned that it was possible to place relations between entities while sketching. These relations can be between arcs, lines, points, or any other entity type. The relations could be parallel, perpendicular, coincident, and so on. Now it is time to take that knowledge and apply it to the 3D world.

What you learned in sketching will be applied almost exactly the same way to parts within an assembly. In a sketch, the entities to be constrained are selected, the Add Relations icon selected, and the relation added. In an assembly, the faces (or other entities) are selected, the mate icon selected, and the appropriate mate applied. The process is the same, but the mechanics are different. The following are the various types of SolidWorks constraints available in the Mate dialog box.

Coincident	Constrains components so that they lie on the same plane.
Perpendicular	Constrains selected objects so that they lie perpendicular to one another.
Tangent	Constrains components so that they remain tangent to one another.
Concentric	Constrains cylindrical surfaces so that they share the same center axis.
Parallel	Constrains components parallel at an unspecified distance from one another.
Distance	Constrains components so that they are a specified distance from one another.
Angle	Constrains components so that they are a specified angle from one another.

Included in the Mate dialog box are the options Aligned, Anti-aligned, and Closest. These are not mates, but alignment conditions for mates. Their meanings are as follows.

Aligned	Places a component's material on the same side of the planes selected for mating.
Anti-aligned	Places a component's material on the opposite side of the planes selected for mating.
Closest	SolidWorks decides on Aligned or Anti-aligned, whichever is closest.

The following illustration shows a simple example of the Aligned and Anti-aligned options. It is not extremely important that you understand these conditions one hundred percent. If the option for alignment is left to Closest, SolidWorks has a 50/50 shot of getting it right. You can normally ignore this setting and just apply or preview the mate. If the mate is incorrect, simply select the opposite alignment condition.

Aligned versus anti-aligned.

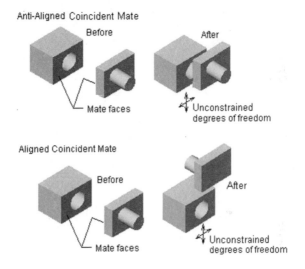

Deleting Mates

Deleting mates in SolidWorks is as easy as deleting anything else. Select the Mate from FeatureManager MateGroup and press the Delete key. To delete a mate, perform the following steps.

1. Select the mate from the FeatureManager MateGroup.

2. Select Delete from the Edit menu or press the Delete key.

3. Select Yes to continue and confirm the deletion.

The table that follows charts types of mating relationships for various types of geometry. Most mating relationships are fairly obvious. For instance, it would be impossible to add a concentric mate between two planes. For the sake of thoroughness, this table has been included.

	Point/Vertex	**Line/Axis**	**Plane**	**Cylinder**
Point/Vertex	Coincident Distance	Coincident Distance	Coincident Distance	Coincident Concentric
Line/Axis	Coincident Distance	Coincident Distance Parallel Perpendicular Angle	Coincident Distance Parallel Perpendicular	Coincident Tangent Concentric
Plane	Coincident Distance	Coincident Distance Parallel Perpendicular	Coincident Distance Parallel Perpendicular Angle	Tangent
Cylinder	Coincident Concentric	Coincident Tangent Concentric	Tangent	Tangent Concentric

Introducing Fix and Float

All components in an assembly are either fixed in space or are able to freely be rotated or moved with respect to the assembly coordinate system. This is true for all assembly components unless the component has been mated. When a component is able to move or rotate, it is said to be floating. If a component is locked at a particular position, it is fixed.

Fixed components cannot have additional mating constraints placed on the component. If a part is fully defined in an assembly, it will appear without a prefix in FeatureManager. If it has any degree of freedom of movement, a small minus symbol will prefix its name. If the component is fixed in space, the component's name will be prefixed with an "f." The following illustration shows component constraint prefixes.

Component constraint prefixes.

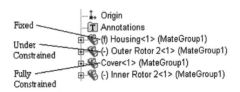

Components that have been mated should not be fixed, and vice versa. However, it is allowable, and sometimes necessary, to temporarily Fix components. If this is done, do not forget to go back later and Float the component in question. The following section provides a more detailed explanation of fixing and floating.

Modifying Assemblies

After components have been added to an assembly, SolidWorks allows the components to be edited and repositioned, and new configurations to be defined. Depending on the defined mating constraints, assemblies can also simulate assembly motion.

Fix Component

Assembly components can be fixed within an assembly. A fixed component cannot be moved. The first component in an assembly is fixed by default. The Fix attribute can be removed by using the Float function. Right click on the component in FeatureManager and select Float. The component can now be moved. To fix the location of a component, perform the following steps.

1. Right click on the component in FeatureManager.

2. Select Fix.

To remove a component's fixed location attribute, perform the following steps.

1. Right click on the component in FeatureManager.

2. Select Float.

As mentioned previously, a fixed component will display an "f" in front of the component name in FeatureManager. Mate relationships should not be added to fixed components, and mated components should not be fixed. Usually, a plus sign (+) prefix is added before the part name in FeatureManager if it is overdefined. However, there is one exception to this, and that is if a part is fixed and constrained at the same time. The "f" prefix will be displayed instead of a plus sign, thereby hiding the fact that the part is overdefined. Be aware of this so that it does not catch you off guard.

Move Component

Assembly components can be moved, depending on their active assembly constraints or mating conditions. The assembly component will move depending on its remaining degrees of freedom. For example, if a component is constrained coincident with the *x-y* plane, the component will move only along the *x-y* plane, but not in the *z* direction.

As mentioned previously, the first assembly component is fixed by default (automatically). This component can be moved only after the Float condition has been applied to it. The assembly components added afterward are allowed to move or rotate along any unconstrained axis. Move Component can be used to show assembly motion. The assembly components are allowed to move along the unconstrained degrees of freedom previously mentioned.

As a reminder, these degrees of freedom of movement are translation along the *x-y-z* axes and rotation about the same. This totals six degrees of freedom of movement. This allows for some simple assembly motion studies but does not provide the capabilities of a complete kinematics software program. To move a component, perform the following steps.

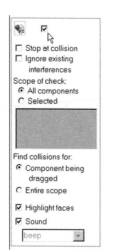

PropertyManager displays collision detection information when Move Component icon is selected.

1. Select the component in FeatureManager.

2. Click on the Move Component icon, or select Tools/ Component/Move.

3. Hold down the left mouse button and drag the component to its new location.

Move Component is similar to AutoCAD's Move command only in its ability to move an object from one point to another. In SolidWorks, unlike AutoCAD, there is no need to enter displacement values or pick base points, and components can be dynamically moved from one position to another.

Another aspect of moving assembly components is collision detection. Once the Move Component icon is selected, the PropertyManager tab is displayed and temporarily takes the place of FeatureManager. A portion of PropertyManager is shown in the following illustration. If the Collision Detection option is checked (see top of previous image), collision detection will occur for selected components.

Collision detection can be performed for all components by checking the All Components option. What happens when a collision takes place can also be controlled, such as a warning beep or face highlighting. These options are self-explanatory and will not need further elaboration here.

Rotate Component

Assembly components can be rotated, depending on their active assembly constraints. The Rotate Component Around Centerpoint (shortened after this point to Rotate Component) function works only with unmated assembly components. Rotate Component cannot be used to show assembly motion because only completely unconstrained parts can be rotated.

Another option, Rotate Component About Axis, allows for rotating components about a selected axis or edge. Unlike its Rotate Component counterpart, this function can show assembly movement very well, as long as the mating relationships have been correctly applied. Take a look at the following illustration to see an example of what might be accomplished with the Rotate Component About Axis function.

Using Rotate Component About Axis.

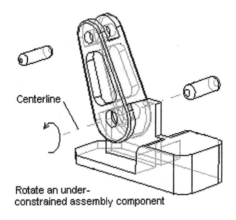

Centerline

Rotate an under-constrained assembly component

To rotate a component that has not been mated, perform the following steps.

1. With the left mouse button, select the component in FeatureManager.

2. Click on the Rotate Component Around Centerpoint icon.

3. Hold down the left mouse button and drag to rotate the component to its new location.

To rotate a component around an axis, perform the following steps.

1. With the left mouse button, select the component in FeatureManager.

2. Hold down the Control key and select an edge or axis of rotation.

3. Click on the Rotate Component Around Axis icon.

4. Hold down the left mouse button and drag to rotate the component to its new location.

When components or assemblies are manipulated onscreen in any way, the system will temporarily turn the display mode to wireframe if Hidden Lines Removed or Hidden in Gray have been selected. This is done because attempting to recalculate hidden lines during the rotation process would result in too many computations to make redisplaying in those modes feasible. Therefore, the recalculation of hidden lines is not completed until you let go of the mouse button. Wireframe and shaded modes will offer the best performance when moving or rotating components, but you will probably want to use shaded mode as it is more aesthetically pleasing.

Edit Part

Editing a part within an active assembly allows you to reference assembly component geometry while creating new features. This allows for other parts to be created in-context to the assembly. Creating geometry using a reference to another assembly member creates an external reference to that assembly component.

The advantage to this methodology is that design features can be logically interrelated. For example, a bearing hole can be created with .001 clearance to the mating shaft. The clearance is maintained between these components any time the shaft dimension changes. The disadvantage is that if the parent part or feature is removed, the referenced feature will fail to find the parent part or feature and fail to rebuild.

Any time a feature or part is created with external references, an arrow is added as a suffix to the part or feature with the external reference. As previously discussed, creating parts or features in the context of an assembly automatically creates an external reference. Take a look at the following illustration to view the external reference symbol at the end of a part in FeatureManager.

External References arrow.

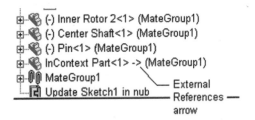

An assembly component can be opened in a new document window. This allows the document to be edited without the visual clutter of the other assembly components and ensures that no unintentional assembly references are made within the document. This is a shortcut method that allows you to open an assembly with two mouse picks: right click on the component and select Open *Filename.sldprt*. This shortcut can actually be shortened to just one mouse click if the right mouse button is held down after right clicking on the part name. Hold down the mouse button and highlight the desired menu option. When the mouse button is released, the desired choice is selected.

➥ **NOTE:** *The right mouse button click, select and release method will work with any right mouse button menu command in any Windows program. This is not a function that pertains only to SolidWorks.*

The alternative to opening the part in its own window is to edit the part within the context of the assembly. Each method is presented in the material that follows. To edit a selected part file in the active assembly, perform the following steps.

1. Right click on the part name in FeatureManager.

2. Select Edit Part. The part will display in a salmon color, and the rest of the assembly will display as gray.

3. Modify the part as desired.

4. To return to the assembly, right click in the graphics window or on the assembly name in FeatureManager and select Edit Assembly.

To open a selected assembly component in its own document window, perform the following steps.

1. Right click on the part name in FeatureManager.

2. Select Open PartName.sldprt. The part file will be opened independent of the assembly.

3. To return to the assembly, click anywhere in the assembly graphics window to make it active or select the assembly name from the Window pull-down menu.

Assembly Information

Find References in the File menu displays the file location of assembly components. This is an ideal way of seeing just what files are in the assembly and where those files are located. Optionally, you can copy these files to a new location using the Copy Files button. External file references can be lost when a file is moved or deleted using the Windows Explorer. When the SolidWorks software cannot find an external file it will prompt you to locate the file. More specifically, it will ask you if you want to Browse for the file.

The advantage to using SolidWorks to copy a document is that any existing external references are kept intact. Otherwise, if the file was copied using Windows Explorer, the external file references may be missing. The following are some of the actions that will create an external file reference.

- Components (parts and subassemblies) within an assembly (these are referenced by the assembly).

- Parts and assemblies used in drawings (these are referenced by the drawing).

- In-context assembly features. The referenced feature becomes an external reference for the part.

- The Base Part function. The base part becomes the external reference for the new part.
- The Mirror Part function. The original part becomes the external reference for the mirrored part.

To display the file location of components used in an active assembly, perform the following steps.

1. Select Find References from the File menu.

2. Select Close to exit the dialog box when finished.

To copy the components used in an active assembly to a new location, perform the following steps.

1. Select Find References from the File menu.

2. Select Copy Files to copy the assembly and related component files to a new drive or directory.

3. Select the location of the new documents. The copied files will have the same file name as the originals.

4. Check Preserve Directory Structure to maintain the entire directory structure for every part listed in the dialog box. This is optional. If this is not selected, all assembly files will be placed in the location you choose.

5. Select Close to exit the dialog box.

The Copy Files function can be used to make changes to a released document in a temporary working directory. It can also be used to create a new assembly similar to an existing assembly or to explore experimental changes without affecting the original assembly. The system will prompt you prior to overwriting any files, although when copying an entire assembly it is best to create a directory specifically for that purpose.

Component Properties

Component Properties can be used to control and define specific characteristics of the parts in an assembly. These properties are used to suppress or hide a component, change the component's source file or location, exchange the component with a different part or assembly, or specify a named configuration for use in the assembly.

It is possible to control what configuration is being used in specific assemblies. For instance, a part may have many configurations, but one configuration in particular is needed for the assembly. It is possible to control, from the assembly, exactly which configuration of a part is being used within the assembly. This feature is extremely important in reducing the number of files needed to create multiple assemblies that require different revisions of the same part. The part does not have to be copied every time it needs to be placed in a different assembly.

External file references can be lost when a file is moved or deleted. When the system cannot find an externally referenced file, it will prompt you to Browse for the file. This occurs when, for example, you attempt to open an assembly that references components (part files) that are missing from the hard drive. There is a wrong way and a right way to rename or move Solid-Works files.

The Component Properties of an assembly component allows for replacing one component with another. Component replacement usually works beautifully when performed on a component that has simply been mated in an assembly. You should be aware, however, that problems may arise with geometry references if a part with externally referenced geometry is replaced. These lost geometry references must either be removed or redefined (in essence, reattached to the new component).

Through the use of Component Properties, the following characteristics of an assembly component can be controlled.

- Component location
- Component name (if "Update component names when documents are replaced" is left unchecked in the External References tab of the Tools/Options dialog box)
- Visibility
- Suppression state
- Ability to use or override visibility properties for parts contained within subassemblies
- Display of defined configurations for individual parts or subassemblies

- Component color

To change a component's properties, perform the following steps.

1. Right click on the component and select Component Properties.

2. Set or change the desired properties.

3. Select OK to continue.

Not all of the properties in the Component Properties dialog box are adjustable. Some are hard coded by the software. These are pointed out in the list that follows, which contains the Solid-Works component property types and descriptions of each.

Component Name	The component name. This name is derived from the component name and cannot be changed unless "Update component names when documents are replaced" is left unchecked in the External References tab of the Tools/Options dialog box. When that option is unchecked, the component name will be independent of the actual part file name of the component.
Instance ID	Indicates the number of the component in the assembly when more than one instance of the component is present. This is not the total number of instances. It is SolidWorks' way of identifying the same components as being individual instances within the assembly. If an instance is deleted, SolidWorks will not reorder the ID to begin at number 1 for the remaining instances. This item is maintained entirely by the system.
Full Name	Component Name plus the Instance ID number.
Model Document Path	Displays the path to the component.
Browse...	Select this button to view the Open (file) window in order to locate a replacement component or specify a new path to the component if it has been moved. This is only available for assembly parts or subassemblies, not individual subassembly components.
Use visibility properties specified in configuration "*<configuration name>*" of "*<subassembly name>*"	This option is only available for subassembly components. Selecting this option uses the visibility settings specified in the subassembly.
Override visibility properties specified in configuration "*<configuration name>*" of "*<subassembly name>*"	This option is only available for subassembly components. Selecting this option uses the visibility settings of the assembly.

Hide Model	Check to hide the component and the component's design tree. When the component is not shown, the component's icon will appear as an outline in FeatureManager.
Suppression State	This section allows for setting a component to either Suppressed or Resolved. Suppressing removes the component from memory. The term *Resolved* is used when referring to assembly components, whereas the term *Unsuppressed* refers to part features. Both terms have essentially the same meaning.
Color	Click on this button to change the component's color. This information will not revert back to the original part. It will only take affect in the current assembly. In addition, this is a great way to show interior workings of an assembly by making outer components transparent. Use the Advanced button after clicking on Color to change advanced color characteristics.
Use component's "in-use" or last saved configuration	Select this option to use the active or last saved configuration of the selected part.
Use named configuration	Select this option to use a named configuration. Select a configuration name from the drop-down list. The list will be directly connected to the configurations created in the part file.

Assembly Structure Editing

There are a number of aspects of an assembly the SolidWorks user has control over. This includes the ability to reorder components in FeatureManager, form new subassemblies from existing components, or dissolve subassemblies into separate components. This section will take a look at these options and will instruct you in how to accomplish these functions.

Component Reordering

This is one of the easier aspects of an assembly over which the user has control, so you take a look at it first. Reordering was first examined in the last chapter with regard to features (see the section in Chapter 6 titled "Parent/Child Relationships"). Reordering assembly components doesn't require any rebuilding, so this function is easier on the computer than reordering features. The process is exactly the same as reordering features, and is accomplished using the following steps.

1. Position the cursor over the component you want to reorder.

2. Drag the component to a new position in FeatureManager using your left mouse button. Drop the component being moved to the component you want to see it appear after.

That is all there is to it, but it doesn't answer the question as to why you would want to reorder assembly components. Consider a bill of materials. SolidWorks uses the order of the components in the assembly when creating a bill of materials (BOM). Because the BOM is always sorted by the order in which components were added to an assembly, you may find it useful to reorder those components, thereby having control over the order of components in the BOM.

Restructuring Components

What is meant by restructuring? Restructuring is nothing more than dissolving subassemblies into separate components or forming new subassemblies out of existing components. The user has control over these aspects of an assembly, but it should be noted that restructuring may adversely affect the assembly. Before doing any restructuring of assembly components and subassemblies, consider the following.

- Components created in the context of an assembly cannot be moved to another subassembly. These include components such as weld beads.

- Any assembly features will be deleted if components affected by the features are restructured.

- Explode steps will be deleted if components involved with the explode steps are restructured.

- Equations can be affected due to component names changing. Specifically, the instance numbers at the end of a component name may change if that component is moved, and this will have an impact on existing equations.

In the case of assembly features or exploded view steps, nothing will be deleted without warning you first. The user will have an opportunity to cancel out of any restructuring effort if he or she changes their mind. To dissolve an existing subassembly and break it down into individual components, perform the following steps.

1. Right click on the subassembly to be dissolved.

2. Select Dissolve Subassembly.

Any mates present in the subassembly will be transferred to the overall assembly. To create a subassembly out of existing components, perform the following steps.

1. Select the components that will belong in the new subassembly. This is most easily accomplished by selecting the components in FeatureManager.

2. Select Insert/Component/Assembly from [Selected] Components.

3. Give the new subassembly a name.

4. Click on Save to complete the process.

Any mates between components used to form the new subassembly will automatically be transferred to the Mate Group of the new subassembly. Feel free to reorder the newly created subassembly if desired.

Advanced Assembly Features

This section discusses many of the advanced features available within a SolidWorks assembly document. Certain operations must be completed within an assembly because those operations require more than one part to be open at one time. Remember, unlike AutoCAD, a SolidWorks part can only contain one contiguous solid model. Operations that require multiple components, such as the Join operation or the Cavity routine, must have two or more components to be completed.

Another requirement that often comes into play in operations such as these is that a new part will be created. Therefore, it is necessary to begin a new part and save it first, much like the sequence needed when inserting a new component into an assembly.

Assembly Component Pattern

This function creates a pattern of a selected component using a pattern defined in another part or by defining a new pattern within the assembly. Component patterns can be defined using either a linear or circular pattern. Assembly patterns can be used to create a copy of an assembly component without having to insert and mate the same component repeatedly. This pattern has parametric values that can be edited just as part feature patterns can be edited.

As previously mentioned, the assembly pattern can be based on an existing component feature pattern, or the pattern can be created locally within the assembly. If the pattern is based on a component feature pattern, assembly components will be

Assembly component pattern.

dependent on the component feature pattern. If the feature pattern changes, the assembly components update accordingly. The illustration at left shows an assembly pattern. To create an assembly pattern based on an existing feature pattern, perform the following steps.

1. Insert the component containing the pattern into the assembly.

2. Insert and mate the component to be patterned into the assembly.

3. Select Component Pattern from the Insert menu.

4. Select the "Use an existing feature pattern (Derived)" field.

5. Select Next to continue.

6. Pick inside the Seed Component field and select the component to be patterned from FeatureManager.

7. Pick inside the Pattern Feature field and select an existing pattern feature from one of the components listed in FeatureManager.

8. Select Finish to continue.

To create an assembly pattern based on a locally defined pattern, perform the following steps.

1. Insert and mate the component to be patterned into the assembly.

2. Select Component Pattern from the Insert menu.

3. Select the "Define your own pattern (Local)" field.

4. Select either the Linear or Circular option.

5. Select Next to continue.

6. Pick inside the Seed Component field and select the component to be patterned from FeatureManager.

7. Pick inside the Along Edge/Dim field and specify an edge, axis (if a circular pattern), or dimension that will define the direction of the pattern.

8. Specify the Spacing distance (angular if a circular pattern, linear distance if a linear pattern).

9. Specify the number of Instances.

10. Specify Reverse Direction if needed.

11. If a linear pattern was specified and a second direction is required, select Second Direction and repeat steps 7 through 10.

12. Select Finish to continue.

It is possible to delete instances after creating a pattern. If this is done, the deleted items will appear in the Items to Skip field when editing the definition of the Component Pattern. To delete an instance, select a surface on the desired instance and press the Delete key. SolidWorks will ask if you want to delete the entire pattern or just an instance. Make sure you specify just an instance. If at a later time you want to undelete the instance, edit the definition of the pattern and delete the instance name from the Items to Skip field.

Assembly Features

An assembly feature is used to add features to an assembly that will not revert to the original components. This function can be used to show cutaway views of an assembly for display purposes. Assembly features can only be made as cuts, and cannot add material to the assembly. Assembly features can be used to add important design features that would normally be machined into the assembled model. They can also be used to create some very nice demonstrations.

The Feature Scope function is used to define which assembly components an assembly feature will affect. The feature scope can be changed after an assembly feature has been defined. Therefore, just like most anything else in SolidWorks, you are not locked into a decision. Multiple assembly features in the same part can have different feature scopes. The following illustration shows an assembly feature in which a 90-degree section of an assembly has been removed.

Two components were not included in the feature scope.

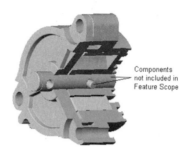

Components not included in Feature Scope

The desired feature scope should be defined prior to creating an assembly feature. Otherwise, the assembly feature will affect all parts in the assembly and the feature scope would more than likely need to be edited. This is not a problem, but it is easier to set the feature scope beforehand. To specify the scope of an assembly feature, click on Feature Scope in the Edit menu. Any changes to the feature scope will affect only new assembly features. To define the scope of an assembly feature, perform the following steps.

1. Select Feature Scope from the Edit menu.

2. Select items to add to the Feature Scope by selecting components in FeatureManager.

3. To delete items from the Feature Scope, highlight the item in the Feature Scope list box and press the Delete key.

4. Select OK to continue.

When the assembly feature is added, only those items listed in Feature Scope will be affected. Assembly features can be only cuts or holes, as previously stated. To modify the scope of an existing assembly feature, perform the following steps.

1. Right click on the assembly feature in FeatureManager.

2. Select Feature Scope.

3. Select components in FeatureManager to be added, or highlight the component in the Feature Scope list box to be deleted and press the Delete key.

4. Select OK to continue.

At this point, you could probably figure out for yourself the steps used to create an assembly feature. However, to make things as straightforward as possible, the following steps will guide you through the entire process of creating an assembly feature. It is assumed you know how to define the feature scope. To create an assembly feature using a cut, perform the following steps.

1. Define the feature scope for the assembly feature (Edit/Feature Scope).

2. Select a plane or planar face.

3. Select the Sketch icon or select Sketch from the Insert menu.

4. Sketch a profile to use as the cutting profile.

5. Dimension and geometrically constrain the profile as necessary.

6. Select Insert/Assembly Feature/Cut, and specify Extrude or Revolve.

7. Select the End Condition and Depth. Select any other option as required.

8. Select OK to complete this function.

To create an assembly feature using a hole, perform the following steps.

1. Define the feature scope for the assembly feature (Edit/Feature Scope).

2. Select on the face where the hole will be created.

3. Select Simple from the Insert/Assembly Feature/Hole menu.

4. Specify the End Condition and the Diameter.

5. Select OK to complete this function.

After creating the hole, you will more than likely need to edit the sketch for the hole in order to position the sketch at a particular location. Edit the sketch in the usual manner and position it using dimensions or constraints, depending on your design intent.

In summary, the following guidelines should be used when creating an assembly feature.

- The scope of an assembly feature should be defined prior to creating an assembly feature.
- An assembly feature will affect only components listed in the Feature Scope.
- The scope of the assembly feature can be edited to include or exclude assembly components after the assembly feature has been created.
- Assembly features will not revert to the original components.

Interference Detection

Interference checking is used to determine if two or more components interfere with one another. This is a useful feature because assemblies can be checked for problems prior to physical prototypes. Assembly interference can be identified and corrected early in the design process. Interference detection can only be performed in an assembly.

The system will display the interference by creating a rectangular boundary box around the interference. This box is a representation of the interference and illustrates the smallest possible bounding area the interference could exist in. This helps illustrate the height, width, and length of the area of concern. Objects to be included in the interference check must be selected. These selected assembly components will highlight in the usual green color if using shaded mode. There may be more than one area of interference. The following illustration shows interference detection.

Interference detection.

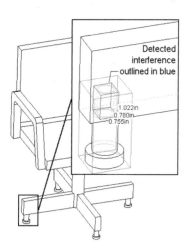

Detected
interference
outlined in blue

1.022in
0.780in
0.755in

To check for component interference in an assembly, perform the following steps.

1. Select Interference Detection from the Tools menu.

2. Click in the Selected Components list box and select the components to be checked from FeatureManager.

3. Click on the Recheck button.

4. Pick the interferences listed in the Interference list box to display each occurrence of interference. New components can be selected by holding down the Control key and selecting or deselecting components. Select Recheck to start an interference check with the selected components.

5. Select OK to exit, or select Recheck to recheck interference with newly selected components.

Each listed interference can be viewed graphically by selecting it from the Interference list. The system will graphically display where and how much interference occurred. The fields Component1 and Component2 will display the component names for the selected interference. Components can be selected before or after starting this function, as is the case with most SolidWorks dialog boxes.

AutoCAD allows for checking interference and actually creating a solid from the result. The solid can then be used in a Boolean operation to modify components in the assembly. SolidWorks does not permit this. It is up to you to find the interference and then modify the assembly as needed.

As you learned earlier, it is possible to simulate motion in an assembly. It would be ideal if the interference detection could be run at the same time motion was being applied to the assembly components. However, each position must be defined manually and checked individually within SolidWorks. Concurrent function is the domain of high-end modelers and analysis software and is outside the limitations of SolidWorks without special add-on packages.

Insert Cavity

Assembly components can be used to create cuts in other assembly components, very similar to an AutoCAD Boolean subtraction. A common use of this function is to create a cut in a mold base (or mold base insert) that defines the mold cavity. This method allows you to enter part shrinkage that takes into account the cooling of the material, which is something AutoCAD only offers through the use of the Scale command.

The mold cavity is derived from the design part geometry, and any changes to the design part will be updated in the mold cavity parts. A cavity block or insert could be defined and reused to create two or more pieces of a mold. Care should be taken to create a copy of the cavity block because the mold creation process will change the part. Save As can be used to rename the original cavity block part, along with some other options explained next.

Derive Component Part can be used to break a mold base into multiple mold parts. The cavity function is performed first to cut the design component from the mold base. The following are differences between Base Part, Cavity, and Derived Component Part. The illustration that follows shows a cavity.

- Base Part is used when a single part can be used to create a new part. An example would be a casting and the machined finished part. The new "machined" part references the base part.

- Derived Component Part is used when a part needs to be created from an assembly component part. An example would be a mold base. Once the mold base is created, the top and bottom of the mold can be derived from it using Derived Component Part.

- Cavity is used to create a cut in an assembly component while applying a shrinkage value. This also requires assembly references to the cavity block and the design component. This function is used typically before the Derived Component Part function. Whereas Cavity would be used to create the cut itself, Derived Component Part would be used to create both halves of the mold.

Results of the Cavity routine.

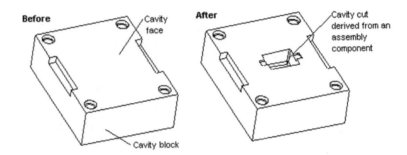

The following are the required documents when you create a cavity cut.

- Create the cavity block or mold base. This will be the target for the cavity cut.

- Create the design component used to define the cavity cut in the cavity block or mold base.

- Create an assembly that contains the cavity block and the design component. The assembly should be mated to ensure proper placement of the design components.

To create a cavity cut, perform the following steps.

1. Create the mold base.

2. Create the molded part.

3. Create a new assembly.

4. Insert both parts into the assembly.

5. Mate the mold base to the component. Distance and coincident mates are commonly used for this. Ideally, the design component should be centered within the mold base.

6. Select the Save icon or select Save from the File menu to save the assembly.

7. Right click on the mold base component in FeatureManager and select Edit Part.

8. Select Cavity from the Insert/Features menu.

9. Select the Design Component.

10. Specify a Scaling Factor, if required.

11. If a scaling factor is used, specify the Scaling Type.

12. Select OK to continue.

You might know some of the Cavity routine properties. The following expand on these properties.

Design Component	Displays the name of the design components used to define the cavity. More than one design component can be selected. Click in the list box before selecting components.
Scaling Type	Select About Component Centroids, About Component Origins, or About Mold Base Origin. This is analogous to AutoCAD's Base Point when scaling geometry.
Scaling Factor in %	Enter the scaling (shrinkage or expansion) factor for the material being used (+/– 20%).

Some important guidelines for performing the Cavity routine follow. This summary is provided to help new users through this somewhat advanced topic.

- Create a backup copy of any standard parts (i.e., mold cavities) so that if they are accidentally overwritten, the original part can be retrieved.

- It may be convenient to create a directory with standard mold base sizes for common use.

- Multiple components may be placed in the mold cavity.

- The mold base normally would totally enclose the component parts, but this is not mandatory.

- The part should have a datum plane that can be used as the parting line. This will allow for easy alignment and cutting the mold components. This is a suggestion; it is not mandatory.

- Create derived component parts to split the mold into multiple parts, or use Base Part to create the halves of the mold.

Derived Component Part

A derived component part is used to easily create a new part from an assembly component. A new part is derived from the original assembly component. Any changes to the original assembly component will be reflected in the derived component parts. The example described in this section is a mold base created via the derived component part method. The following illustration shows a derived component part.

Possible results of using a derived component part.

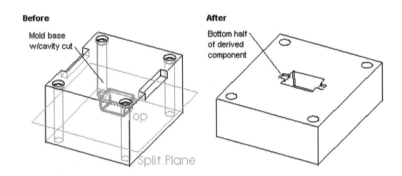

Before

Mold base w/cavity cut

op

Split Plane

After

Bottom half of derived component

To create a component derived from an existing assembly component, perform the following steps.

1. Using the left mouse button, select the desired assembly part in FeatureManager.

2. Select Derive Component Part from the File menu.

3. A new window will open with the derived part file.

4. Change or add features to the new part. In the previous illustration, a cut was performed to chop the derived part in half (discarding the top half).

5. Save the part under a new name.

Remember, any features that reference another feature in another part or assembly is called an in-context feature. In-context features are shown in the FeatureManager design tree using the in-context arrow. This also holds true when creating parts based on other parts, such as using the Base Part or Derived Component Part functions.

Explode

Explode expands the active assembly configuration, revealing how the assembly would be put together. This image can be used on drawings to describe assembly sequences or images for user manuals. Each assembly step can be defined within the exploded view. Only one exploded view can be defined within a configuration.

Exploded views are controlled using ConfigurationManager. This is a FeatureManager area used for setting, defining, and editing configurations.

➥ **NOTE:** *Configurations were discussed in Chapter 6.*

Each configuration can have its own exploded view. To set the exploded view to a specific assembly configuration, activate the desired configuration prior to creating the exploded view.

Explode views can be defined to show the assembly by exploding each step of the assembly process one by one. The step editing tools (shown in the illustration of Assembly Exploder further in this chapter) allow you to define and modify multiple assembly steps with the current exploded view. Assembly steps can be added, edited, or deleted. The following illustrations show an exploded assembly and the objects within ConfigurationManager.

An exploded assembly.

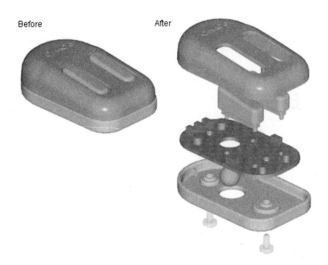

ConfigurationManager objects.

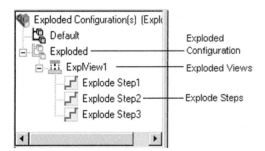

To add an exploded view to an active configuration, perform the following steps.

1. Open ConfigurationManager by selecting the Configu-rationManager tab at the bottom of FeatureManager.

2. Right click on a configuration to create the exploded view and select New Exploded View.

3. Select Auto Explode to allow SolidWorks to explode the assembly automatically.

4. The exploded view may be edited as desired (see the next procedure).

5. Select OK to continue, or select Cancel to quit.

To edit an exploded view or create one manually, you would perform the following steps. The illustration that follows shows the Explode dialog box.

1. Open ConfigurationManager by selecting the Config-urationManager tab at the bottom of FeatureManager.

2. Right click on the configuration for which to create the exploded view and select New Exploded View; or if an existing exploded view, select Edit Definition.

3. Select the New (Step) icon.

4. Specify an explode direction.

5. Specify a component to explode. Steps 4 and 5 can be inter-changed. Just make sure you have the proper list box acti-vated (salmon colored).

6. Specify an explode distance. Reverse the explode direction if necessary.

7. Click on the green check icon.

8. Repeat steps 3 through 7 as required.

9. Click on OK when finished.

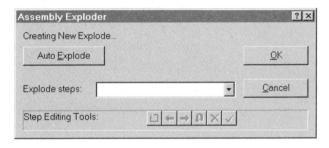

Assembly Exploder dialog box prior to adding an explode step.

Adding the exploded view is easy, but usually SolidWorks does not place the components in the desired location. This is because Auto Explode is really only meant for very small assemblies (maybe half a dozen parts or less). If the Auto Explode button is not convenient, you can add steps on your own. As soon as you click on the icon for adding a step, the rest of the dialog box (shown in the following illustration) opens up to show additional attributes you can use to add your own steps in the explode process. The following are the attributes in the SolidWorks Assembly Exploder dialog box.

Auto Explode	Pick to explode the current configuration.
Explode Steps	Displays the current explode step. Pick from the pull-down list or select the step editing tool arrows to show an explode step.

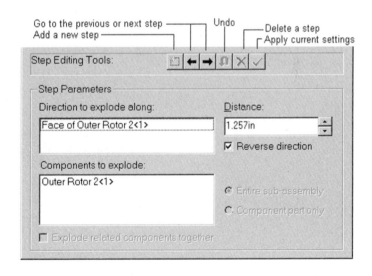

Assembly Exploder dialog box after adding an explode step.

The following are the step editing tools.

New	Select to define a new explode step. Select Apply to add the explode step.
Previous	Select to list the previous explode step.
Next	Select to list the next explode step.
Undo	Select to reverse a change.
Delete	Select to delete the current explode step.
Apply	Select to apply (and show) the current step parameters onscreen.
Direction to explode along	Pick an edge that defines the explode direction. To select a new edge, pick inside the field and select the desired edge. A cylindrical surface whose axis defines the explode direction, or a planar face whose normal defines the direction, may also be used.
Distance	Enter the distance to explode the selected components.
Reverse direction	Select to reverse the explode direction.
Components to explode	Displays the components to be exploded. Components can be added by selecting within this field and then selecting the components to be added. Components can be removed by selecting the component and pressing the Delete key.
Entire subassembly	Check to explode the entire selected subassembly. This option is only available when a subassembly component has been selected.
Component part only	Check to explode only the selected component of a subassembly. This option is only available when a subassembly component has been selected.
Explode related components together	Check to explode related (parent/child) components together. This option is only available when selecting components with external references.

Once the exploded view has been obtained, expanding and collapsing it is only a right mouse click away. To collapse an exploded view, perform the following steps.

1. Right click on the exploded view name in ConfigurationManager, or anywhere in the work area.

2. Select Collapse.

Theoretically, because you can have an infinite number of configurations, it is also possible to have an infinite number of exploded views. Because only one exploded view can be active at one time, being specific about which view to collapse is not necessary. Exploding a view, on the other hand, is a different story. To activate an exploded view, perform the following steps.

1. Right click on the configuration name in ConfigurationManager.

2. Select Explode.

In Chapter 8, you will learn how an exploded assembly can be shown in a drawing layout. Multiple exploded views can be defined, but only one can be defined per configuration. New explode steps are placed at the end of the explode step list. The name defaults to "Explode Step x," where x is the explode step number.

To edit an exploded view, the parent configuration must be active. If the exploded step is selected in ConfigurationManager, a dashed green line will be shown. Dragging the small green triangular "handle" on this green line is a very easy way of changing the explode distance of the exploded component. The component will be dynamically moved onscreen as the handle is dragged. Think of these small green handles as AutoCAD grips while in the "move" grip mode, but while working with solids and with physical constraints in place.

AutoCAD does not have explode capabilities. An AutoCAD assembly file is a drawing file with a collection of entities within it. In this respect, the individual entities are not really separate. It is not possible to create various configurations because the separate components are actually all part of the same file. About the closest workaround for creating configurations would be to make duplicates of all entities in the file and place them on a different layer, where the entities can be moved to create the appearance of an exploded state. This is at best a cumbersome workaround. Without configuration capabilities, exploded states cannot exist.

Join   Join creates a new part out of multiple assembly components, joining them into one component. This could be used to simulate a welded assembly, or to create a part for use with analysis software. The Join command does not affect the original parts. The join operation would fail if the parts that defined the join were not available, whether because they were moved or deleted. This is because the parts are needed to define the joined component. An external reference is established to the defining components. The following illustration shows three joined parts.

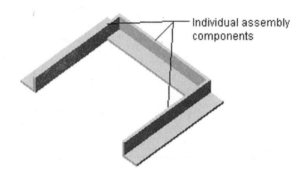

Similar to the Cavity routine, Join requires that you create a new part in the assembly. This new part file will become the joined part. To join two or more parts from an assembly into one part, perform the following steps.

1. Create an assembly and insert the parts to be joined.

2. Constrain the parts as necessary.

3. Click on Insert/Component/New.

4. Select a face on which to sketch. This is not overly important, as a sketch will not actually be required. However, the plane should still be selected. You can safely exit the sketch afterward, because it is not necessary to be in sketch mode.

5. Click on Insert/Features/Join. The Join dialog box will appear.

6. Using the left mouse button, select the parts to be joined in FeatureManager.

7. Select OK to continue.

8. To return to editing the assembly, right click on the assembly name in FeatureManager and select Edit Assembly.

AutoCAD's Union command is similar to Join. Union adds regions or solid parts together, creating new regions or solid. Unlike SolidWorks, creating the new solid does not result in a new part file. SolidWorks creates a completely new part file that externally references the originals. Making changes to the originals results in the joined component being updated. This type of associativity is not present in AutoCAD. AutoCAD's unioned part could be Write Blocked out to a separate file, but there is no associativity to the originals.

Another command that follows closely behind the Join command is the ability to add weld beads to an assembly. This is discussed in the next section.

Weld Beads

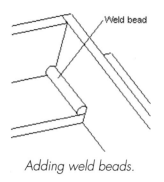

Adding weld beads.

Weld beads can be defined in an assembly to represent the weld beads used to weld parts together. These can be used to create detail drawings and show an added aspect of reality. The engineering characteristics of the feature (weld beads) are defined at the time the beads are created. This can be easier than trying to document these characteristics when the drawings are produced, but you do have to create the weld beads. The illustration at left shows a weld bead.

When a weld bead is created, SolidWorks creates a new part file that contains the weld bead. The bead becomes another part file in the assembly. After creating weld beads, it is not necessary to join the components. It is common to do so, but this is not a requirement. To add a weld feature between two part faces, perform the following steps. The illustration that follows shows the Weld Bead wizard.

1. Select Weld Bead from the Insert/Assembly Feature menu.

2. Select the weld type and click on Next to continue.

3. Select the weld shape, specify the measurements for the bead, and click on Next to continue.

4. Select the weld faces. These will be different, depending on the weld type being created. Click on Next to continue.

5. Enter the name for the weld bead part file or accept the default name SolidWorks assigns.

6. Select Finish to complete the process.

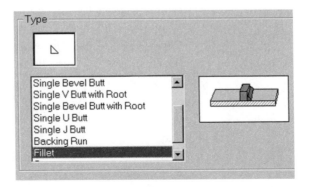

Page 1 of the Weld Bead wizard.

A few more words could be said about weld beads. For instance, adding weld beads certainly will increase the size of the assembly you are working with. You will want to weigh the options of simply calling out the various welds in the assembly layout versus creating the weld beads and thereby increasing the overall size of the assembly. Keep in mind that every weld bead will be another component in the assembly.

Another interesting aspect of weld beads is that there is an option to Float a weld bead, accessible through the right mouse button. However, if you attempt to perform this action, Solid-Works will inform you that weld beads cannot be floated. They must remain fixed, which, of course, is only logical and makes perfect sense. The Float option should probably have been removed from the right mouse button menu for weld beads so as not to confuse new users.

Example Assembly

The robot assembly shown in the following illustration uses an assembly layout sketch. The layout sketch is used to control the angle of components in the assembly. Plane Base_Ang can be changed to rotate the robot arm. The plane Gripper_Ang can be used to rotate the gripper arm around the attachment point. Some of the components in this assembly were created using the top-down design process discussed early in this chapter. Other components were already built and simply inserted into the assembly. It is not intended that you attempt to create this assembly, as many steps have been left out regarding the actual design process.

The assembly layout is used to define the linkage between the robot components. The sketch is aligned to the center of the base and each sketch entity has a defined length. The angle of each sketch leg can be defined with an angular dimension or dragged within the sketch to the desired position. The following series of illustrations depicts a typical assembly procedure. The completed assembly is shown in the final illustration.

Step 1:
Define the function
of the assembly.

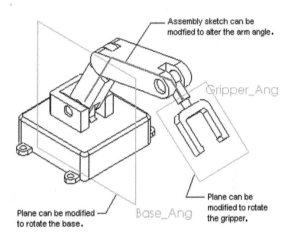

Defining the function of an assembly is something you would do in your mind's eye or through the use of notes. The image shown in step 1 is not what would appear on your screen at this point. It is just being shown for reference purposes.

Step 2: Create the
assembly skeleton and
assembly control layout.

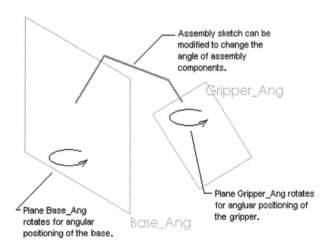

Step 3: Incorporate the base unit into the assembly, and mate the assembly and base unit origins.

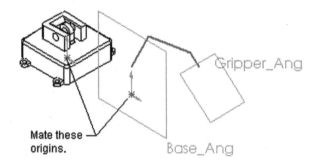

Mate these origins.

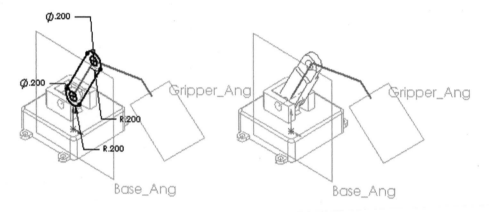

Step 4: Create Arm 1 in context with the assembly.

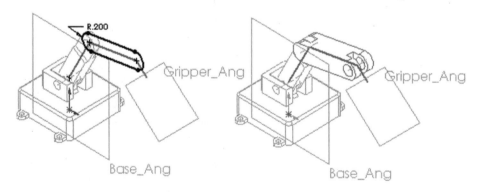

Step 5: Create Arm 2 in context within the assembly.

Step 6: Insert the gripper brace and mate it to Arm 2.

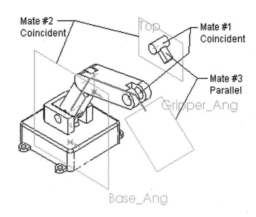

Mate #2
Coincident

Mate #1
Coincident

Mate #3
Parallel

Step 7: Insert the gripper finger into the assembly and mate it to the gripper base and the plane Gripper_Ang.

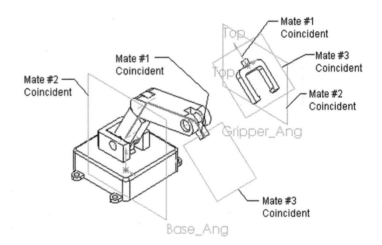

Mate #1
Coincident

Mate #2
Coincident

Mate #1
Coincident

Mate #3
Coincident

Mate #2
Coincident

Mate #3
Coincident

Completed assembly.

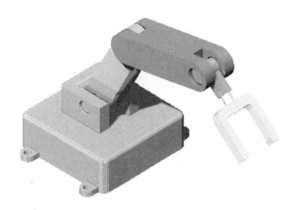

As mentioned previously, this section shows just one example of how a SolidWorks user might go about creating an assembly. The options of whether or not you actually create parts in the context of an assembly is up to you, but may also be dictated by the situation at hand. If the only way to create a part is to build it within a partially completed assembly, that is what you must do. Do not create in-context parts unnecessarily, though. Doing so creates unwanted external references that could hamper you farther down the line. Creating components separately limits external references and will make your assembly more flexible.

Summary

The purpose of an assembly is to create a real-world type of working model that can be used to define, drive, and document the design of a product. The assembly can provide information that can help define the form, fit, and function of the design. Assembly mates should be defined in a manner that allows for the simulation of the actual assembly method. If a door hinges on a pivot and can be opened, the assembly should be constructed to mimic this functionality. The assembly can be reviewed in a number of different positions to determine the clearance and fit to other assembly components.

There are two main assembly types. These are bottom-up assemblies, where parts are brought into the assembly, and top-down assemblies, where parts are created within the assembly. In the latter, parts created will contain external references to geometry in the assembly.

Mating relationships play a big role in assemblies. Without mates, parts would be free to move without rhyme or reason. Add appropriate mates that result in a model in which components move as they would in the real world. Use the Assembly toolbar icons to move or rotate assembly components individually. Interference detection can be used to check for component interference, and Explode can be used as a visual aid and to show assembly instructions within a drawing layout.

8 Drawings

Introduction

This chapter discusses how to create 2D drawings based on existing parts and assemblies. The automated process used to create drawings with SolidWorks is very user friendly and eliminates many of the tedious aspects normally associated with drafting. The ability to alter aspects of certain views, such as section lines and detail circles, can make what used to be hours of painstaking drafting nothing more than a tweak of a line or circle.

Dimensions created in a part do not have to be recreated. All of a part's dimensions can be easily inserted into a drawing with a few mouse clicks. The amount of detail and type of dimensioning or geometric tolerances can vary greatly between companies; therefore, you should establish a standard before embarking on a full-scale project. Many standards can be set in SolidWorks, such as ANSI, ISO, and JIS, to name a few.

Drawings are a unique type of document within SolidWorks, just as parts and assemblies are unique. The pull-down menu headings remain constant, as they did with assembly documents, which helps reduce confusion for the new user. What commands are contained within the pull-down menus depend on what SolidWorks document type is active at the time (i.e., part, drawing, or assembly). You will find that some of the pull-down menu commands remain the same and that the other menus display drawing commands.

Prerequisite   Because this chapter deals primarily with engineering drawings and drafting terminology, you should have at least a basic understanding of drafting. In addition, many of the topics and the step-by-step procedures covered will assume that you have completed the previous chapters.

Content   The "Creating a Drawing" section describes drawing formats and sheets. These are the basic components used when creating a drawing. Drawing views, dimensions, and symbols are added after a drawing format has been established. This chapter will attempt to follow as closely as possible the actual sequence of events found when creating a drawing.

The section on creating views describes the various drawing views. Following that, the "Adding Dimensions" section describes how to display existing part or assembly annotations or create new dimensions to document a design. The section on drawing symbols describes these symbols (i.e., notes, balloons, and so on) and how they are used to add information to a drawing that dimensions do not provide.

Objectives   When you have finished the "Creating a Drawing" section, you should understand what a drawing format is and how it can be used, be able to define and import a drawing format, and understand drawing sheets and how they can be used to create multisheet drawings all within the same drawing file. Upon completion of the section on creating views, you should understand the various types of drawing views, be able to select a part or assembly for view creation, and understand how to define and manipulate drawing views.

With completion of the "Adding Dimensions" section, you should know the difference between an existing model annotation and a reference dimension, be able to insert existing model annotations from a part or assembly, and be able to modify an existing dimension. When finished with the section on drawing symbols and annotations, you should understand how to add a wide range of model annotations and drawing symbols and be able to modify existing symbols, text, geometric tolerances, and any of the other annotation symbols. When you have completed the "Bill of Materials" section, you should be able to import a bill of material, including balloons.

About SolidWorks Drawings

Drawings are created as separate documents that become associated with existing parts and assemblies. This helps separate the functions of design and drafting into separate documents. The part or assembly documents are used in the design phase, and the drawing documents the finished product for manufacturing. This helps simplify the user interface and allows you to focus on the current task, rather than have the commands needed for parts, assemblies, and drawings all gathered in one document type and one interface.

Drawings are created using existing parts or assemblies. These part documents are created prior to creating the drawing, although the reverse can also be true. Not all that long ago, drafters and designers created 2D layouts that represented a model before the model was ever built. Now, through the use of solid modeling programs such as SolidWorks, a part can be created, checked, and visualized before the layout is ever started.

Most of the dimensions needed for a part are actually added while designing the part. It would be foolish to have to recreate all dimensions for a drawing; therefore, SolidWorks allows the importation of all part dimensions directly into a drawing layout. The dimensions and annotations (i.e., surface finish, geometric tolerance, and so on) can all be extracted from the part or assembly to help produce the detail drawing. The following are a few uses of drawings.

- Detail drawings for parts. Detailed drawings are used to display the dimensions used to create and document a design. The drawing should have sufficient views to display the desired dimensions.

- Design layouts can be created from assemblies. These layout drawings, which display important design-related features or functions, can be produced from a part or assembly.

- Quick reference drawings can be produced to measure a feature or dimension value. The purpose of this type of drawing is similar to the design layout but is simpler in nature. Quick prints can be used to impart information to others without creating a complete layout.

- Exploded assembly drawings can be produced from an assembly. An assembly can be exploded to indicate the assembly direction and order for those manufacturing the assembled product.

- Assembly drawings can be used to produce a bill of materials.

SolidWorks documents are fully associative. This means that any changes made to the part or assembly used to create a drawing will be automatically updated in the drawing. The display within the views is automatically cleaned up for hidden line removal or hidden lines in gray (or dashed) display modes.

The part or assembly is designed in the currently selected unit of measure (i.e., inches, mm, cm, and so on) at a scale of 1:1. This is similar to AutoCAD, wherein parts are normally drawn at full scale. Parts are not normally scaled until they are printed. AutoCAD's use of Paper Space allows for different scaling of viewports. This is similar to SolidWorks, where the drawings are created using views that can be displayed at a different view scale (i.e., 1:2, 2:1, and so on) than the part or assembly. Drawing formats and sheets are created using a selected paper size. The drawing views can be automatically scaled to fit within the selected paper size.

You do not have to be concerned with the paper space versus model space AutoCAD uses to simulate the various document types within SolidWorks. An approach wherein a part is created full size and drawing views are scaled to display the part or assembly is more flexible and easier for most users to work with. A drawing is shown in the following illustration.

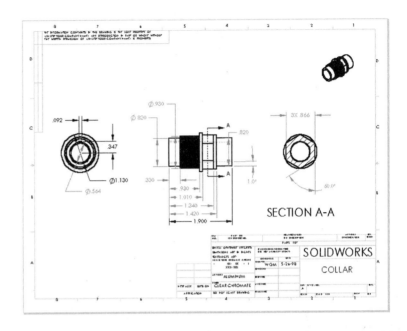

A SolidWorks drawing.

Creating a New Drawing

Drawings are created as a means of documenting a part or assembly. A drawing can contain more than one part or assembly. As a matter of fact, a drawing can contain absolutely any combination of parts or assemblies of any configuration from any file. A drawing is a document that references the part or assembly used to define the drawing views. Because of this, if the original part used to create the drawing is somehow misplaced or deleted, the drawing will fail to open.

Drawings, or 2D layouts as they may be referred to in this chapter, can be used as a design tool by creating drawing views and dimensions that show the critical design dimensions of a part or assembly. As the design progresses, this drawing can be used as a reference to check these features and can even be used to make modifications to the part. Because drawings are bidirectionally associative with the part or assembly contained within them, any change made to a dimension will affect the part or assembly. Some companies do not want the people viewing the drawings to have this capability. For those companies, this function can be disabled during the installation of the SolidWorks program.

Selecting a Template

Starting a new drawing is the same as starting a new part or assembly, but with one extra dialog box (shown in the following illustration). This extra dialog box is used for selecting your template or paper size. To start a new drawing, perform the following steps.

1. Click on File/New, or select the New icon.

2. Specify Drawing from the New dialog box.

3. In the Template to Use dialog box, specify Standard, Custom, or No Template for the template to use.

4. If No Template is selected, specify a Paper Size to use.

5. Click on OK.

Selecting a template to use.

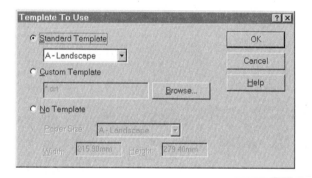

Creating a Standard 3 View Layout

Once a drawing is started, the next item on the agenda is usually bringing the part or assembly into the drawing. There are four methods for bringing a part or assembly into a drawing in order to achieve the standard top, front, and right-side engineering layout. The following are the steps required to create a standard engineering drawing using any of the four methods available.

1. Open Windows Explorer.

2. Drag the part or assembly file into the drawing window.

or

1. Open the part or assembly.

2. From the top of FeatureManager, drag the part or assembly name to the drawing window.

or

1. Click on Insert/Drawing View/Standard 3 View, or click on the Standard 3 View icon.

2. Click in the window of the part or assembly to be inserted into the drawing.

or

1. Click on Insert/Drawing View/Standard 3 View, or click on the Standard 3 View icon.

2. Right click on the drawing sheet and select Insert From File.

3. The Insert Component dialog box appears. Select the desired part or assembly and click on Open.

Any one of these methods will produce top, front, and right-side views of the part or assembly used. The fourth method can also be used to create a Named View or a Relative to Model view. Both of these views will be covered in more detail later in this chapter.

AutoCAD makes use of layers to separate 2D geometry from 3D geometry (if that is what the user desires). In addition, Paper Space can be used to create specific views within AutoCAD. None of the AutoCAD process is automated as it is in Solid-Works. If working with solids in AutoCAD (or even wireframe), it is a chore to create section views, manipulate scales for specific views, organize layers so that the proper entities are visible, and so on. It is certainly possible, and there are methods for all of these routines, but SolidWorks automates all of these processes.

In AutoCAD, inserting a drawing template or title block can be arranged so that it automatically asks you for all of the appropriate title block information during the import process. Templates can be set up in SolidWorks quite easily, but the template information is edited differently in SolidWorks. The material that follows contains a further explanation of the drawing formats (templates) and template options available. The section titled Linking to File Properties discusses how templates can be customized with data that can be inserted similar to the way AutoCAD's attributes can be inserted.

Drawing Formats (Templates)

Templates are used to define drawing formats. These templates can be customized to suit your company's requirements. Drawing formats will be one of the first things that need defining when creating a drawing. Existing drawing formats, logos, and so on can be imported via DXF- or DWG-formatted files.

There are existing formats included with SolidWorks that can be customized to your requirements, or existing title blocks can be inserted via DXF or DWG file formats. These files are located in the \sldworks\data directory. A company logo can be inserted as a DXF, DWG, or OLE object.

The Edit Template function can be used to customize desired formats. Save the formats and then back up the format directory. It is a very good idea to back up formats into another directory and make these files read-only so that they cannot be changed or deleted. SolidWorks will replace standard format names when the software is updated. Having a backup copy ensures that your custom formats are not overwritten. The following is information that could be included in a drawing format. The illustration that follows shows a drawing format.

- Company name
- Revision
- Approved by/Date
- Scale
- Part Number
- Drawn by/Date
- Checked by/Date
- Sheet Number

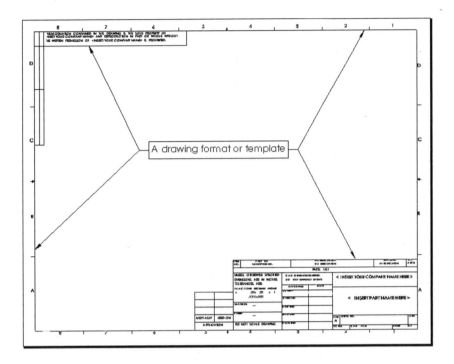

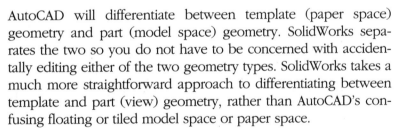

Typical drawing format.

AutoCAD will differentiate between template (paper space) geometry and part (model space) geometry. SolidWorks separates the two so you do not have to be concerned with accidentally editing either of the two geometry types. SolidWorks takes a much more straightforward approach to differentiating between template and part (view) geometry, rather than AutoCAD's confusing floating or tiled model space or paper space.

Any of the sketch tools normally used when editing sketch geometry can be used for editing template geometry with only a few exceptions. You saw how to select a drawing template in the previous section. To edit a drawing format (template), perform the following steps.

1. Right click anywhere in a blank area on the sheet and select Edit Template.

2. Make any desired changes or additions to the format.

3. Select Save Template from the File menu.

4. Select Custom Template and specify a name for the template, or select Standard Template to overwrite an existing standard template (this second option is not recommended).

5. Select OK to continue.

Once again, it is recommended that you make a copy of the new templates using Windows Explorer. These should be kept in a directory outside the SolidWorks directory. To import an existing drawing format from a DXF or DWG file, perform the following steps. The illustration that follows shows the import of a DXF or DWG file as a template.

1. Open the DXF or DWG drawing by selecting the Open icon, or by selecting Open from the File menu.

2. Select DXF or DWG from the Files of Type field.

3. Select the file to be opened.

4. Select OK to continue.

5. The Open DXF/DWG File dialog box will appear. Select the "Import to template" check box.

6. Click on OK.

7. Make any desired changes or additions to the format.

8. Select Save Template from the File menu.

9. Select the new format name by selecting Custom Template, or select Standard Template to overwrite an existing standard template.

10. Select OK to save the template.

Importing a DXF or DWG file as a template.

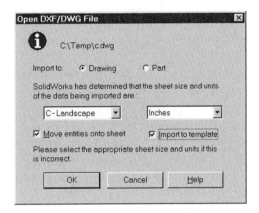

It is typically easier to start with an existing template when making your own templates because this gives you something to work from. Back up all templates into a different directory before creating new templates so that there will be copies of the originals, and back up all new templates into a separate directory to keep them in a safe place. New releases to SolidWorks will overwrite any template changes made to the original files in the SolidWorks\Data directory. All drawing templates will have a .drt file extension.

Linking to File Properties

AutoCAD has a very nice function that allows attributes to be associated with blocks. In this way, a drawing template (or format) can be turned into a block and then have attributes attached to it that describe different aspects of the template, such as Drawing Title, Date, Checked By, Company, Scale, and any other number of attributes commonly associated with a drawing and/or title block. This same functionality can be accomplished through the use of File Properties in SolidWorks. The mechanics are different, but the final result is basically the same.

What does linking to a file's properties actually accomplish? The properties of a file can be thought of as little bits of custom information associated with a file. This information can range from being something as simple as the name of a file to the part number of the part contained within a SolidWorks part file. By linking a specific property to a note, the note displays the value of that property.

Lets look at this another way. Imagine a file property is created that has the name File Name. Now imagine the value of this property has the value of Gasket Seal. A note can be selected to link up to the property field of File Name, and then the note displays the text Gasket Seal.

Linking to a file's properties usually requires that you know how to add a note to a drawing. However, the drawing templates SolidWorks provides already contain notes. It is possible to link a file's properties to these notes using the methods described in this section. In addition, because this area of the book deals with drawing templates, this is a good place to discuss linking to file properties. If you want to read about inserting notes first, feel free to skip ahead to the section titled Notes in this chapter.

You do not have to use SolidWorks to modify the Properties of a SolidWorks document. Windows Explorer will suffice. To access the properties of *any* file, not just a SolidWorks document, right click on the file's name in Windows Explorer and select Properties. This will work in Windows 98 or NT operating systems. The following illustration shows the Custom tab of a SolidWorks file's properties.

File properties of a SolidWorks drawing file.

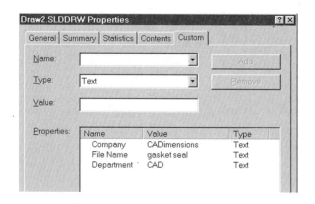

Some Custom Properties have already been added to the file's properties shown in the previous illustration. The Properties dialog box of a file when accessed via SolidWorks (as opposed to Windows Explorer) will look somewhat different, but the information will be the same. Some of the tabs, however, will not be present (see the following illustration). To access the properties of a SolidWorks document, perform the following steps.

1. While in SolidWorks, Select Properties from the File menu.

To go one step further and create your own custom properties, continue with the following steps. Keep in mind that these properties will be able to be linked to notes in the drawing template.

2. Select the Custom tab.

3. In the Name text field, type in a property name. This would be analogous to an attribute tag in AutoCAD.

4. In the Value text field, type in a property value. This would be analogous to an attribute value in AutoCAD.

5. Click on the Add button. You will see the new property added to the Properties list area.

6. Continue steps 3 through 5 until all of the properties are defined.

7. Select OK when finished.

The following illustration shows an example of what the properties list area might look like after adding three custom properties. Your file properties may vary. Notice the dialog box looks somewhat different than the previous illustration. SolidWorks calls this dialog box the Summary Info dialog box. Nevertheless, you would select Properties from the File menu to access it.

*Defining custom
properties.*

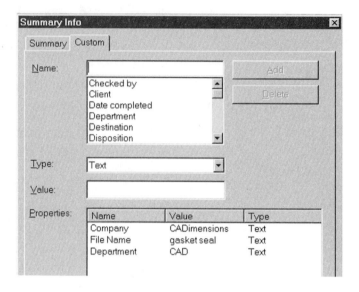

Linking Example

Now that you understand how to create custom properties, lets
tackle how to link those properties to text in a drawing or a
drawing template. In previous sections you learned how to
insert and edit a drawing template. Assuming the reader currently has a new drawing started complete with a drawing template, follow along with these steps to link custom properties to
text in the drawing template. (Additionally, it is possible to link
properties to any note in a drawing, whether or not it is part of a
template.)

1. If the note to be linked is on a template, right click on the
   drawing sheet and select Edit Template.

2. Right click on the note to be linked and select Properties.
   Optionally, you can double click on the note.

Link to Property button.

3. Click on the Link to Properties button, shown in the illustration at left.

4. From the Link to Property window (shown in the following
   illustration), select a property to link to from the drop-down
   list. The value of the selected property will be displayed as
   the note's text when you finally exit from the Note dialog
   box.

5. Select OK to close the Link to Property window. A code will appear in the Note text field. It will appear as $PRP:"*property_name*", where *property_name* is the name of the property that was linked to.

6. Click on OK to close the Note Properties window.

Link to Property window.

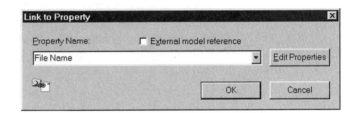

If you have been following along with the illustrations, you probably noticed that the property value was "File Name." In a previous illustration (see the Summary Info dialog box shown earlier) it is also evident that the value of the property name "File Name" is "gasket seal." Therefore, "gasket seal" is the text that will appear as the note.

This functionality has far-reaching implications. For example, it is possible to create many notes in a drawing template that act as placeholders for Custom Properties. For example, the property name "File Name" could have a value of < *file_name* >. Dozens of such properties can be used throughout a template. When that template is used in a drawing, all the user has to do is open up the file's Properties and modify the values for the various property names. The user would never even need to edit the template.

As you can see, from an end user standpoint, this is almost exactly the same as filling in the data for the attribute values when inserting a template block into an AutoCAD drawing. It would more than likely fall on the shoulders of the company CAD administrator to set the drawing templates up in the first place, but once it was done, nobody would have to edit a template again. This makes it very easy to set up standardized templates.

The topic of File Properties will be touched on again near the end of this chapter in the section regarding bills of materials.

There you will see how it is possible to set up a Description property that will automatically fill out the description column of a BOM.

Drawing Sheets

Drawing sheets are used to create a drawing using multiple drawing formats instead of one larger, single-format drawing. Using multiple sheets can allow for more space on individual sheets. This also allows for more information in one drawing file but less crowding on individual sheets. The tabs at the bottom of the drawing allow for easy navigation between different drawing sheets. Sheets can be added or removed depending on your needs. The following illustration shows an example of drawing sheet tabs.

Drawing sheet tabs for moving between sheets.

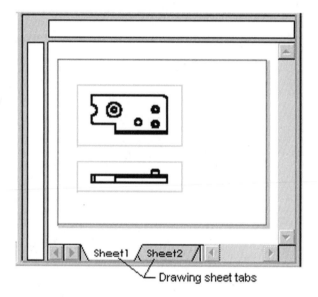

Drawing sheet tabs

Adding Drawing Sheets and Changing Properties

To add a new drawing sheet, perform the following steps.

1. Select Sheet from the Insert menu, or right click on a drawing sheet and select Add.

2. Select the format template, drawing scale, paper size, title, and type of projection (first or third).

3. Select OK to continue.

To delete a drawing sheet, perform the followi[ng]

1. Right click on the drawing sheet tab to be dele[te] Delete.

To change drawing sheet properties, perform the f[ollowing] steps.

1. Select Sheet Setup from the View menu, or right click on [the] desired tab and select Properties.

2. Edit the desired sheet properties.

3. Select OK to continue.

Drawing Sheet Properties

When editing the properties of a drawing sheet, the current template or sheet size can be changed, along with the drawing sheet scale and many other options. The following is a list of the various options available to you.

Name	Name of the drawing sheet. This is the name that appears on the sheet's tab.
Paper Size	Size and orientation of the sheet. Width and Height are only selectable for a custom paper size.
Width	Width of the sheet. Pick inside the field to change.
Height	Height of the sheet. Pick inside the field to change.
Scale	Scale of the entire drawing sheet. Pick inside each field to change. Be aware that individual drawing views can have their scale changed independently of the sheet.
Template	Select the template name to use for the sheet.
Type of Projection	Check first angle to view objects from the front and projected back. Check third angle to view objects from behind and projected forward. This method is primarily used in the USA, Canada, and parts of the UK.
Next Section Label	The next default section letter. Pick to change the default for the next section.
Next Detail Label	The next detail letter. Pick to change the default for the next detail.
Next Datum Label	The next datum letter. Pick to change the default for the next datum.

Activating Drawing Sheets

Because it is possible to have multiple drawing sheets, it would benefit you to know how to alternate between these drawing

. This is a very simple procedure. To make another draw-
...heet active, perform the following.

. Select the sheet name from the sheet tab at the bottom of
the drawing, or select Next Sheet from the View menu.

AutoCAD does not use sheets in the same sense that Solid-
Works does. The closest way to simulate drawing sheets in
AutoCAD would be to have independent layers in AutoCAD's
Paper Space. Layers would have to be turned on and off to view
the specific "sheets." It is much easier to click on a tab to view a
sheet, as in SolidWorks.

Creating Views

Drawing views are created to display different orientations or
snapshots of a part or assembly. The drawing view's alignment,
orientation, and hidden line display are controlled and updated
automatically when any changes are made to the part or assembly. Drawing views can be used to create views of parts or
assemblies. These views display a part or assembly using ortho-
graphic views and can be projected using first- or third-angle
projection. More than one part or assembly can be displayed
within a single drawing.

Drawing views are used to display different view orientations of
the same part or assembly. The scale of the view is independent
of the actual part or assembly. The scale of the drawing views
can be different from the part or assembly, and different views
within the drawing can have different view scales. The illustration at left shows the standard three views.

The three standard views.

AutoCAD does have the ability to create any of the views Solid-
Works can. However, the process is much more cumbersome.
Model geometry must either be duplicated in order to create the
other "view" in model space, or viewports must be created and
set up within Paper Space and then the geometry and layers
manipulated to create the desired view. SolidWorks creates
snapshots of the model, which update as the model changes.
Crosshatch is created automatically, not manually. Projected
and auxiliary views are automatically projected.

Alignments between views are automatic and adjustable. The
list is quite extensive. Without getting into every last difference

between AutoCAD and SolidWorks, suffice it to say that many of the procedures used to create drawing views will probably be a lot easier and much more user friendly than those the AutoCAD user is accustomed to.

Because the methods for inserting the three standard views of top, front, and right side into a drawing layout have already been discussed, the rest of this section is devoted to the remainder of the view types accessible to the SolidWorks user. It will be assumed at this point that you have already created a new drawing and have the standard three-view layout inserted into the drawing.

Activating Views

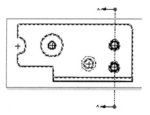

An activated view has a gray shadow border.

For specific geometry to be associated with a particular view, that view must be activated. Activating a view is simple enough. What is most important is that you remember to do it! To activate a view, either double click on the view border, or right click on the view border and select Activate View. If the view is not activated before adding a detail circle or section line, the view options of Detail View or Section View will not be available. The illustration at left shows an activated view.

Activating a view before adding geometry serves another function. If the view is moved to a different location, any geometry added to the view while it was active will move with the view.

➠ **NOTE:** *If the Dynamic View Activation option is set, views will be activated automatically as you move the cursor.*

Projected Views

A projected view.

A Projected View can be created above, below, or to either side of an existing view. To create an orthographically projected view from an existing view, perform the following steps. The illustration at left shows a projected view.

1. Select the view to project from.

2. Click on the Projection icon or select Projection from the Insert/Drawing View menu.

3. Select the side and location for the new view. For instance, clicking above the selected view will project a new view above it.

Relative to Model View

One of the view types you can specify is the relative-to-model view. This takes a little bit of input on your part, but can be used when a precise orientation needs to be created normal to a particular face. To create a view oriented relative to model faces, perform the following steps.

1. Select the Relative to Model icon, or select Relative to Model from the Insert/Drawing View menu.

2. Use the Window menu to activate the part or assembly window.

3. Click on a face of the part or assembly.

4. Select the orientation for the selected face in the dialog box that appears (see the following illustration).

Drawing View Orientation dialog box.

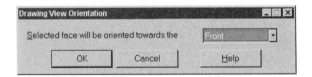

5. Click on a second face.

6. Select the orientation for the second face.

7. Make the drawing the active window.

8. Select the location for the new view on the drawing. (The illustration that follows shows a view oriented relative to model faces.)

View relative to model.

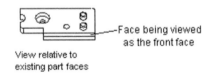

It should be noted that steps 5 and 6 in the previous procedure are optional. It is not mandatory that the second face be selected, in which case SolidWorks will only use the first face and orientation. The user should realize that the model in the view may or may not be fully oriented in a particular position, but at least the first face orientation condition will be satisfied.

Named Views

A named view allows you to bring any previously saved view into the drawing. This can be a system view already defined by SolidWorks, a view named by you, or the current model view. First, take a look at how to save a specific view within a part. Open a sample part to use in this example and follow along with these steps.

1. Rotate the model until the desired view is obtained.

2. Click on View/Orientation to open up the Orientation dialog box (or press the spacebar).

3. Click on the Add icon (first icon to the right of the pushpin).

4. Name the view.

5. Click on OK to complete the procedure.

Notice that the new view with the name you gave it is listed in the View Orientation list box. Double click on the view name to go to that view, just like you would to access any other view listed in the Orientation dialog box.

Now that you know how to create a user-defined named view, take it one step further and import that named view into the drawing. Of course, you do not have to create a new user-defined view first if you do not want to. Any view listed in the Orientation window (i.e., isometric) can be brought into a drawing layout. To create a view from a named part or assembly orientation, perform the following steps. The illustration at left shows a named view.

Named view (Iso)

Sample of a named view; in this case, an Isometric view.

1. Select the Named View icon, or select Named View from the Insert/Drawing View menu.

2. Use the Window menu to activate the part or assembly window.

3. Click once anywhere in the part or assembly graphics area.

4. Select a view from the Orientation menu list. A single click on the view is all that is needed.

5. Use the Window menu to activate the drawing window.

6. Select the location for the new view on the drawing.

Auxiliary Views

An auxiliary view is used to show a planar face straight on that otherwise would not appear normal to the screen. Sometimes there are details that are difficult to distinguish on the surface of a part unless that surface is being displayed in a plan view. Many times the only way to see the surface straight on is to use the relative-to-model view or auxiliary view.

In SolidWorks, selecting a model edge allows a view to be projected perpendicular from that model edge so that the face of that model edge can be viewed parallel to the screen. To create an auxiliary view from a selected model edge, perform the following steps. The illustration at left shows an auxiliary view.

1. Select the edge to project the auxiliary view from. The view will be defined perpendicular to the selected edge.

2. Click on the Auxiliary icon, or select Auxiliary from the Insert/Drawing View menu.

Auxiliary views must be from an angled edge. If the edge is horizontal or vertical, SolidWorks thinks you are trying to create a projected view. Projected views (covered in a previous section) are created slightly differently than auxiliary views.

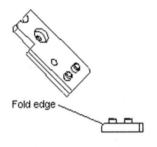

Fold edge

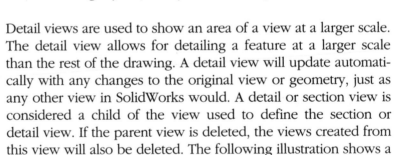

Auxiliary view.

Detail Views

Detail views are used to show an area of a view at a larger scale. The detail view allows for detailing a feature at a larger scale than the rest of the drawing. A detail view will update automatically with any changes to the original view or geometry, just as any other view in SolidWorks would. A detail or section view is considered a child of the view used to define the section or detail view. If the parent view is deleted, the views created from this view will also be deleted. The following illustration shows a detail view.

Detail view.

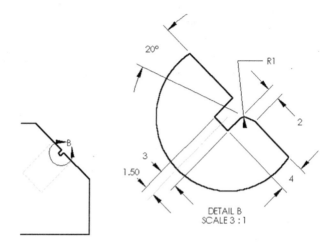

DETAIL B
SCALE 3 : 1

To create a blown-up detail of an existing view, perform the following steps.

1. Activate the view to create a detail view.

2. Create a circle that defines the boundary of the detail view.

3. Select the circle.

4. Select Detail from the Insert/Drawing View menu or select the Detail View icon.

5. The detail view will be created. Select the detail view border with the left mouse button and drag it to the desired location.

To change the detail view label, perform the following steps.

1. With the left mouse button, double click on the Detail label.

2. Enter the new detail letter. The detail view name will update automatically.

When creating a detail view, remember that detail view notes are associated with the detail letter after insertion. Therefore, both values can be modified and the other value will update automatically. To change the name of a detail view, double click with the left mouse button and type in the new name. These properties can also be changed by right clicking on the detail view and selecting Edit Properties, or by selecting Properties from the Edit menu. To reposition the detail view letter, select the detail letter and drag it with the mouse to the new location. It is also possible to drag the position and size of the detail circle itself. The detail view will automatically update, but may need repositioning.

Detail and section views have something in common as far as the mechanics involved with creating the views. Geometry must be added to the view in each case. This is not too surprising, considering that detail circles are needed for detail views and section lines are needed for section views. What is significant is how SolidWorks keeps track of the geometry that needs to be created. This is the reason for activating a view prior to adding geometry.

Section Views

A section view is used to show a cutaway view showing the interior detail of a part or an assembly. This view does not affect the part or assembly components sectioned for the drawing view. The drawing section is used only for visualization purposes within the drawing.

A section view differs from a part section view and an assembly feature cut. A part section view is used only within a part to view a cutaway. A part section cannot be displayed within another document. An assembly feature cut can be used to create a cutaway, typically for visualization purposes. An assembly feature can be displayed within a drawing, but the section lines are not created and managed automatically like a drawing section view would be. Actual assembly section views can be created within a drawing for this purpose.

Creating a Section View

Activating a view and sketching a line that defines the section line is the prerequisite to creating a section view. This line can be one segment or can be a number of contiguous (joined) segments. These line segments are then turned into a section line, and the view can be defined from the section line. Both a detail or section view is dependent on the view used to define the section or detail view. If the parent view is deleted, the views created from this view will also be deleted. The illustration at left shows a section view.

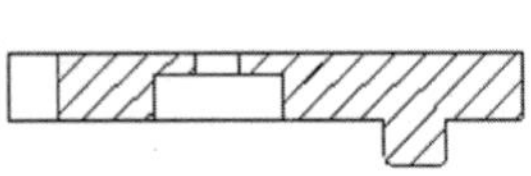

SECTION B-B
Scale 1 : 1

Basic section view.

As with detail views, section views require geometry to be added to the view. Before this is done, the view must be activated so that SolidWorks can associate the geometry with the correct view. Section views require lines, arcs, or centerlines, which can then be turned into section lines or used to create section views directly. To create a section view, perform the following steps.

1. Activate the view.

2. Sketch lines, arcs, or centerlines across the geometry to be sectioned. The sketch geometry should extend all the way across the part or assembly view geometry for a complete section view.

3. Select a linear sketch entity to project from when creating the section view.

4. Select the Section View icon, or select Section from the Insert/Drawing View menu.

5. Skip this step if creating a section view of a part. If creating an assembly section view, select the assembly components, if any, to be excluded from the section view. This is known as the Section Scope. Click on OK when done.

6. Select to position the section view at the desired location.

Crosshatching

Crosshatching properties for section views can be changed. Crosshatching is created automatically when a section view is created. Unlike AutoCAD, crosshatching cannot be created as a separate entity. The crosshatching pattern can be changed when the default crosshatching characteristics are not sufficient for a section view due to the size or pattern type. The illustration at left shows section view crosshatching. To modify an existing section's crosshatching, perform the following steps.

1. Right click on the crosshatching to be modified and select Crosshatch Properties.

2. Edit the values as desired (see the following illustration of this dialog box).

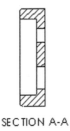

SECTION A-A

Section view crosshatching.

3. Select OK to continue. Select Apply to make the change but keep the Crosshatch Properties dialog box active for further changes.

Crosshatch Properties dialog box.

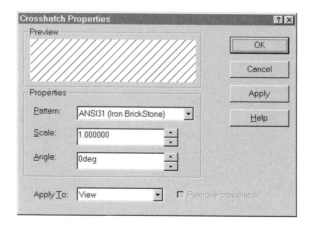

The following are adjustable pattern properties and other aspects of the Crosshatch Properties dialog box.

Preview	Shows a preview of the crosshatching with the specified changes.
Pattern	Pick different crosshatch patterns from this list.
Scale	Crosshatch scale. Enter a higher value to make the crosshatching smaller. Enter a new value in this field or select the up/down arrows to increment/decrement the scale value.
Angle	Crosshatch angle. Enter a new value in this field or select the up/down arrows to increment/decrement the angular value.
Apply To:	Select the scope of change the crosshatch will apply to from the following:
View	• All faces in the active section view.
Component	• All faces of an assembly component.
Region	• Selected faces only.
Remove crosshatch	Check to remove crosshatch modifications from this section view.

The default crosshatch type is ANSI31 (Iron Brick Stone). The default crosshatch pattern can be defined for the current drawing or for all new drawings by setting the values in the Crosshatch tab in the Tools/Options menu.

Aligned Section Views

Aligned section views are very similar to section views but with one difference. The section line itself must contain two segments that should not be orthogonal. During the creation of the aligned section view, SolidWorks "unfolds" the section line so that the view is perpendicular to each section line segment. The illustration that follows shows an aligned section view.

Aligned section view.

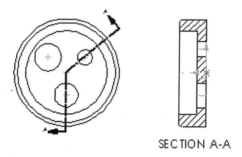

SECTION A-A

Take a look at the illustration that follows to get a better idea of the various section lines that can be created in SolidWorks.

Section lines that can be created in SolidWorks.

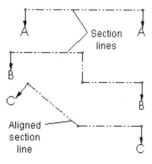

Creating an Aligned Section View

Aligned section views are used most often for cylindrical parts, but an aligned view can be used anywhere a two-segment angled section line needs to be created. To make an aligned section view, perform the following steps.

1. Activate the view.

2. Sketch a centerline across the geometry to be sectioned. The centerline should extend all the way across the geometry for a complete section view and should contain exactly 2 segments. If the part is cylindrical, the point where the section lines meet should usually be at the center of the part. However, this is not a requirement.

3. Select the linear sketch entity you want to project the section view from.

4. Select the Aligned Section View icon or select Aligned Section from the Insert/Drawing View menu.

5. Pick a position to position the section view.

To modify the direction of the section line arrows, double click on the section line. To change the properties of a section line, perform the following steps.

1. Right click on the section line and select Properties.

2. Change the desired values (see the list that follows).

3. Select OK to continue.

Section Line Properties

A common mistake made by new users when attempting to edit a section view is to edit the properties of the section view, when in reality it is the section line itself that must be addressed. Right clicking on the section line and selecting Properties will bring up a dialog box with the following options. The illustration that follows shows the Section Line properties dialog box.

Label	Section line text label.
Change direction of cut	Check to change the direction of the section cut. This can also be accomplished by double clicking on the section line.
Scale with model changes	Check to make the section scale change with the drawing sheet.
Partial section	A section view shows only the geometry cut by the section. If the section line does not pass completely through the geometry, check this option.
Display only surface cut	Only the portion of the model cut by the section line will be shown. Geometry behind the cut will not be shown.
Font	Uncheck the Use Document's Font field to make the Font button selectable. Select the Font button to change the section line text font. This works well for setting the size of the section line labels.

Section Line properties dialog box.

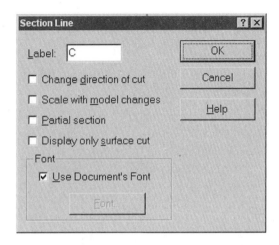

Sometimes it may be desirable to create a simple section line without creating a section view. To do this, follow the same procedure for adding a section view up to the point where the view is actually inserted. Instead of adding a section or aligned section view, click on Insert/Make Section Line. This will change the sketch entities into a section line without creating a new view.

One other comment on section views should be made. If the scale of a section view is modified, the text associated with the view will alter to display the new scale. The reverse is also true. By modifying the text associated with a section view, the scale of the view is changed as well.

Broken Views

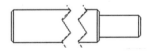

Broken view.

Broken views are used to display one or two ends of a view with the middle portion removed. This type of view can be used to show details on long parts that would normally not fit on the paper. A typical application would be a tube or pipe with detail on both ends and the center removed to fit the part on a smaller drawing. This also allows the part to be shown at a greater scale than would be possible if the entire geometry was required on the drawing. A broken view is shown in the illustration at left.

Broken views can be created in AutoCAD, but once again this is a manual procedure that must be done by cutting and trimming entities, adding break lines, and modifying dimension values to read correctly. SolidWorks automates this process in a way you should be coming to expect. To create a broken view, perform the following steps.

1. Select the view.

2. Select Horizontal Break or Vertical Break from the Insert menu. Break lines will appear on the view.

3. Position each of the breaks by selecting a break line with the left mouse button and dragging the break line to the desired location.

4. Right click on the view and select Break View. The view will be broken at the break line positions.

To remove a broken view, perform the following.

1. Right click on the view and select Unbreak View.

To change break line types or move the break lines, perform the following steps.

1. Select the break line with the left mouse button and drag it to a new position.

2. Right click on one of the break lines and select Straight Cut, Curve Cut, or Zig Zag Cut to change the break line type.

Only one broken view can be created per view. However, more than one broken view can exist on a drawing. Dimensions will read accurately when applied to the broken view.

Empty View

Empty views are just what they sound like: views without geometry. SolidWorks includes this function so that you can create a view, make it active, and add geometry that will move with the view border. In this respect, you can "create" your own views from scratch. With all of the functionality SolidWorks provides, this function does not seem very useful. It does, however, make for a convenient way to group geometry by way of associating it all with one specific view.

Modifying and Aligning Views

Preexisting views can have their attributes (i.e., scale and alignment) changed. The alignment of views is defined by the creation method of the view. Alignment properties can be removed or added to existing views. Views that have other

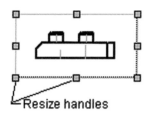

Resize view handles.

drawing views projected or created from them can also have their alignments broken. Moving a view requires that you select the view's border and drag the view by its border. Sometimes increasing or decreasing the size of the view's border is needed to make moving the view easier. To move a view or resize a view's borders, perform the following steps. (The illustration at left shows view handles used for resizing the view.)

1. Select the view by selecting on the view boundary with the left mouse button.

2. Select one of the view's resize handles with the left mouse button. These handles appear at each corner of the view boundary and at the center of each boundary segment.

3. Drag the handle to the desired size or position. The view border cannot be resized to the point where the geometry will not fit in it. Dragging the border from any point other than a handle will move the entire view.

To change a view's properties, perform the following steps.

1. Right click on the view and select Properties.

2. Change the drawing scale or other parameters if desired.

3. Select OK to continue.

In the last example, you may also have noticed the ability to change a drawing view's configuration. This is possible only if other configurations already exist in the part or assembly. Any configuration present in the part or assembly can be viewed within a particular view. To access a particular configuration, perform the following steps.

1. Right click on the view and select Properties.

2. In the Configuration Information section, select Use Named Configuration.

3. Specify the configuration name from the drop-down list.

4. Select OK.

It is sometimes necessary to break the alignment of a view so that it can be moved to a different location. To break an existing

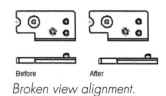

Before   After

Broken view alignment.

Horizontal view alignment.

Vertical view alignment.

view alignment, perform the following steps. The illustration at left shows a broken view alignment.

1. Right click on the view whose alignment you want to break and select Break Alignment.

To align a view horizontally, perform the following steps. The illustration at left shows a view aligned horizontally.

1. Select the view to be aligned horizontally.

2. Press the right mouse button and select Align Horizontal.

3. Select the target view to set the alignment.

To align a view vertically, perform the following steps. The illustration that follows shows a view aligned vertically.

1. Select the view to be aligned vertically.

2. Press the right mouse button and select Align Vertical.

3. Select the target view to set the alignment.

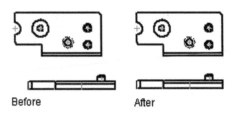

Before   After

To reset the view alignment to its original default alignment state, perform the following. The illustration that follows shows a reset view alignment.

1. Right click on the view whose alignment should be returned to its default state and select Default Alignment.

Reset view alignment.

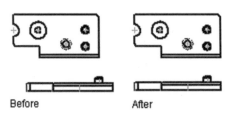

Before   After

Rotating Views

Sometimes aligning views is just not enough. SolidWorks gives you the ability to rotate views to any angle you see fit. Use the Rotate icon just like you would to rotate a part on the screen. Additionally, the user can type in the angle of rotation. The Rotate Drawing View dialog box is shown in the following illustration. Here are the steps you would perform to rotate a view.

1. Select the Rotate View icon.

2. Click on the view to rotate. The Rotate Drawing View dialog box will appear.

3. Drag the view to rotate it using the left mouse button, or type in a rotational value in the Drawing View Angle field.

4. Uncheck "Dependent view update to change in orientation" if you do not want other dependent views to be affected by the parent view's rotation.

5. Click on Apply to rotate the view if a value was entered in the Drawing View Angle field.

6. Click on Close when finished.

Rotate Drawing View dialog box.

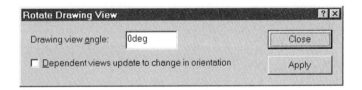

Hiding a View

It is often convenient to hide a drawing view. One example where this is an advantage is in the case of showing a large section view on one sheet by itself. Typically, you would have to have the parent view and section view on the same sheet because the section view is dependent on the view it was created from. It is possible to hide the parent view, leaving the section view behind. To hide a view, perform the following.

1. Right click on the view to be hidden and select Hide View.

When a view is hidden, its border is still selectable. By selecting the view's border, it is possible to restore a hidden view. To restore a hidden view, perform the following.

1. Right click on the border of the hidden view and select Unhide View.

There are a few things before closing this section on drawing views that should be mentioned. Drawing views can be copied and pasted to other drawing sheets. View borders are shown with a green border when they are selected, and a selected view can be moved, deleted, or activated. Their borders are shown with a light gray edge when they are not selected or activated. These borders will not print, but the borders can be turned off, if you prefer, in the Drawing tab of the Options dialog box. Remember to activate a view before adding geometry. Otherwise, geometry will not be associated with the view and will not move with the view. When activated, a drawing view has a gray shadow border.

Layers and Lines

The ability to define custom layers is an option that AutoCAD users will be very familiar with. Layers are a crucial part of AutoCAD, but they are not any real necessity in SolidWorks. SolidWorks layers are more of a convenience than a requirement.

Discussing layers and line format options has been placed at this area of Chapter 8 because it is a function of SolidWorks the reader should be made aware of before going forward with annotations and dimensions. Many of the annotation dialog boxes, for instance, have a layer option built right in. An example would be the Note dialog box, as shown in the following illustration.

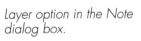

Layer option in the Note dialog box.

Layers

Layers in SolidWorks are similar to those in AutoCAD. Layers can basically be used for the same purpose; that is, to separate or sort various entity types. For example, you may want to put notes or dimensions on a particular layer so that they can easily be turned off. SolidWorks layers do not have quite all the bells and whistles that AutoCAD layers do, but they serve the purpose.

The Layers dialog box, shown in the following illustration, shows a number of example layers typical of what a user might create. As with AutoCAD, layers can be turned on or off by clicking on the respective light bulb associated with any particular layer. In the example, two layers, notes and dimensions, are currently turned off. The active layer is the section layer as shown by the yellow arrow. This yellow arrow will always point to the current layer, which can be selected by simply clicking just before the layer name. This will position the yellow arrow before the layer name to show that it is current.

Layers dialog box.

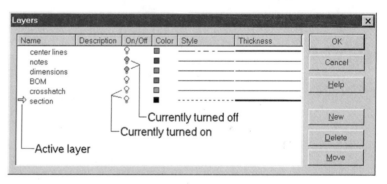

Layers can only be created in a drawing document. Once a new layer is created, you should assign the desired color, line style, and line weight characteristics to the layer. This is a simple procedure and is outlined in the steps that follow.

1. Click on the Layer Properties icon on the Line Format toolbar.

2. To create a new layer, click on the New button.

3. Type in a name for the new layer and press Enter to accept the layer name. This new layer will automatically become the current (active) layer.

4. To change the color from black, click on the small square to the right of the layer name under the Color column heading. This will open the Color palette window and allow you to choose a color. Click on OK when done.

5. To choose a different line style, click on the current line style to the right of the layer name below the Style column heading. This will open a selection window. Choose a new line style and it will be shown in the Style column.

6. To choose a different line thickness, click on the current line thickness to the right of the layer name below the Thickness column heading. Once again, a selection window will be displayed. Choose a new line thickness and it will be shown in the Thickness column.

7. Click on OK when finished to exit the Layer dialog box and save your changes.

Layers will of course be saved with the drawing, just as you would expect. You will have noticed there are two additional buttons in the Layer dialog box. These buttons are Delete and Move. To delete a layer, select it and click on the Delete button. The Move button is for moving entities to a different layer. For example, you can select entities, click on the Move button, and the selected entities will be moved to the current layer.

Line Formatting

When it is desirable to change the appearance of lines in a drawing, the user has a choice between modifying individually selected lines, or by placing entities of a certain type on a layer in which a line type has been specified for that layer. Specifying a line type for a particular layer was mentioned in the previous section. Here, modifying the line type of individually selected lines will be explored.

A number of aspects of lines can be changed. This includes changing the thickness (often referred to as line weight), color, and line style (such as phantom or dashed). These line characteristics are accomplished through the use of the Line Format toolbar, shown in the following illustration. Line characteristics can only be altered in a drawing. It is not possible to alter edges of model geometry in this same respect.

Line Format toolbar.

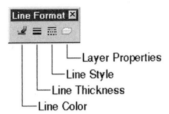

Whether you are changing the color, style, or thickness of a line or set of lines, the procedure is the same. The steps for performing this procedure are as follows. These steps assume that you have a drawing already opened that contains at least one drawing view or added sketch geometry.

1. Select the lines (model edges or sketch geometry) to have color, line style, or thickness changed.

2. Select the appropriate icon from the Line Format toolbar. This will either be the Line Color, Line Thickness, or Line Style icon.

3. Choose the appropriate color, thickness, or style. The changes will take affect immediately.

It is not currently possible to change more than one characteristic of the selected entities at once. Therefore, if another aspect of the selected entities must also be changed, the entities must be selected again and the previous steps performed an additional time. However, an additional option is to create a layer that contains multiple attributes, such as a heavier line weight, different color, and specific line style, and then place selected entities on that particular layer. In this way it is possible to change the attributes of as many entities as desired in one fell swoop.

Adding Dimensions

Dimensions are added to a drawing to communicate the design intent of the part and to identify the critical functions and inspection dimensions. The dimensions discussed within this section are discussed within the context of drawing creation.

⊸ **NOTE:** *Chapter 5 discusses creating 2D sketch dimensions to create part features.*

AutoCAD's dimensioning commands seem fairly straightforward and easy to implement. However, if you have had to change the characteristics of just a few dimensions, you know the difficulties that can arise in changing a dimension's properties. There are many system variables that must be modified manually or through the use of a dialog box. In addition, a dimension style must then be saved and a dimension updated in order to be able to use the specified style. SolidWorks does away with all that by incorporating a global setting for all dimensions and the ability to modify any dimension's properties to achieve the desired result.

The dimensions used to create the 2D sketches can be used to annotate the drawing. These dimensions are referred to as model dimensions. The Dimension, Horizontal Dimension, and Vertical Dimension commands can be used in sketch mode to create model dimensions. These modifiable dimensions are used to drive the part feature size.

The Dimension, Horizontal, Vertical, Baseline, and Ordinate Dimension commands can be used to create dimensions within the drawing. These dimensions are referred to as reference dimensions. These dimensions cannot be used to drive the model geometry. They show the size of the feature and will change value only if the geometry changes. Model dimensions are shown in black. Reference dimensions are shown in gray, with parentheses being an optional trait.

Model Dimensions Versus Reference Dimensions

The following are points to keep in mind when comparing model dimensions and reference dimensions.

- Model dimensions reuse sketch dimensions from a part, eliminating the time required to redefine layout dimensions.

- Model dimensions can be changed within the drawing and the corresponding drawing views and part will update automatically.

- Dimension parameters and annotations can be defined within a model during feature creation. These parameters could be tolerance values, appended or prefixed text, geometric tolerances, and so on. The advantage to defining these parameters during feature creation is that a designer can capture design intent when the feature is defined and not have to go back later and try to remember the intent for the feature. For companies that employ engineers, designers, and draftspersons, the person who creates the model can also define the engineering attributes for the design.

- Reference dimensions can be added when and where it makes sense in creating a drawing.

- Reference dimensions cannot be altered and cannot change model geometry.

AutoCAD does not differentiate between model dimensions and reference dimensions. It has no need to, because there are no parametric relationships to be concerned with. Dimensions cannot alter model geometry; therefore, all of AutoCAD's dimensions might be thought of as reference dimensions.

Model Dimensions and Annotations

Model dimensions are the sketch dimensions used to define part geometry. Annotations are additional attributes that can be defined within a part or assembly and displayed in a drawing. Model dimensions shown on a drawing can be modified, and the model geometry will update after the Rebuild command has been issued. Model dimensions are shown in black, and reference dimensions are shown in gray. The following are types of model annotations.

- Cosmetic Threads
- Datum Targets
- Geometric Tolerance
- Notes
- Surface Finish

- Reference Dimensions
- Feature Dimensions
- Welds
- Datums

Model dimensions and annotations can be added to an entire view, feature, or assembly component. For this to occur, the proper object must be selected from the drawing's FeatureManager. The following list includes a brief description of each optional object that can have annotations added within a particular view. The illustration that follows shows model annotations.

View	Select the view or views to display annotations. The selected types of annotations are displayed within the selected views.
Feature	Select the feature or features to display annotations. The selected types of annotations are displayed for the selected features.
Component	Select the assembly component or components to display annotations. The selected types of annotations are displayed for the selected assembly components.

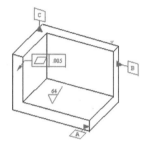

Model annotations.

To display model annotations by view, perform the following steps.

1. Select the views to display model dimensions. To select multiple views, hold down the Control key and select the desired views. If no views have been preselected, annotations will be added to all views.

2. Select Model Items from the Insert menu.

3. Select the types of annotations to be inserted. All Types can be selected to display all model annotations within the selected view. The default selection is Dimensions.

4. Select OK to continue.

To display model annotations using only selected features, perform the following steps.

1. Select the specific features to display model annotations. This is usually accomplished via the drawing's FeatureManager. Multiple features can be selected by holding down the Control key and selecting the desired features.

2. Select Model Items from the Insert menu.

3. Select the types of annotations to be inserted. All Types can be selected to display all model annotations within the selected view. The default selection is Dimensions. The Import From field displays selected features.

4. Select OK to continue.

To display model dimensions using only selected assembly components, perform the following steps.

1. Select the specific assembly components to display model annotations. Multiple components can be selected by holding down the Control key and selecting the desired features.

2. Select Model Items from the Insert menu.

3. Select the types of annotations to be inserted. All Types can be selected to display all model annotations within the selected view. The default selection is Dimensions. The Import From field displays Selected Component.

4. Select OK to continue.

Creating Reference Dimensions

Reference dimensions are added to a drawing as individual dimensions, as opposed to importing them from the model, as in the previous procedures. This type of dimension is not parametric and cannot be used to drive model geometry. The default dimension type is point-to-point, point-to-line, and line-to-line, as shown in the following series of illustrations. When selecting a single line to dimension, the dimension lines are inserted parallel to the selected line. Generally speaking, adding reference dimensions is done the same way as when dimensions are added to sketch geometry. The mechanics are the same.

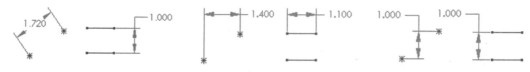

Reference dimensions.   *Horizontal reference dimensions.*   *Vertical reference dimensions.*

The Dimension command will place the dimension value in a direct line between the two points. A dimension can be placed between two vertices, a single sketch entity, or two sketch entities.

➥ **NOTE:** *If you require a refresher course on how to insert dimensions, see Chapter 5, regarding inserting dimensions.*

Once again, adding dimensions in a sketch is exactly the same as adding reference dimensions. However, there are a few extra dimension types that can be added to a drawing as reference dimensions. These are covered next.

Baseline Dimensions

A baseline dimension is a reference dimension that uses a common endpoint and displays dimensions in the same manner as a horizontal or vertical dimension. A baseline dimension will follow any changes to geometry used to create the dimension. To create a baseline dimension referencing one start point, perform the following steps. The illustration at left shows baseline dimensions.

1. Select Baseline Dimension from the Tools/Dimensions menu, or press the right mouse button and select the dimension type.

2. Select the edge to be used as the baseline origin.

3. Select each additional edge or vertex to be dimensioned.

4. Click on the first dimension's text location to place all dimensions.

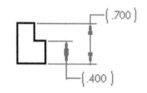

Baseline dimensions.

Ordinate Dimensions

An ordinate dimension is a reference dimension that uses a common endpoint and displays dimensions with a single extension line. An ordinate dimension will follow any changes to geometry used to create the dimension. The type of ordinate dimension (point-to-point, horizontal, or vertical) would be selected based on the same reasoning as the other dimension types. To create parallel point-to-point ordinate dimensions referencing one start point, perform the following steps. The illustration at left shows ordinate dimensions.

1. Select the Dimension icon.

2. Select Ordinate Dimension from the Tools/Dimensions menu, or press the right mouse button and select the dimension type.

3. Select the edge to be used as the ordinate origin.

4. Select the text location.

5. Select additional edges or vertices to be dimensioned.

6. Press the Escape key to exit the command when finished.

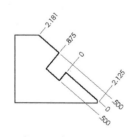

Ordinate dimensions.

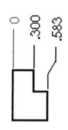

Horizontal ordinate dimensions.

To create horizontal ordinate dimensions referencing one start point, perform the following steps. The illustration at left shows horizontal ordinate dimensions.

1. Select the Dimension icon.

2. Select Horizontal Ordinate Dimension from the Tools/ Dimensions menu, or press the right mouse button and select the dimension type.

3. Select the edge to be used as the ordinate origin.

4. Select the text location.

5. Select additional edges or vertices to be dimensioned.

6. Press the Escape key to exit the command when finished.

To create vertical ordinate dimensions referencing one start point, perform the following steps. The illustration at left shows vertical ordinate dimensions.

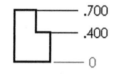

Vertical ordinate dimensions.

1. Select the Dimension icon.

2. Select Vertical Ordinate Dimension from the Tools/Dimensions menu, or press the right mouse button and select the dimension type.

3. Select the edge to be used as the ordinate origin.

4. Select the text location.

5. Select additional edges or vertices to be dimensioned.

6. Press the Escape key to exit the command when finished.

If new features are added to the model, or if you simply forgot to add a dimension, it is possible to insert extra ordinate dimensions after exiting the command. To add new ordinate values to an existing set of ordinate dimensions, perform the following steps.

1. Right click on an ordinate dimension and select Add to Ordinate.

2. Select the new item to be dimensioned.

3. Select additional edges or vertices as necessary.

4. Press the Escape key to exit the command when finished.

Once the ordinate dimensions have been added, they are often too crowded together. Sometimes individual ordinate dimensions need to be jogged or have their alignment broken. First, take a look at breaking the alignment of an ordinate dimension. To break the alignment of an ordinate dimension, perform the following.

1. Right click on the ordinate dimension and select Break Alignment.

To align ordinate dimensions in a drawing that have previously had their alignment broken, perform the following steps.

1. Right click on the ordinate dimension to be aligned and select Align Ordinate.

2. When ordinate dimensions are to be aligned, it does not matter which dimension in the ordinate string is selected. All dimensions in the string will be aligned. If any particular ordinate dimensions must have their alignments broken, this must be done on an individual basis.

Jogging an ordinate dimension can be done in conjunction with breaking alignment, or without. Use whatever combination of jogging and breaking you need to get the desired results. To jog an ordinate dimension, perform the following steps.

1. Right click on the ordinate dimension and select Jog.

2. Drag the text to jog the dimension. Use the green handles to drag the jog line angle and position.

3. When an ordinate dimension is jogged, its alignment is broken out of necessity. You can unjog a dimension by right clicking on that same dimension and unchecking the Jog option. The leader for the dimension will straighten out, but its alignment will still be broken. To realign unjogged ordinate dimensions, follow the procedure shown previously regarding aligning ordinate dimensions.

Linked Dimensions

Linked dimensions are usually added in a part to control design intent, but they can also be added in a drawing. A linked dimension allows a single dimension name to drive many dimensions. Equations could be used for the same purpose, but the link function can be used to do the same thing more quickly when a number of dimensions are required to be the same value. If an equation is used, all of the dimensions need to be added to separate equations.

Equations are useful when mathematical operators are required. Linked values can be used for equalities. When dimensions are linked, they all have the same name. If one of the dimension values is changed, all of the linked dimensions will change. Reference dimensions cannot be linked because they are driven. Linked dimensions require that they drive the geometry of the part. To create a linked dimension, perform the following steps.

1. Right click on a dimension and select Link Value.

2. Specify a name for the Link Value.

3. Select OK to continue.

4. Right click on the next dimension to be linked and select Link Value.

5. Type in the same name or select the link name from the drop-down list.

6. Select OK to continue.

7. Repeat steps 4 through 6 for every dimension to be linked.

It is possible to have multiple sets of linked values all within the same part file. To link values within the same sketch, however, is a poor tactic. Consider the Equal constraint if there are two or more circles (for example) you want to have equal diameters.

This ability to link dimension names is strictly a SolidWorks feature. AutoCAD has nothing that can be compared with linked dimensions. AutoCAD dimensions do not have names with which to link. Because of this lack of underlying intelligence within the dimension entities, the ability to create equations or linked values does not exist.

Modifying Dimensions

Existing dimensions can have any attribute changed through their dimension properties. Model dimensions are created within the part and are used to drive sketch and part geometry. Reference dimensions are typically created in a drawing and do not drive geometry, but will automatically update when geometry is modified.

The option to disable or enable changing model dimensions in a drawing to update model geometry is selectable during installation and can be optionally disabled. When this option is disabled, model dimensions cannot be changed within the drawing. This is usually a decision made by management and is not something you can easily change without reinstalling SolidWorks.

Moving dimensions between views can be accomplished while holding down the Shift key. Hold the Control key down while dragging dimensions to be copied to other views. Only orthographically correct views are allowable. In other words, if Solid-Works cannot make the dimension look reasonably correct in a view, such as a horizontal dimension on edge, it will not place it there. The following are methods used to modify a dimension.

Double click	Double clicking on a model dimension with the left mouse button will bring up the Modify dialog box. This dialog box allows you to change the dimension value and rebuild the feature. Double clicking on a feature will display its dimensions.
Properties	Right click on a dimension and select Properties. Dimension attributes (i.e., font, decimal precision, tolerance, and so on) can be changed. Multiple dimensions can be selected for this function by holding down the Control key, selecting the dimensions, and then right clicking on any one of the dimensions.
Dimension handles	Selecting a dimension will display drag points (handles) that can be used to change the location of the dimension, dimension text, extension line length, or slant.

A dimension's handles can be picked and dragged to the desired location. Handles are similar to AutoCAD's grips in the way they are selected and dragged to a new position. The following illustration shows dimension handles.

Dimension handles located at either end of each extension line.

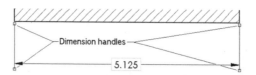

A model dimension can be changed within any type of Solid-Works document (i.e., part, assembly, or drawing). The Modify dialog box appears when you double click with the left mouse button on a model dimension. One advantage to using the Modify dialog box is that different dimensional values can be tried out on the fly by using the built-in Rebuild icon. If multiple configurations are present, it is also possible to modify a dimension for only the current configuration (see the following illustration). It should be noted that the two configuration options will not be present if multiple configurations do not exist. To change a model dimension value, perform the following steps. The following illustration shows the Modify dialog box.

1. Using the left mouse button, double click on the dimension.

2. Enter the new value in the value field.

3. Select from the optional functions available within the Modify menu.

Modify dialog box and its icons.

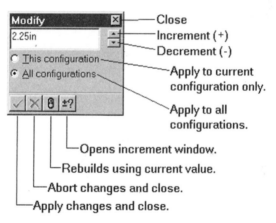

If the increment window is opened, the user can type in a new value for the increment/decrement arrows. Make sure the Enter key is pressed after typing in a new incremental value, or your change will not take effect. Checking the Make Default option (see the illustration at left) makes the new incremental amount the default value when changing any dimension. This will only affect the current document, though, and does not become the system default for new documents.

Increment dialog box.

Dimension attributes can be changed or redefined for individual or multiple dimensions after dimensions have been created. The attributes for dimensions can be changed via the Properties function. Setting dimension properties within the Options dialog box may affect existing dimensions, and in other cases will only affect new dimensions. This takes time to learn and should be dealt with on a case-by-case basis. The following is an example.

There is an existing drawing that, for whatever reason, should have the witness line gap and extension characteristics altered. You click on Tools/Options, go to the Detailing tab, and make the changes. Every witness line in the drawing immediately reflects the changes. Also imagine there are quite a few reference dimensions in this drawing, and they are all surrounded by parentheses. Once more you access the Detailing tab and turn off the "Show parenthesis by default" option. This setting, you will discover, only affects new dimensions.

Let it be said that *most* of the settings that can be changed in the Detailing tab will affect existing dimensions. These are global settings, meaning these settings will affect the entire document.

➡ **NOTE:** *This topic is covered in greater detail in the "Preferences Settings" section in Chapter 3.*

To change an individual dimension's properties, perform the following steps.

1. With the left mouse button, select the dimensions to be changed. Hold down the Control key to select multiple dimensions. Multiple dimensions can also be selected by picking a corner of a rectangle and dragging it to define the opposite corner. All dimensions that fall within the rectangle will be selected.

2. Press the right mouse button and select Properties.

3. Change the desired parameters.

4. Select OK to continue.

Dimension Properties

When the previous procedure is implemented, the dialog box that opens contains specific properties for various dimension types. Some of the property values in the list that follows may not be present in all Dimension Properties dialog boxes. The following are the more common properties.

Value	The dimension's value. Select to change the value of a model dimension. Select Rebuild to update the model once the dialog box is closed. This option is selectable only for model dimensions.
Name	To change the name of a dimension, pick inside the field and enter the new name. The new name must be unique to the part.
Full Name	Full name assigned by the system. This name can be referenced in design tables, equations, or notes. The full name's syntax could be *<DimensionName>@<SketchName>*, *<DimensionName>@<FeatureName>*, or *<DimensionName>@<FeatureName>@<PartName>*, depending on where the dimension name originates (such as in an assembly).
Arrows	Select an arrow type from the pull-down menu. The arrow will automatically be placed outside the extension line when Smart has been selected.
Display Precision	Uncheck "Use document's precision" and press the Precision button to set the primary, dual, and angular dimension decimal precision.
Font	Uncheck "Use document's font" and press the Font button to select a different dimension text font.
Driven	Checked when other dimensions or parameters drive a dimension.
Read Only	Check to disallow any modification of model dimensions in the drawing.
Display with Parenthesis	Check to display parenthesis on drawing-created dimensions.
Modify Text	Select to prefix and append text strings and symbols to the dimension. The dimension remains fully associative with the part/assembly geometry.
Tolerance	Select to define a tolerance type and values for the selected dimension.
Display as Dual Dimension	Check to display a dimension formatted with a dual unit of measure.

To edit the decimal precision of individual dimensions, perform the following steps.

1. Right click on the dimension and select Properties.

2. Uncheck "Use document's precision."

3. Select the Precision button.

4. Select the decimal precision for the Primary Units Value and Tolerance, and for the Alternate Units Value and Tolerance. The value refers to the dimension value (text).

5. Select OK to return to the Dimension Properties menu.

6. Select OK to continue.

SolidWorks uses Windows fonts, which offers many fonts, sizes, and styles for SolidWorks documents. To edit a dimension's font type, perform the following steps.

1. Right click on the dimension and select Properties.

2. Uncheck "Use document's font" to set the dimension's font properties.

3. Click on the Font button.

4. Select the desired font characteristics.

5. Select OK to return to the Dimension Properties menu.

6. Select OK to continue.

The visibility of the extension and dimension lines is definable within SolidWorks. To edit witness line visibility, perform the following steps.

1. Right click on the dimension and select Properties.

2. Click on the Display button.

3. Select the items to be displayed or suppressed. The dimension graphic will display the changes as they are made.

4. Select OK to return to the Dimension Properties menu.

5. Select OK to continue.

Text can be added before, after, above, or below a dimension. This function can be used to add symbols to the dimension. The dimension value is displayed in the form <DIM>. Any symbol is also displayed with a less than and greater than symbol (<symbol name>). Anything displayed with the less than and greater than symbols is being controlled by SolidWorks. If the "<>" symbols are removed, the dimension value (text) will not read correctly. To add or change prefixed or appended text to a dimension, perform the following steps.

1. Right click on the dimension and select Properties.

2. Click on the Modify Text button.

3. Position the flashing cursor where you want text to be added.

4. Type in the desired text, add symbols, or modify existing text. The Preview field will display the changes as they are made.

5. Select OK to return to the Dimension Properties menu.

6. Select OK to continue.

It is possible to modify a dimension's properties in AutoCAD, but it is just not as user friendly as it is in SolidWorks, where there are no messy variable names to deal with, such as DIMALT, DIMASO, or DIMLFAC. SolidWorks takes care of all of these settings, and lets you make modifications by right clicking on a dimension and changing its properties.

Dimension Text Properties

This section expands on the previous procedure for modifying text and adding symbols. For reference, the illustration that follows shows the Modify Text of Dimension dialog box.

Center Text	Checking this option centers the text between the extension lines if the text is inside the extension lines. If outside the extension lines, the text remains stationary.
Dimension Text: (Field 1)	Text entered in this field will be inserted above the dimension.
Dimension Text: (Field 2)	Text entered in this field will be inserted on the dimension line.
Dimension Text: (Field 2)	Text entered in this field will be inserted below the dimension.
Preview	Dynamically updates to show what the dimension text will look like.
Add Symbol	Opens up the Symbol dialog box.
Add Value	Select to add the dimension value back into field 2. The dimension value cannot be changed in this dialog box. The dimension value can only be displayed in field 2. The dimension value will appear as <DIM>. This option is only available if for some reason the value has been removed from field 2.

*Modify Text
of Dimension dialog box.*

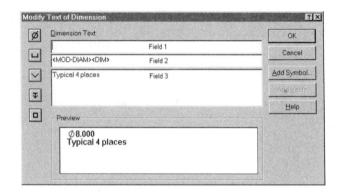

Tolerances can be defined on dimensions. When a model dimension is modified, the tolerance is also displayed on the dimension when selected within the part. To change a dimension's tolerance properties, perform the following steps.

1. Right click on the dimension and select Properties.

2. Click on the Tolerance button.

3. Select the desired tolerance characteristics. The preview field in the lower right-hand corner of the dialog box will display the changes as they are made.

4. Select OK to return to the Dimension Properties menu.

5. Select OK to continue.

Tolerance Properties

The Dimension Tolerance dialog box, shown in the following illustration, has a number of options that should be elaborated upon. Use the illustration that follows this list to reference the various tolerance options described in it.

Dimension Tolerance dialog box.

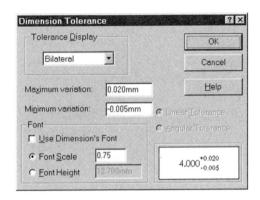

Tolerance Display	Select a tolerance type from the pull-down menu. Select None to display no tolerance. The preview field in the lower right-hand corner displays the selected tolerance type and values.
Maximum Variation	Display the upper tolerance limit. Pick inside the field to change the value. Symmetrical tolerance types are defined by the value in this field.
Minimum Variation	Display the lower tolerance limit. Pick inside the field to change the value.
Use Dimension's Font	Uncheck to set font scale and height of the tolerance value. The Font Scale and Font Height fields are not selectable with this field checked.
Font Scale	Check to select a tolerance font scale value. Pick inside the field to change the value.
Font Height	Check to select a font height for the tolerance font. Pick inside the field to change the value.
Linear Tolerance	Field checked for linear dimensions (not editable, for reference only).
Angular Tolerance	Field checked for angular dimensions (not editable, for reference only).

Tolerance types.

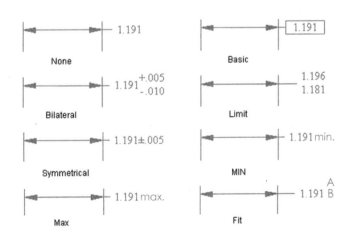

Drawing Symbols

Drawing symbols are 2D entities added to a drawing to document the design intent. These symbols communicate the additional information (i.e., note, surface finish, weld symbol, and so on) that dimensions cannot convey. Model annotations can be defined within the part or assembly and displayed on the detail drawing. The following are annotation features that can be defined on a part or assembly. The illustration that follows shows model annotations.

- Cosmetic Threads
- Datum Targets
- Geometric Tolerances
- Surface Finish Symbols

- Datums
- Dimensions
- Notes
- Weld Symbols

Samples of model annotations.

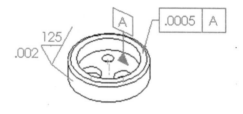

Drawing symbols in AutoCAD are created through a much different process than in SolidWorks. There are third-party programs that will allow an AutoCAD user to insert symbols, but if that type of software is not available, other means must be devised. For instance, it is possible to create symbol libraries within AutoCAD and then insert those symbols as needed into a drawing. This is usually accompanied by specifying x and y scaling, rotation angle, and so on. This is really just a workaround because AutoCAD does not have symbol creation methods of its own. There is the ability to create a few symbols, such as the degree symbol, by using the %% codes, but this is inconvenient.

Notes   Notes are used to insert text, labels, and balloons on a SolidWorks drawing. The note can also be made a hyperlink to a World Wide Web (Internet) page or to another document located on the local computer or over the company network or intranet.

In SolidWorks, notes can be produced in a model or drawing. You can also use an external program (e.g., Microsoft Word) to cut and paste notes via the clipboard as a text string into the drawing. This allows for functions such as spell checking to be performed prior to a text note being inserted into a drawing. The following illustration shows examples of a note, balloon, and hypertext link.

Samples of a note, balloon, and hypertext link.

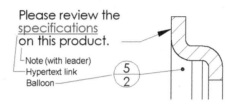

It is important to remember to select a location for the note prior to executing the command. The command will not be accessible otherwise. To insert a text note into a drawing, perform the following steps.

1. With the left mouse button, select a location for the note. This need not be exact, as it is easy to move a note afterward. If the note will contain a leader, the selected location will define where the arrow will point to.

2. Select the Note icon, or select Note from the Insert menu.

3. Enter the text in the Note Text field. Define any additional options.

4. Select OK to continue.

AutoCAD's text commands have come a long way from the simple TEXT command. SolidWorks lacks the capabilities of AutoCAD release 14, such as its formatting of text and specification of line justifications, page anchor points, and overflow type. Changing text styles and colors within the same paragraph is outside the ability of SolidWorks at this time.

Note Properties

The following list describes the various properties found within the note Properties dialog box (shown in the following illustration).

*Note Properties
dialog box.*

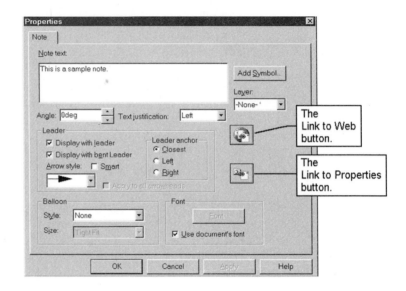

Note Text	Displays the text as it is entered.
Angle	Enter the text angle or pick the up/down arrows to increment/decrement the angle value.
Text Justification	Select left, right, or center justification. Default setting is left.
Add Symbol	At any point in the note, the Add Symbol button can be selected to insert standard drafting symbols.
Layer	Use the drop-down list to specify what layer the note is to be on.
Display with Leader	Check to add a leader to the note.
Display with Bent Leader	Leave this field unchecked to use a straight leader for the note. Check this field to place a tail before the note text.
Arrow Style	Select an arrow style from the pull-down menu when Display with Leader has been checked.
Link to Web button	Pick to enter a file or URL (i.e., *http://www.solidworks.com*) link to jump to when selected.
Link to Properties button	Pick to enter a file property to associate the note to.
Balloon Style	Select the balloon style from the pull-down list.
Balloon Size	Select the balloon size from the pull-down list.
Use document's font	Uncheck "Use document's font" to set dimension font properties. Select the Font button to define a new font style.

Editing a Note

To change the contents of an existing note, perform the following steps.

1. With the left mouse button, double click on the note.

2. Modify the parameters of the note as previously described.

3. Select OK to continue.

Symbol Properties

Symbol Properties are very self-explanatory, but have been included in the list that follows for reference. The Symbol dialog box appears when the Add Symbol button is selected. Describing all of the various symbols is outside the scope of this book. The following illustration shows the Symbols dialog box.

Symbols dialog box.

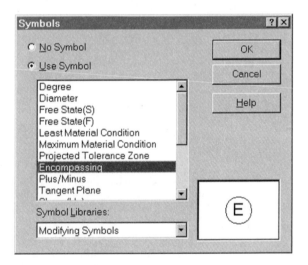

No Symbol	Check to disallow the display of symbols.
Use Symbol	Check to allow for the display of symbols.
Symbol Field	Select the desired symbol from the scrolling list. The list displays symbols from the selected symbol library. The field in the lower right-hand corner will display a preview of the selected symbol.
Symbol Libraries	Select a symbol library from the pull-down menu.

Various Note Options

This section contains the steps for adding notes using various options, such as with balloons or bent leaders. They are listed here for reference purposes and for the reader's convenience. To insert a note with the leader option, perform the following steps.

1. Select the leader attachment point (where the arrow will point to).

2. Select the Note icon, or select Note from the Insert menu.

3. Enter the note text.

4. Check the Display with Leader option. Also check Display with Bent Leader to add a short line to the end of the leader near the text.

5. To specify an arrow type for the leader, uncheck the Smart option to override the system defaults. Otherwise, Solid-Works will choose what it thinks is the best arrow type.

6. Select OK to continue.

7. Drag the text to a new location if necessary.

To insert a note with the balloon option, perform the following steps.

1. Select the balloon leader attachment point.

2. Select the Note icon, or select Note from the Insert menu.

3. Enter the note text. Specify the Balloon Style and Size.

4. Select OK to continue.

5. Drag the balloon to a new location if necessary.

To relocate the attachment point of a leader, perform the following steps.

1. Using the left mouse button, select the annotation. This can be a note, weld symbol, or any annotation with a leader.

2. Drag the arrow's handle (at the tip of the arrow) to a new position. The leader's arrow can be attached to a vertex, edge, or face. The cursor will display system feedback as the cursor passes over these entities.

3. Drop the leader's arrow onto its new attachment point.

4. Drag the note to the desired location, if necessary.

Notes can be cut or copied onto the clipboard and pasted into the current drawing or other drawing and applications. Simply paste any notes copied to the clipboard into the text box when inserting a note. Drawing notes can also be associated with a view. Activate the view or select a model edge or face and insert the note. When the view or model reference moves, the note will move.

Balloons

This command inserts a balloon in a drawing. Balloons can be used to label features and relate them to items on a bill of material. Various types of balloons can be defined. The item in the balloon can be an item number, quantity, or custom value. The following illustration shows some of the balloon styles.

Balloon styles.

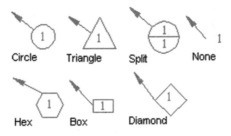

Circle Triangle Split None

Hex Box Diamond

To insert a note with the balloon option, perform the following steps.

1. Select the balloon attachment point.

2. Select the Note icon, or select Note from the Insert/Annotations menu.

3. Enter the note text. Specify a balloon Style and Size.

4. Select OK to continue.

5. Drag the balloon with the left mouse button to the desired location.

To insert a balloon using the balloon icon, perform the following steps.

1. Select the Balloon icon, or select Balloon from the Insert/Annotations menu.

2. Select the balloon attachment point.

3. Drag the balloon with the left mouse button to the desired location.

To edit a balloon, perform the following steps.

1. Double click on the balloon with the left mouse button to open the Note dialog box.

2. Make the desired modifications to the balloon's definition.

3. Select OK to continue.

The following are two points to remember when creating balloons.

- Balloons can be edited using the same methods as notes.

- Default balloon characteristics can be defined under the Detailing tab in the Tools/Options dialog box.

Center Marks

Center marks show the centerpoint of an arc or circle. The illustration at left shows a center mark. To insert a center mark, perform the following steps.

Center mark.

1. Select the Center Mark icon, or select Center Mark from the Insert/Annotations menu.

2. Select any arcs or circles on which to add center marks.

To change the attributes of a center mark, perform the following steps. The illustration that follows shows the Center Mark dialog box.

1. Double click on the center mark with the left mouse button.

2. To change the default properties, uncheck the "Use document's defaults" field.

3. Edit the display attributes.

4. Select OK to continue.

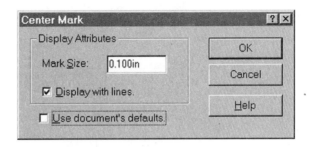

Center Mark dialog box.

Geometric Tolerances

Geometric tolerance symbols and datum symbols can be added to a drawing or part to define the position, shape, or form of a feature. These symbols can be added to a drawing or model directly using the Geometric Tolerance annotation function. If added to a model, these same symbols can then be inserted and displayed on a drawing. The illustration at left shows geometric tolerance symbols. To insert a geometric tolerance symbol in a drawing, perform the following steps.

Geometric tolerance symbols.

1. With the left mouse button, select the location for the geometric tolerance symbol.

2. Select the Geometric Tolerance icon, or select Geometric Tolerance from the Insert/Annotations menu.

3. Click on the GCS button and select the type of tolerance. The preview field will display the symbol chosen.

4. Select OK to return to the Geometric Tolerance menu. The preview will update as the geometric tolerance is created.

5. Enter the value for Tolerance 1, and set whether or not the diameter symbol is to be included.

6. Enter whether any material conditions apply (i.e., MMC or LMC).

7. Repeat for additional tolerances if necessary.

8. Set whether the tolerance is to apply to a projected tolerance zone if required.

9. Enter a Datum identifier, if applicable.

10. Select the Options button if font or leader characteristics need to be modified. This will open a dialog box. Click on OK when finished.

11. Select OK to close the Geometric Tolerance Properties dialog box and continue.

12. Drag the tolerance value to the desired location.

Geometric Tolerance Properties

The following list describes the various properties within the Geometric Tolerance dialog box. Use the following illustration for reference.

Geometric Tolerance dialog box.

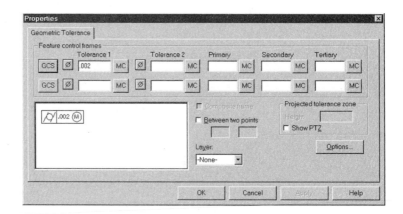

GCS	Pick the GCS button to change or define the GD&T symbol field. This can be defined for the first and second lines of the geometric tolerance symbol.
Tolerance 1	Pick inside the field to enter the first tolerance value. Select the diameter symbol if applicable. Select the MC button to define the material condition if applicable. This field can be defined for the first and second lines of the symbol.
Tolerance 2	Pick inside the field to enter the second tolerance value. Select the diameter symbol if applicable. Select the MC button to define the material condition if applicable. This field can be defined for the first and second lines of the symbol.
Primary	Pick inside the field to enter the primary datum letter. Select the MC button to define the material condition if applicable. This field can be defined for the first and second lines of the symbol.

Secondary	Pick inside the field to enter the secondary datum letter. Select the MC button to define the material condition if applicable. This field can be defined for the first and second lines of the symbol.
Tertiary	Pick inside the field to enter the tertiary datum letter. Select the MC button to define the material condition if applicable. This field can be defined for the first and second lines of the symbol.
Composite Frame	Control both the form and orientation of a profile feature. The top specifies the locating tolerance zone and the bottom the form and orientation.
Between Two Points	Check to add a symbol used to define two locations to constrain the tolerance. Enter the two point names in the field below the check box.
Show PTZ	Check to specify a tolerance zone located at true position extending away from the primary datum. A projected tolerance zone is used to control perpendicularity of threaded or press-fit hole features with the mating part.
Height	Projected tolerance zone height. Pick inside this field to enter a new value. This field is selectable only with Show PTZ checked.
Options...	Pick to select leader and font characteristics for the symbol.
Layer	Use the drop-down list to specify a layer for the geometric tolerance.

Optional Geometric Tolerance Properties

The Options button, found within the Geometric Tolerance dialog box, opens another dialog box. The characteristics described in the list that follows is found in this dialog box, shown in the following illustration. To set optional properties for a geometric tolerance symbol, perform the following steps.

1. Insert the geometric tolerance symbol as previously described.

2. Select the Options button.

3. Select the desired optional symbol characteristics.

4. Select OK to return to the geometric tolerance menu.

5. Select OK to continue.

6. Drag the tolerance note to the desired location if Display with Leader was checked.

The Options button displays this Geometric Tolerance Options window.

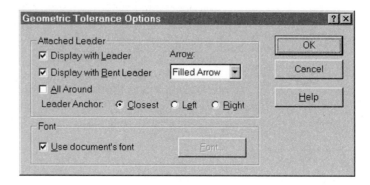

Display with Leader	Check to display the symbol with a leader attached to the selected origin.
Display with Bent Leader	Check to add a leader tail before the symbol.
All Around	Check to add a circle around the intersection of the leader and the leader tail.
Leader Anchor	Select the leader tail placement direction.
Arrow	Select an arrow style from the pull-down menu when Display with Leader has been checked.
Use document's font	Uncheck "Use document's font" to set the dimension font properties. Select the Font button to define a new font style.

To edit a geometric tolerance symbol, perform the following steps.

1. Right click on the geometric tolerance symbol and select Properties, or double click with the left mouse button to automatically open the Geometric Tolerance dialog box.

2. Edit the symbol's definition as required.

3. Select OK to continue.

To relocate a geometric tolerance or datum feature symbol, perform the following steps.

1. With the left mouse button, select the geometric tolerance symbol.

2. Hold the left mouse button down and drag the symbol to its new location.

The geometric tolerance symbols are inserted, by default, using the document's font. For a more complete description of the principles of geometric tolerancing and dimensioning, refer to *ANSI Y14.5M, Dimensioning and Tolerancing.*

Annotations can be turned off globally within any SolidWorks document. If drawing annotations are not visible, right click on the Annotations feature in FeatureManager and select Display Annotations.

Datum

ANSI geometric tolerance datums, datum point symbols, and datum points can be created within SolidWorks. A datum symbol describes the alpha reference (i.e., A, B, C, and so on) for a datum plane. A datum point is used to aid in the measurement of geometric tolerances. A datum point is used to define specific locations for the definition of a reference datum plane. An example would be to define three points that define a reference plane.

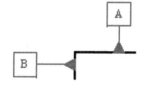

Datum feature symbol.

Inserting a Datum Feature Symbol

A datum feature symbol is used to define properties for datum points. These symbols can be added to a drawing or model directly using the annotation function. These symbols can also be displayed within a drawing. The illustration at left shows a datum. The following illustration shows a datum point and datum point symbol. To insert a datum feature symbol, perform the following steps.

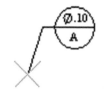

Datum target (point) and datum target symbol.

1. Select the Datum Feature icon, or select Datum Feature Symbol from the Insert/Annotations menu.

2. Select the origin of the datum feature symbol. Add as many datum feature symbols as necessary.

3. Press the Escape key to exit the command.

4. Drag the datum feature to the desired location.

To edit a datum feature symbol, perform the following steps.

1. Double click on the Datum Feature icon, or right click on the Datum Feature icon and select Properties.

2. Enter the desired datum feature symbol characteristics. The preview field in the upper left-hand corner will display the specified datum feature symbol.

3. Select OK to continue.

Inserting a Datum Target Symbol

To insert a datum target symbol (as opposed to a datum *feature* symbol), perform the following steps.

1. Select the geometry on which to place the datum target.

2. Select the Datum Target icon, or select Datum Target from the Insert/Annotations menu.

3. Enter the desired datum target characteristics. The preview field will display the specified datum target symbol characteristics, but not the target area.

4. Select OK to continue.

To edit a datum target symbol, perform the following steps.

1. Double click on the datum target symbol, or right click on the datum target symbol and select Properties.

2. Enter the desired datum target characteristics. The preview field will display the specified datum target symbol characteristics, but not the target area.

3. Select OK to continue.

To relocate a datum target symbol or datum target point or area, perform the following steps.

1. With the left mouse button, select the target area or text of the symbol.

2. Hold the left mouse button down and drag the datum target symbol or datum target area to its new location.

Datum Target Symbol Properties

Refer to the following illustration for the options present in the Datum Target Symbol dialog box. The list that follows describes the properties contained in this dialog box.

Datum Target Symbol dialog box.

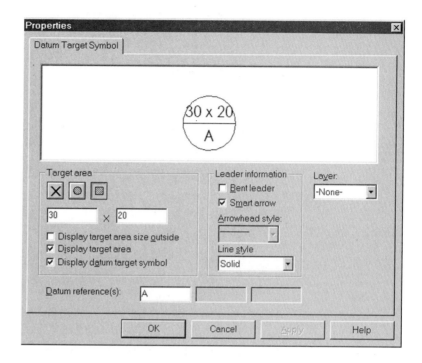

Target area size	Enter the target size for measuring the target area. The Diameter symbol can be selected to indicate a diameter target area.
Display target area size outside	Check to display the target area outside the symbol. Typically this is done when the target area note is too long or large and would be more easily read outside the symbol.
Datum reference(s)	Enter the datum reference letters.
Bent leader	Leave this field unchecked to use a straight leader for the note. Check this field to place a tail before the note text.
Arrowhead style	Displays the arrowhead style. Select a new arrowhead style from the pull-down menu.
Line style	Displays the leader line font style. Select a new line style from the pull-down menu.

Weld Symbols

Weld symbols are used to denote the manufacturing method and processes used to fasten parts using welding processes. These symbols can be added to a drawing or model directly using the annotation function. These symbols can also be displayed within a drawing. The following illustration shows weld symbols.

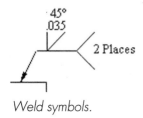

Weld symbols.

Inserting a Weld Symbol

To insert a weld symbol in a drawing, perform the following steps.

1. Select the weld symbol attachment point.

2. Select the Weld Symbol icon, or select Weld Symbol from the Insert/Annotations menu.

3. Enter the desired weld characteristics.

4. If a weld is to be applied to the opposite side of where the leader arrow is pointing, click on the Arrow Side option and enter the desired characteristics.

5. Select OK to continue.

6. With the left mouse button, drag the weld symbol to the desired location.

Welding Symbol Properties

There are numerous welding symbols. Describing all of them is outside the scope of this book. However, the list that follows describes the most commonly used properties. Refer to the following illustration of the ANSI Weld Symbol dialog box.

*ANSI Weld Symbol
dialog box.*

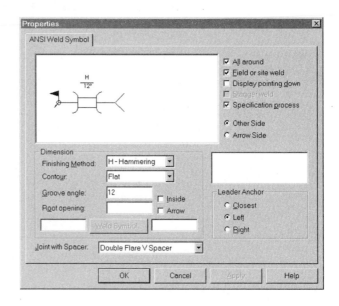

Finishing Method	Specifies the finishing method for the weld. Default is None.
Contour	Specify whether the weld should be convex, concave, or flat. Default is None.
Groove angle	Angular opening for groove weld. Pick inside the field to change the groove weld opening angle.
Root opening	Distance between weld members. Pick inside the field to change the root opening distance.
Weld size (text field)	The text field preceding the weld symbol is used to indicate weld size. Pick inside the files to enter a new weld size.
Weld Symbol	Click on the Weld Symbol button to select a symbol type. Select a weld symbol from the Symbol field. Select OK to continue.
Pitch or length of weld (text field)	The text field following the weld symbol is used to indicate the pitch or length of the weld. Pick inside the text box area to enter a new pitch or weld length.
Joint with Spacer	Select from the pull-down list to specify a joint with spacer. Default is None.
All around	Check to indicate a weld completely around the specified joint.
Field or site weld	Check to display the field weld symbol. Field welds designate a weld done at the work site.
Display pointing down	Only available when "Field or site weld" is checked. Flips field arrow to the opposite side of weld symbol.
Stagger weld	Only available under certain conditions. Check to specify a stagger weld.
Specification process	Check to enter a text string in the tail section of the weld symbol. This area is used to specify a weld process or number of welds. Enter the text string in the text field underneath this option.
Leader Anchor	Defaults to Closest, but leader arrow anchor point can be forced to the left or right side of the weld symbol.

Surface Finish Symbols

Surface finish symbols are used to describe the manufacturing methods, surface roughness, and machining direction for a part surface. These symbols can be added to a drawing or model directly using the annotation function. These symbols can also be displayed within a drawing. The illustration at left shows surface finish symbols.

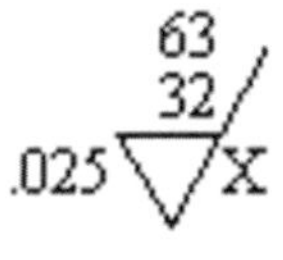

Surface finish symbols.

Inserting a Finish Symbol

To insert a surface finish symbol, perform the following steps.

1. Select the surface finish symbol attachment point. This should be an edge or surface.

2. Select the Surface Finish Symbol icon, or select Surface Finish Symbol from the Insert/Annotations menu.

3. Enter the desired surface finish characteristics. The surface finish symbol is displayed as changes are made.

4. Select OK to continue.

5. With the left mouse button, drag the surface finish symbol to the desired location.

Surface Finish Properties

The properties that can be added using the Surface Finish Symbol dialog box are described in the list that follows. The following illustration shows the Surface Finish Symbol dialog box.

Surface Finish Symbol dialog box.

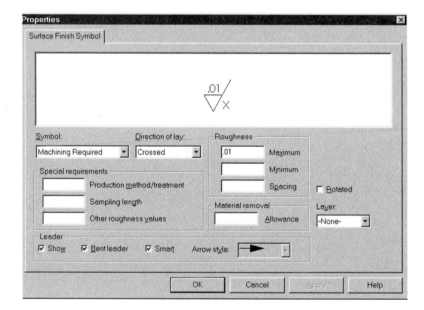

Symbol	Display the current surface finish symbol type. Select a new symbol from the pull-down list. Basic, Machining Required, or Machining Prohibited can be selected.
Direction of lay	Displays the surface finish direction or pattern.

Roughness	Defines height deviation and spacing between the peaks and valleys of the surface.
Minimum	The maximum arithmetic-average height deviation of the surface from the mean line in the profile. Pick inside the field to change this value.
Maximum	The minimum arithmetic-average height deviation of the surface from the mean line in the profile. Pick inside the field to change this value.
Spacing	The maximum allowable distance between repetitive features of the surface pattern. Pick inside the field to change this value.
Special requirements	Defines any necessary requirements.
Production method/ treatment	Specify the production method or process. Pick inside the field to change this value.
Sampling length	Specify the length of surface to be measured. Pick inside the field to change this value.
Other roughness values	Specify the width cutoff value. Pick inside the field to change this value.
Material removal	Specifies the amount of material to be removed by machining. Pick inside the field to change this value.
Leaders:	
Show	Check to show a leader pointing to the surface and to enable the other options in this section.
Bent leader	Check to specify bent leader type.
• Smart	Lets SolidWorks decide on the most suitable arrow type.
• Arrow Style	Displays the currently selected arrowhead type. Select an arrowhead type from the pull-down list.
• Rotated	Rotates the symbol 90 degrees.
• Layer	Specify the layer on which to place the symbol.

Cosmetic Threads

Cosmetic threads are used to display the major or minor diameter of a thread. They appear as dashed lines indicating threads within a hole or on the exterior of a cylinder. These symbols can be added to a drawing or model using the annotation function. The following illustration shows cosmetic threads.

Cosmetic threads.

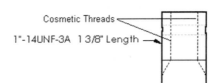

Cosmetic Threads

1"-14UNF-3A 1 3/8" Length

Cosmetic threads can be placed in a part or a drawing. The process is the same in either case. To insert a cosmetic thread into a part or drawing, perform the following steps.

1. On the part or in the drawing view, select a circular edge on which to start the cosmetic threads. It does not matter if you are viewing the edge from the side or straight on.

2. Select Insert/Annotations/Cosmetic Threads.

3. Specify the thread parameters.

4. Enter the thread callout if desired.

5. Select OK to continue.

Hole Callouts

Hole callouts are used to annotate circular features. These features can be cylindrical cuts, or simple, countersunk, counterbored, or tapped holes. These symbols can be added to a drawing directly using the annotation function.

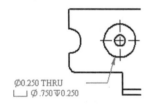

∅0.250 THRU
⌴ ∅.750 ▽0.250

Hole callout.

The hole callout feature uses the additional information defined by the Simple Hole or Hole Wizard commands to help automate detail drawing creation. The illustration at left shows a hole callout. Hole callout information can also be automatically obtained from an extruded cut. To add a hole callout, perform the following steps.

1. Select the Hole Callout icon, or select Hole Callout from the Insert/Annotations menu.

2. With the left mouse button, select the circular feature to be dimensioned.

3. The Modify Text dialog box will appear with the recommended hole callout displayed in the preview box. Add additional text or symbols as required.

4. Select OK to continue.

5. Select additional holes to call out, or press the Escape key to exit the command.

Bill of Materials

A bill of materials (BOM) is created on assembly drawings to describe component names, part number, and the number of instances for each component in the assembly. This BOM can be created in part from the individual component properties of the parts within the assembly. AutoCAD does not currently have functionality similar to SolidWorks' BOM, but there are third-party programs that may provide this feature.

The item number is derived from the creation order of the assembly. This order is displayed within the FeatureManager design tree from top to bottom. The BOM will display only unique assembly components. Multiple assembly components are noted within the Quantity field. Any changes to the assembly are updated within the BOM. A BOM can be added only to an assembly. The following illustration shows a bill of materials.

Bill of materials.

ITEM NO.	QTY.	PART NO.	DESCRIPTION
1	1	Mounting Plate	
2	1	Housing	
3	1	Control Mount	
4	2	Round Cover Plate	
5	1	SubAssy	
	1	Worm Gear Shaft	
	1	Worm Gear	
6	1	Cover Plate	
7	1	Offset Shaft	

Inserting a Bill of Materials

To insert a BOM, perform the following steps.

1. With the left mouse button, select a drawing view. The view should contain the parts in the assembly to be included in the BOM.

2. Select Bill of Materials from the Insert menu.

3. Specify the BOM template to use and select Open. The default BOM template file is titled bomtemp.xls.

4. Select the desired options, which are described in the section that follows.

5. Select OK to continue.

After following the steps listed, SolidWorks will create the bill of materials. There are a couple of issues that immediately come to mind, such as the order of the components listed in the BOM. Because this is determined by the order of the parts as specified in the original assembly's FeatureManager, reordering the components in the assembly will also reorder the components listed in the BOM.

⊷ **NOTE:** *To review this process, read through the section "Assembly Structure Editing" in Chapter 7.*

Editing a Bill of Materials

Another issue that arises regarding a BOM is how to take advantage of the Description column. This can be done one of two ways. The first is to edit the BOM and simply type in a description for the desired components. New components you would not typically have created an actual SolidWorks part for can be added in a similar manner. Examples of such components would be paint or glue. To edit a BOM, perform the following steps.

1. Double click on the BOM. The interface (menus and toolbars) will change to an Excel interface.

2. Click in the cell you want to add a description to and type in the desired description. Make sure this is done under the Description column heading.

3. To add another row to the BOM, click anywhere on the row containing the $$END tag and select Insert/Rows. A new row will be inserted above the selected row.

4. Type in the desired information.

5. When finished, click once outside the BOM to return to SolidWorks.

Another option the user has available when it comes to filling in the Description column is to use the File Properties of the individual part files. This requires some work up front, but the Description column is filled in automatically when the BOM is inserted into the assembly layout. Specifically, the user must create a description property and give that property a value, which is what actually becomes the description. Here are the steps for performing this function.

1. Open the file you want to create the description for (such as a part) and select Properties from the File menu.

2. Click on the Custom tab of the Summary Info dialog box.

3. Click once in the Name field and type in the word Description. Your spelling must be correct, but it is not case sensitive.

4. Leave the Type field set to Text.

5. Click in the Value field and type in whatever you want for the description. Use any characters you wish, including symbols or punctuation.

6. Click on the Add button and the item will be added to the Properties list (see the following illustration for an example).

7. Click on OK to close out of the Summary Info dialog box.

8. Save your SolidWorks file.

Creating custom properties.

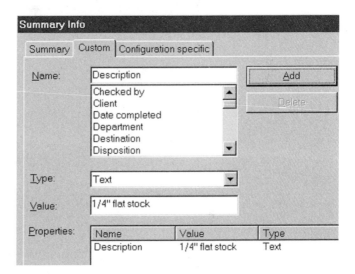

Now when you insert a BOM into an assembly that uses this particular part (the one in the example, whatever it may be), the Description field will automatically display a description for the part. The only issue that arises using this type of functionality is getting all of the SolidWorks users in a particular company to make sure they fill out the properties when a part file is created. This would be information that could be disseminated to the employee population by the CAD administrator.

There are other ways in which one can edit a BOM. Read on to discover some of the other features found when creating a bill of materials.

Bill of Material Properties

The Bill of Materials Properties dialog box contains options described in the list that follows. Use the following illustration as reference for the options described in the list.

Bill of Materials Properties dialog box.

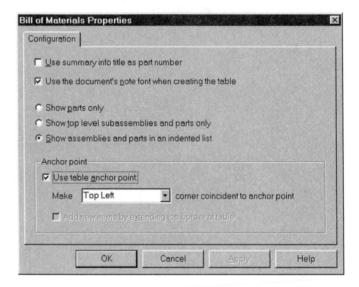

Use summary info title as part number	Check to use a component's Summary Info title field for the part number.
Use the document's note font when creating the table	Uses the font set up in Tools/Options/Detailing tab under the Note font button. Otherwise, uses the default Excel font, which is more than likely not the same font or size as your drawings note font.
Show parts only	Check to show only parts in the BOM. Subassembly components will be listed separately.
Show top level subassemblies and parts only	Check to show subassemblies and parts in the BOM. Subassemblies will not have their parts listed separately.
Show assemblies and parts in an indented list	Check to show subassemblies and parts in the BOM. Subassemblies will have their parts listed separately, but the subassemblies' components will not be issued item numbers.
Use table anchor point	Allows anchoring the BOM to a point on the drawing template.

Make *(corner_name)* corner coincident to anchor point	Pick from the drop-down list to make a particular corner of the BOM coincident with the anchor point.
Add new items by extending top border of table	Check to add new items to the top of the table. This is only available when "Use table anchor point" is *not* checked.

Using Anchor Points

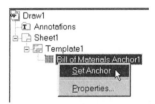

Setting an anchor point for a bill of materials.

An anchor point can be specified if using a template in your drawing layout. This anchor point can then be used to anchor the BOM to a specific location on the sheet. If, for example, you want to anchor the BOM to the top right corner of the sheet, you can then specify Top Right as the corner to anchor on the BOM. Before you can specify which corner to anchor, you must create the anchor point. Perform the following steps to create an anchor point. Use the illustration at left for reference.

1. In the drawing's FeatureManager, expand the Template item to show the Bill of Materials Anchor item (see the illustration at left).

2. Right click on the Bill of Materials Anchor item and select Set Anchor.

3. Pick a location for the anchor point using your left mouse button. This is usually an endpoint of a line on the template.

4. Right click on the sheet and select Edit Sheet when finished.

To take advantage of this newly created anchor point, specify "Use table anchor point" in the BOM dialog box (see earlier illustration in this same section). You must then pick one of the anchor point options from the drop-down list, such as Top Left or Bottom Right. There are a total of four options, one for each corner of the BOM.

Updating, Moving, and Resizing BOMs

Updating a BOM can be an automated process, or you can also choose to manually update the BOM. In the Drawings tab of the Options dialog box, you will find an option that states Automatic update of BOM. When this option is checked, adding or removing components from the assembly is automatically reflected in the drawing layout containing the BOM. If Automatic update of BOM is not enabled, you must manually update the BOM to see the changes reflected. To manually update a BOM, perform the following steps.

1. Select the drawing view used to create the BOM.

2. Select Bill of Materials from the Insert menu. The existing BOM will update to show any additions or deletions.

With regard to moving a BOM, you will find that your options are severely limited if an anchor point is being used. This is an understatement, because actually you will not be able to move the BOM. This makes sense, considering the fact that it is anchored! However, you can edit the properties of a BOM and turn off the anchor point. To move a BOM, perform the following steps.

1. If no anchor point is being used, select the BOM (one mouse click). This will show the green handles at each corner of the BOM.

2. Hold the left mouse button down and drag the BOM to a new location.

If an anchor point is being used and you change your mind and decide the anchor point needs to be different, or you want to "float" the BOM, perform the following steps.

1. Right click on the BOM and select Properties.

2. Change the anchor point, or change whether or not an anchor point is being used.

3. Select OK when finished.

The BOM will automatically move to a new location if the anchor point is enabled or changed. If you have tried this particular set of steps on your own, you will have noticed a couple of extra tabs when accessing the BOM's properties. Specifically, there is an additional tab titled Contents, and another titled Control. The Contents tab allows hiding items in the BOM. By removing the green check placed before the item in the BOM, it is possible to keep that item from appearing in the BOM. This decision can be reversed at any time, and the BOM will automatically update accordingly.

Another aspect of the Contents tab is the ability to place the column headings on the bottom of the BOM as opposed to on the top. Doing this also reverses the order of the components in the BOM. This option is called "Display labels on top," and unchecking it in essence flips the entire BOM. The Contents tab is shown in the following illustration.

Contents tab of a BOM's properties.

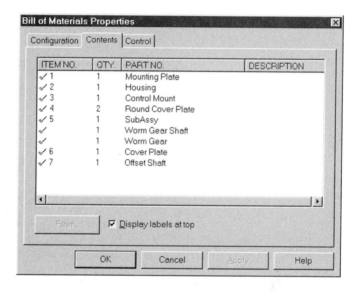

With regard to resizing a BOM, that is really dependent on the font being used and the width of the columns as defined in Excel. If the note font being used in the drawing is changed, and you specified "Use the documents note font when creating the table" as you were first inserting the BOM, the BOM will grow or shrink in size to accommodate the new font. Solid-Works will even ask you if you want it to automatically refit the column widths to the new font size.

It is possible to change the width of the columns manually, simply by editing the BOM and changing their widths. See the earlier section "Editing a BOM" to see how this is done (or just double click on the BOM).

Splitting a BOM

It does not take a very large assembly to find that a BOM is getting too long for the assembly layout sheet. This is where the Control tab of the BOM's properties comes into play. By right clicking on the BOM and selecting Properties, then selecting the Control tab, you will see an option for splitting the BOM, as shown in the following illustration. The actual steps for splitting a BOM are listed here for convenience.

Splitting a BOM.

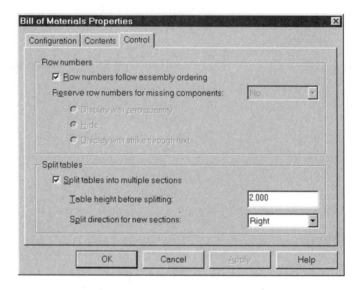

1. Right click on the BOM and select Properties.

2. Click on the Control tab.

3. Place a check in the "Split tables into multiple sections" option.

4. Specify the height the BOM should be allowed to extend before a new section is split off.

5. Specify the direction new sections of the table should be split off in, either to the right or left.

6. Click on OK when finished.

In this same Control tab used to split a BOM, there are also some options available for controlling the row numbers. Both the Row Numbers and Split Table options are disabled by default. Concerning the Row Numbers option, this means that it is possible to edit a BOM and change the values under the Item No. heading as listed in the BOM. If the "Row numbers follow assembly ordering" option is left unchecked (see previous illustration), it is possible for the user to modify the item numbers. If an Item No. is modified, the BOM will continue to display the users modification even if updated. This is not the case if "Row numbers follow assembly ordering" is checked. The following are a few other notes regarding BOMs that merit mention.

- Microsoft Excel 97 is required to use this function. It is also highly recommended that Service Pack 2 for Excel 97 is installed (or for Office 97, whatever is appropriate).

- The default template for BOM creation is Bomtemp.xls and is located in the SolidWorks/LANG/<language> directory (where <language> is English by default).

- It is possible to have more than one BOM on a drawing sheet. Theoretically it would be possible to have a BOM for every view on the sheet, though this scenario is unlikely.

- It is also possible (though unlikely) to insert a BOM for a part, in which case you will wind up with a BOM with only one item.

Summary

Drawing formats can be imported from your existing CAD systems using a DWG or DXF file format. SolidWorks allows for multisheet drawings within the same drawing file. Drawing views are created to display different orientations of parts and assemblies. These drawing views are associative to the part or assembly used to create them. Any changes to the part or assembly are automatically updated in the drawing. Changes made to the drawing's model dimensions will update the part or assembly. This is a function that can be disabled during installation.

Drawing views also benefit from the ability to automatically display them with hidden lines removed or hidden lines in gray or dashed. SolidWorks allows you to define the views without spending time cleaning up views for hidden line removal or updating geometry changes. Dimensions can be added to 2D sketch geometry for a solid model and then reused on drawings, or recreated inside the drawing as reference dimensions. The advantage to using model dimensions is that these dimensions do not have to be recreated in a drawing.

Annotations can be defined within parts and assemblies to assign engineering attributes. The advantage to defining these attributes during the design phase is that the designer or engineer that created the design can assign an attribute (i.e., surface finish, geometric tolerance, and so on) when design criteria are first established. This can be easier than performing this task later or having someone else less familiar with the design add these attributes.

SolidWorks' ability to create all of the standard views from a solid part file and have that geometry update when design changes are made results in significant time savings. The ability to drag detail circles and section lines to obtain best overall views can result in countless hours saved on the drafting floor.

9 Rendering

Introduction

This chapter discusses the lighting and shading capabilities of SolidWorks. Models can be created and edited while a wide array of light sources and color options are active. Other uses of implementing these features include such venues as marketing and design reviews, customer presentations, advertising and promotion, and production of manuals and documentation. The PhotoWorks module described at the end of this chapter has more capabilities and allows for the creation of professionally rendered images.

Prerequisite

Chapters 6 and 7 should be understood. It is not necessarily essential to comprehend how every feature is created, nor is it mandatory that every aspect of assemblies and mating relationships be completely understood. What is important, however, is the theory behind feature-based modeling and knowing that a part consists of individual features and that an assembly consists of individual parts. This will be important when discussing the properties of these individual features and parts.

Content

Within this chapter, you will find procedures on lighting, making use of colors, modifying colors, lighting properties of individual features, and saving images. The section on lighting describes the types of light sources available and how light can be added, changed, or deleted. Light sources alter the way a model appears within SolidWorks. Lighting can add a quality to a part that makes it look as though it was made of certain materials, such as, for example, glossy plastic or transparent resin.

The section "Feature Colors" explains how to modify a part in numerous ways with regard to color options. Specifically, you will see how to make certain colors apply to certain types of features. In addition, a section on feature properties explains how individual features can be altered to display specific properties and how this can apply to individual faces.

PhotoWorks is an optional module that produces photorealistic images of parts and assemblies. A comparative description, including reasons a SolidWorks user might want to purchase PhotoWorks, is provided at the end of the chapter.

Objectives

With completion of the "Lighting" section, you should be able to define, change, or delete a light source, and should understand how these light sources are saved with a document. The section "Advanced Shading Characteristics" steps you through the process of modifying light properties. These properties can be light emission qualities, transparency, and so on.

When you have worked through the "Feature Colors" section, you should be able to define or change a feature's shaded color properties and understand how these color settings are stored within documents. You should have a perspective on global settings versus color settings for independent features. You should also understand how assembly parts could have their colors changed and how to control whether these changes are propagated to the part.

Lighting Sources

Lighting can be used for any number of reasons in a CAD program. The different forms of light sources can be combined to effectively highlight the shaded part or assembly. Working within shaded display mode can make working with complex parts or assemblies easier. The other reason for working within shaded display mode is that parts and assemblies will display faster. Lightweight assemblies use shaded display mode to quickly open and display large assemblies. Lightweight assembly features are discussed within the chapter on assemblies.

Lighting may also be used for other, more business-oriented reasons. Because AutoCAD is more suited to the architectural market, lighting may be used to enhance an image of a building, perform sun and shadow studies, and display light source options for interior decorating possibilities. The list is endless. AutoCAD requires an add-on package to complete these tasks. It should be noted that with the release of version 14, a rendering package is now included with AutoCAD.

SolidWorks makes use of lighting for various purposes. Because SolidWorks is suited to the mechanical design market, its lighting capabilities are somewhat different from those required for an architectural program. For example, sun and shade studies are admittedly not very important for a mechanical part that may very well wind up buried inside an assembly.

Where lighting can be important is during the design process. When viewing a part in a more realistic manner, editing the part not only looks more visually appealing, it becomes much more efficient. Finding possible flaws in the design is easier than if the same part or assembly were being displayed in a simple wireframe. When working with complex parts or assemblies, it is usually easier to understand exactly what you are looking at when the part or assembly is shaded.

AutoCAD's shading capabilities are not comparable to Solid-Works' out-of-the-box functionality. Even with release 14, it is still not possible to dynamically rotate a shaded part in AutoCAD. However, as nice as AutoCAD's 2D renderings may be, they are still static images that cannot be rotated, panned, or zoomed during the design process. AutoCAD's AutoShade feature allows for walkthrough animations, but that is a different topic that goes beyond the scope of this book. If photorealistic renderings are needed, PhotoWorks is the solution for Solid-Works users.

One more piece of important information should be noted. Any and all of the lighting and color changes discussed in this chapter are saved with the part or assembly. If an assembly part's color or lighting is changed in any way, and the document is saved, expect to see the changes the next time the part is opened.

Light Sources

Lighting
— Ambient
— Directional1
— Point1
— Spot1

Light Source objects within the FeatureManager design tree.

Light Sources is actually a name of a dialog box available in SolidWorks. Light sources are found within both parts and assemblies. The Lighting icon is displayed at the top of the FeatureManager design tree and is shown in the illustration at left.

The light sources Point1 and Spot1 are not added by default, and must be added by the user if desired. Steps outlining how to employ this option are provided in material to follow. The light source styles (ambient, directional, point, and spot) can be combined to produce a realistic lighting effect. These styles are defined as follows.

Ambient light	A light source applied evenly from all sides of the object. Most of the settings in the Ambient tab section are grayed out because they do not apply to ambient lighting.
Directional light	A light source applied from a distance. The sun would be a perfect example of a directional light source. Light is radiated from a specific direction, but shines uniformly on the part from its specified location.
Spotlight	A tightly focused directional light source. Like directional light sources, spotlights have a source position, but also have a direction the spotlight is pointing in. This creates a definite spot on the part that gradually diminishes as the light cone expands. An example of this type of light source would be a flashlight or an automobile's headlights.
Point light	A light source that applies light in all directions from a single point in space. An example of this type of light source would be a light bulb.

Modifying Light Source Properties

This section discusses the types of light sources previously mentioned, and describes the functions the various settings in each of the light source dialog boxes perform. If you are working along with this book in SolidWorks, open a part to use as a subject for the lighting experiments contained in this section.

The first and most straightforward of the light sources is Ambient. This lighting source will be discussed first, followed by the Directional, Point, and Spotlight light sources.

Camera part with default lighting characteristics.

Basic Lighting Options

All light sources will have a Basic tab that contains the following options: Intensity, Edit Color, "Use as default" (see illustration at left), and On. Descriptions of these options and their functionality are provided in the following list.

Intensity	With this option, the intensity of the light source is defined. To change a value, move the slider to the left to decrease, or right to increase, the value. These options are found within each light source, with the exception of an Ambient light source. An Ambient light source only has an Ambient setting. All values described below have a range of 0 to 1. The Intensity options are as follows: • *Ambient.* Sets the amount of ambient light that comes from all directions. • *Brightness.* Defines the intensity of the light source. • *Specularity.* A specular light source tends to bounce off an object in a particular direction, giving the object a shiny quality. A glossy surface will have a high specularity value.
Edit Color	This option sets the color of the light source. White is generally best for ambient light sources. This setting does not change the color of the part, only the light source that shines on the part. Depending on the display settings determined in the Windows Display dialog box and your graphics card memory, the number of colors available to you may be different. This option is found under, and has the same functionality in, all light source tabs.
Use as default	This option sets whatever changes are made to the Ambient or Directional tabs as the default settings for future parts. This option, however, is not available under the Point or Spot Light tab. Be careful using this option, as it may be difficult to get back to the "factory set" settings if you decide changes are not acceptable. This option is found under, and has the same functionality in, all light source tabs, but is only selectable for the Ambient and Directional light source types.
On	This turns the light source on or off. This option is found under, and has the same functionality in, all light source tabs.

Light Source Position Options

The Position tab is used to define the location of a directional, point, or spot light source. An ambient light source does not have a position; it is applied from all sides. The position options are described in the following list.

Position	This option is used to define the direction of the light source, using spherical or Cartesian (x-y-z) coordinates. However, these adjustments will give you a real-time update of how changes affect a part. Drag the Light Sources dialog box out of the way so that you can see changes occur as you use the slider controls.
Spherical	Define the light source using latitude and longitude values defined in degrees and a distance from the object. This option is available for spot and point light sources.
Cartesian	Define the light source using x, y, and z coordinates based on the part origin. This option is available for directional, spot, and point light sources.
Lock to Model	Set to lock the light source so it will always point at the object. This option is available for all directional, spot, and point light sources.
Target	Define the light source target using x, y, and z coordinated based on the part origin. This option is available only for a spot light source.

Spot Light Advanced Options

A Spot Light has an additional tab used to define the following options. The illustrations that follow show a camera part with a specularity setting of about 60 percent and a spot light source added to the camera part.

Camera part shown with specularity set at about 60 percent.

Spot light source added to the camera part.

Exponent	Sets the degree of beam focus. Move the slider to the right to increase the beam's focus. Think of this as the fuzzy quality around the light where it strikes the object. The Cutoff option, described in material that follows, controls the beam's cone angle.
Attenuation	This property decreases the intensity of the light as the light's distance from the light source increases. Note that the light source itself is not increasing in distance. Rather, it is the distance the light is from the source. It might help to think of the light waves as diminishing in intensity the farther they get from the light bulb. Attenuation is the speed of this diminishing intensity. The distance of the light source itself is determined according to the formula involving Distance. Attenuation is determined according to the following formula: (where d = Distance) $1/(a+b*d+c*d2)$.
Cone Angle	Used to set the spread of the light cone for a spot light source. The default value for the setting is 45, the range of which is 0 to 180. Experiment with this to find an acceptable value for your specific part. The cone angle will dynamically update when this value is changed.

Take a look at the following illustration to see what settings were used for the Spot Light settings. The Spot Position and Direction settings were left at their default values.

Settings used for the Spot Light source in the previous illustration.

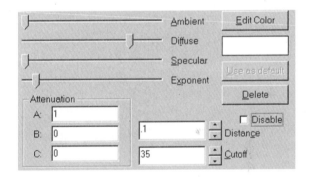

Adding a Light Source

Adding, editing, or deleting a light source is done by selecting the Lighting icon on the FeatureManager design tree. If a "+" appears before the icon, the light sources are not shown. To expand the display to show the defined light sources, press the "+". The display of the FeatureManager design tree should change to display the defined light sources. A minus sign ("−") will now appear before the Lighting icon. To add a light source, perform the following steps.

1. Select the Lighting icon on the FeatureManager design tree.

2. Press the right mouse button and select the desired light source (Directional, Point, or Spot).

3. Edit the lighting properties. Use the definitions previously listed for reference if needed.

4. Continue adding light sources as needed, or select OK to close the dialog box and accept changes.

Changing a Light Source

All that is required to access functionality for adding a light source is to select the Lighting icon on the FeatureManager design tree. To change an existing light source, perform the following steps.

1. Select the light source to change from under the Lighting icon within the FeatureManager design tree.

2. Press the right mouse button and select Properties.

Removing a Light Source

Any defined light source, with the exception of the Ambient light source, can be deleted. To remove a light source, perform the following steps.

1. Select the light source to remove from under the Lighting icon within the FeatureManager design tree.

2. Press the Delete key and confirm the removal.

Colors

Now that you can change the light sources for a part or assembly, it is time to move on. At this stage of the book, you should have a sense of what modifying the color of a part entails. This has been mentioned in a few places, but for the sake of convenience, the steps for changing a part's color follow. To change the color of a part, perform the following steps.

1. Click on Tools/Options.

2. Select the Color tab.

3. In the System area, select Shading.

4. Click on the Edit button and select a color that compliments the part.

Okay, so maybe you do not care about choosing a color that compliments the part. If you prefer purple bearings, who is to judge? In seriousness, though, the color of parts plays an important role when creating an assembly. For example, if all parts were left as their default shade of gray, you can imagine how difficult it would be to differentiate the components in an assembly. In some cases, it would be a nearly impossible task. Consider the idea of color coding by using particular colors for particular part types.

Color can have two main functions: to view and shade the part using a realistic appearance or to help differentiate the part when working on the assembly. A good practice is to get part colors within SolidWorks to aid visualization with the assembly, and to use PhotoWorks to set and define realistic material properties for rendering.

AutoCAD generally associates colors with layers. SolidWorks does not have the need for layers, but colors still need to be managed. This can be done on a global level, per part, or for individual features or faces. AutoCAD is not feature based; therefore, when two or more solids are combined, the lesser components take on the traits of the master solid. For this reason, contiguous solid models in AutoCAD can be one color only when shaded.

Advanced Shading Characteristics

Individual parts can have any one of six settings adjusted that control the lighting characteristics of the part. These settings are found in the Color tab of the Options dialog box. To open the Material Properties dialog box, shown in the illustration that follows, perform the following steps.

1. Click on Tools/Options.

2. Select the Color tab.

3. In the System area, select Shading.

4. Click on the Advanced button to open the Material Properties dialog box.

Material Properties dialog box.

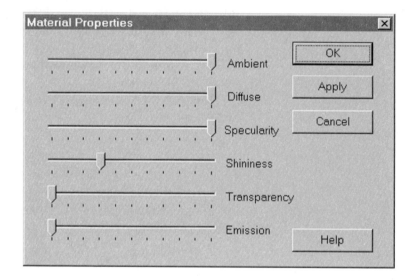

There are slider bar adjustments for each of the six lighting characteristics available, described in the following table.

Ambient	Light reflected and scattered by other objects
Diffuse	Light reflected from the surface of a part and scattered equally in all directions
Specularity	An object's ability to reflect light
Shininess	Determines the reflective nature of a part
Transparency	Determines the amount of light that can pass through an object
Emission	Determines how much light is projected from the surface of a part

By using these settings, it is possible to create some very interesting-looking parts and assemblies. The transparency setting alone is excellent when working with an assembly whose inner workings need to be shown. The use of these settings, combined with SolidWorks' ability to show movement using mating tools, makes for some very impressive visuals.

The Material Property adjustments will update a part onscreen automatically. If the dialog boxes are in the way, drag them to the side or pan the part to the opposite side so that the adjustments can be observed on the part. The illustration at left shows the camera part with a transparency set at about 75 percent.

Camera part with transparency set at about 75 percent.

When a part is to be shown with transparency within an assembly, create two configurations within the part. One would be the normal part color, and the other would be a transparent setting. Two assembly configurations can then be created for displaying the part with the solid and transparent color settings. Configurations allow different versions of a part or assembly to be defined and named.

Feature Colors

In addition to being able to make modifications to an entire part, SolidWorks allows you to specify colors for individual features. One way this is accomplished is using the bottom half of the Colors tab, shown in the following illustration.

Colors can be associated with individual features.

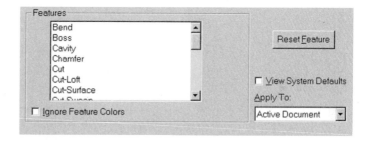

More than twenty features are listed in the Features section. Using the Apply To option, any changes made can apply to the current drawing, future drawings, or both. As always, it is wise to use caution when changing default settings. To specify that certain features are to have specific colors, perform the following steps.

1. Click on Tools/Options.

2. Select the Colors tab.

3. In the Features list box, select the feature whose color you want to change.

4. Click on the Edit button to modify the color of the selected feature.

5. Continue this process until the desired feature colors are modified.

6. Use the Apply To function to specify whether the new settings should be applied to the current drawing (Active Drawing), future drawings (System Defaults), or both (All Possible).

7. Click on OK when finished.

The other two options to note are Ignore Feature Colors and Reset Feature. Ignore Feature Colors will ignore any feature color settings that have been made because this box has been checked and is active. This is a desirable feature when you want to temporarily disable feature colors. Reset Feature is a button that resets any selected feature to the same color the part is currently set at. The current part color is set by the Shading option in the System list box, previously discussed.

Feature Properties

There is another way to change the color of part features that allows for a much greater degree of flexibility. Part features can have any color of the rainbow, independent of part color or feature color. These features can also have independent lighting characteristics, as described earlier in this chapter. Again, the key to accessing this functionality is the right mouse button. The following steps outline this process.

1. In FeatureManager, right click on the feature whose color you want to change.

2. Select Properties.

3. Click on the Color button. This opens the Entity Property dialog box.

4. Click on Change Color to use a color other than the feature's current color settings.

5. Click on Advanced to modify the Material Properties (lighting characteristics) of the feature.

6. Click on OK to exit out of each dialog box and accept changes.

This can be a very convenient alternative to having every feature of a certain type be the same color. By accessing a feature's properties, color and lighting can be controlled on a much more selective scale. It is also possible to change the properties of a single face, instead of an entire feature. The process is very sim-

Camera part with face properties changed.

ilar to the previous one. The following steps outline this process. The illustration at left shows the camera part with some of its face properties changed.

1. In the Document window, right-click on the face whose color you want to change.

2. Select Face Properties.

3. Click on the Color button. This opens the Entity Property dialog box.

4. Click on Change Color to use a color other than the face's current color settings.

5. Click on Advanced to modify the Material Properties (lighting characteristics) of the face.

6. Click on OK to exit out of each dialog box and accept changes.

Assembly Parts

All of the functionality described in this chapter can be applied to SolidWorks parts and assembly features. There are no differences between the two. There is only one other aspect regarding changing the color of a part inside an assembly that should be addressed. Depending on how the part's color is changed will make a difference as to whether or not those changes propagate back to the original part, and possibly any other assemblies the part may be in. The following are the possible scenarios for changes made to a part's color.

Scenario 1: Color changes made to the part are intended to affect the original part and any associated assemblies. To change the color of the part, perform the following steps.

1. In FeatureManager, right click on the part.

2. Select Edit Part. The part is now being edited, not the assembly.

3. Click on Tools/Options.

4. Select the Color tab.

5. Select Shading and make changes, as described earlier in this chapter.

Scenario 2: Color changes will affect only the part in the current assembly. To change a part's color in the current assembly only, perform the following steps.

1. In FeatureManager, right click on the part.

2. Select Component Properties.

3. Click on the Color button.

4. Make desired color changes, as described earlier in this chapter.

If you use this second method, you do not have to worry about how this color change will affect other documents. This is because any changes made to the color of the part with the Component Properties Color button will change the color of the part within the current assembly and will not change the original part.

Saving Images

Once all lighting and color settings have been implemented, it might be nice to save an image or two. SolidWorks allows the current document to be saved as a TIFF image. This functionality can be accessed using the Save As function. Select the Tif Files option from the Save As Type pull-down menu. Photo-Works, however, discussed in the next section, will save to a number of file types of any resolution.

How do you save an image in SolidWorks? If the TIFF image format does not do what you want, you can go out and buy a screen capture utility. That is how many of the images in this book were created. There are many such products on the market. These products provide additional image editing and the ability to save files under numerous file formats.

The second option is built into your computer. Windows 98 and NT both allow copying of the entire screen to the clipboard by pressing the Alt+Print Screen keys on your keyboard. This takes the entire screen; therefore, the image needs to be cropped. It can then be pasted into any program.

PhotoWorks

SolidWorks' PhotoWorks™ is analogous to AutoCAD's AutoShade. PhotoWorks allows for the creation of photorealistic renderings, complete with procedural or texture bitmapped material properties. It also gives you the ability to modify and create custom material libraries, which is a function very similar to AutoShade.

Procedural material properties go one step further than standard bitmapped images. By using mathematical algorithms, specific material properties can be obtained. This process works especially well for materials such as metals, glass, and even water or diamond.

AutoCAD's AutoShade gives you tools for creating elaborate scenes, complete with trees and shrubs, as well as a variety of other options. PhotoWorks, on the other hand, is more suited to the mechanical market instead of the architectural market. Settings can be created and backgrounds displayed to give the impression, for example, that items are sitting on a tabletop surrounded by walls or in an open environment with wispy clouds in the background. Photographs can be used to complete the effect. Shadows can be turned on to heighten the realistic effect.

Full ray tracing and anti-aliasing are available in PhotoWorks for those individuals who are very discriminating and want the highest quality renderings. File output can be in the common TIFF or TARGA formats, as well as BMP and Postscript. If rendering to a file, the resolution is limited only by your machine's hardware and the length of time you are willing to wait for the process to take place. Very high resolutions can be keyed in, and large, E-size drawings can then be plotted with excellent results.

Another feature of this package is the material properties library. This library offers a wide range of preset material settings. You can choose to customize these settings into a user defined material library. A complete selection of predefined light sources and scenes are included in this package.

SolidWorks has excellent shading and lighting functionality, but if you use rendered images on a regular basis, PhotoWorks is the way to go. Applications include professional-looking catalogs, sales brochures, presentations, advertising, and any other number of possibilities. Talk to your vender or reseller to obtain more information on PhotoWorks.

Summary

Adding light sources or modifying colors and lighting characteristics are all options that will increase the SolidWorks user's efficiency by making a model easier to understand. By diversifying the colors of various part features or assembly components, not only is the model more aesthetically appealing, it simplifies the editing process because of the greater degree of contrast these changes invoke. Adding transparency to an assembly or placing a spotlight on a part are just a few ways that can help get a point across or convey information about a solid model.

Any modifications made to light sources, colors, advanced color options, or part properties are saved with the part or assembly. Expect to see the changes the next time the part or assembly is opened. Use the Alt+Print Screen keys on the keyboard to save an image of the screen to the clipboard, where it can then be pasted into another program, such as the Windows Paint program or a text document. A screen capture program can also be used to automatically crop images, save the image to different file formats, and so on.

PhotoWorks is an add-on package that creates photorealistic images that can be used in a variety of ways, such as for business promotions and advertising. The material libraries and advanced rendering capabilities offer the ability to easily create professionally rendered images.

10 Printing

Introduction

This chapter discusses how to print a SolidWorks document using any printer available within the Windows operating system. Output can also be directed to a file if you wish to save a copy of the output, or send the file to another user, customer, and so on. Parts, assemblies, and drawings can be printed using this utility.

Prerequisite

You should have at least a basic understanding of the Windows 95, Windows 98, or Windows NT 4.0 operating system. Almost all of the printing parameters and the actual job of printing are handled by the operating system, not SolidWorks. For this reason, it helps if you are somewhat familiar with Windows so that any problems that arise can be addressed and the causes of such problems can be more easily tracked down.

Content

The sections of this chapter follow the logical order in which printer settings should be defined. The "Printing Basics" section gives you background information needed before using the print function. The section on defining a printer or plotter describes how a new printer or plotter can be added to Windows. The "Page Setup" section describes how to define paper size, margins, and header or footer values for a SolidWorks document. Line weights are addressed in the "Print Settings" section. The section on printing a document describes how to preview and print any type of SolidWorks document.

Objectives

When finished with the "Printing Basics" section, you should be aware of the fundamentals of printing and how printing relates to the hardware, the operating system, and SolidWorks. Upon completing the section on defining a printer or plotter, you should be able to identify what printers are installed and the steps required when defining a new printer. You should be able to change page setup values (i.e., paper size, margins, header, and footer) at the end of the "Page Setup" section. At the conclusion of the section on printing, you should be able to preview and print any type of SolidWorks document.

Printing Basics

Printing in SolidWorks is compatible with all printers and plotters defined under the Windows 98 or Windows NT operating system. The drivers for printers or plotters are obtained from the manufacturer, or from the Windows operating system setup disks or CD-ROM. The same procedure is used whether the output device is a printer or plotter.

The actual task of sending print information to the printer or plotter is the domain of the operating system. SolidWorks does not have any part in this. If the print preview can be displayed, SolidWorks has upheld its end of the bargain.

Loading or installing print drivers should be done before any printing is even attempted. Again, this is done through the operating system, not SolidWorks. AutoCAD users will probably be well aware of the difficulties in setting up and configuring a printer. Whether or not to use an AutoCAD driver or a Windows driver was a common dilemma when initially setting up an AutoCAD printer or plotter.

These decisions do not concern the SolidWorks user. The Solid-Works program was written specifically for the Windows operating system and takes full advantage of this fact. If a printer or plotter is set up correctly through Windows, SolidWorks can print to it.

The Argument for Hard Copy

Obtaining hard-copy output is still an important, often necessary, task when working with engineering drawings. Few companies are operating in a completely electronic environment. Some companies are still using drawing boards. SolidWorks offers a number of methods for distributing documents electronically.

Even when a document can be distributed electronically, there are still advantages to producing a hard copy of a drawing. It is easier to read the entire drawing with a large, formatted plot than on a computer monitor. Often it can be difficult to see the entire drawing on a monitor. The plot can produce a much larger image of the same drawing, and mistakes or design changes can be more easily recognized. The SolidWorks document viewer does not have red lining or markup capabilities. Therefore, a printed drawing can be a way of checking or marking changes on a drawing.

Electronically Sharing Documents

To share a drawing electronically with other anuser can be done using a couple of methods. This allows the person to view and print a document, eliminating the need for a paper copy. The advantage to this is that the most current copy of a document can be made available on a network drive, but a paper copy may or may not be out of date.

The SolidWorks viewer allows any user with a PC to view and print SolidWorks documents (i.e., parts, drawings, and assemblies). It is not necessary to have SolidWorks installed. For that matter, the user or company does not even have to own a copy of SolidWorks. This is a free document viewer that can be obtained from the SolidWorks web site at *http://www.solidworks.com*.

Another method that can be used is a neutral format like DXF or Adobe Acrobat. These formats can be produced and viewed using commercial document viewers. The disadvantage to this approach is that the file must be made and maintained.

Quick Printing

Quick prints can be made of parts or assemblies without a drawing being made. There are many cases where a print of a screen shot of a part or assembly can be made to effectively communicate or document a design element. This can be performed with or without showing the related dimensions. This can be done if scale is not a factor, because there is no way to set the scale of a part or assembly when printing the image onscreen. This is obviously not the case with regard to drawings.

Performing a quick print of the screen in AutoCAD is difficult if you are trying to print a shaded part. AutoCAD automatically regenerates the model during the print process, thereby placing the model back to its wireframe state. This is not an issue in SolidWorks.

If it is a shaded image you desire, you would print the screen from a part or assembly. Shaded views cannot be placed in a drawing using the usual methods described in Chapter 8, but objects can be embedded into a drawing. This might be a company logo, a shaded picture of a part, an AVI movie file, or any other object. The list goes on and on.

This would be accomplished using the Windows clipboard. Inserting objects can be done in SolidWorks using the Insert/ Object command. This capability uses standard Windows functionality and is beyond the scope of this chapter. Any Windows OLE compliant program should be able to perform such tasks.

The quality of the prints obtained from printing a screen shot of your SolidWorks model are only limited by your artistic eye for color and the quality of the printer's output. If the print is being sent to a bubble-jet printer, the results are certainly not going to be as dramatic as what might be seen sending the print to a thermal wax printer.

→ **NOTE:** *See Chapter 9 for options on changing the color or lighting characteristics of a part or assembly.*

Printing SolidWorks Documents

The three types of SolidWorks documents (parts, drawings, and assemblies) have different characteristics when printing or plotting. The following list describes these characteristics for the three types of documents. The routines for printing these document types are discussed later in the chapter.

Parts	Parts are printed using the current zoom scale and view orientation. The document will be printed as it is currently shown onscreen. The active display mode (i.e., wireframe, hidden in gray, hidden removed, or shaded) will be used to create the output file.
Assemblies	Assemblies are printed using the current zoom scale and view orientation. The document will be printed as it is currently shown onscreen. The active display mode (i.e., wireframe, hidden in gray, hidden removed, or shaded) will be used to create the output file. Assembly printing options and functionality are exactly the same as for parts.
Drawings	Drawings will print a full-size copy of the active sheet regardless of the current zoom scale. Scale for drawings is determined by the Page Setup dialog box and through the drawing itself. Multiple-sheet drawings (i.e., Sheet1, Sheet2, and so on) can be printed individually or singly using the print range function in the File/Print menu, as described later in the chapter.

Defining a Printer or Plotter

SolidWorks uses the Windows print system for producing printed output. Whether a printer is going to be used for AutoCAD, SolidWorks, or any other program makes no difference. Defining a printer is done the same way, which is explained in material to follow. There are advantages to using the standard Windows print system, two of which are:

- More printer and plotter drivers are available. Device manufacturers no longer have to produce application-specific drivers. The manufacturer produces a driver for Windows 98 or Windows NT and all applications that run in Windows can then make use of these drivers. Many printer drivers are included with the Windows operating system and more can be obtained by contacting your vender or from the Internet.

- Any printer, local or remote, can be accessed using standard Windows functionality. You do not have to relearn a new interface to print a document. The same printer setup dialog box used to produce a printed document from a Microsoft Office application can be used to print a drawing from SolidWorks.

Printing devices are defined by selecting Printers from the Windows Start button and clicking on Settings to open the printer setting's folder. The location, driver, port, and printer name must be supplied to set up a printer. Using the left mouse button, double click on the Add Printer icon to start the new printer wizard (see the illustration at left).

Adding a printer in the Printers folder.

A series of dialog boxes will ask you questions concerning the location, type, name, and driver to define the printer. Whenever a new printer is added, make sure you respond with a Yes when asked if you want to print a test page during the new printer installation process. This will let you know that everything is functioning correctly.

Printer Configuration

A printer can be attached directly to your computer (local) or attached to another computer on a network (a remote computer). When accessing a remote printer, your computer must be connected to the same network and the printer must be accessible by remote users. For a further explanation, use the Windows help index to search for "printer sharing." This can be found by performing the following steps.

1. Double click on My Computer.

2. Click on the Help pull-down menu.

3. Select Help Topics.

4. Select the Index tab.

5. Type in the words *printer sharing*.

6. Select the option you want to see help for, and click on the Display button.

If you have more than one printer, the printer you use most often can be defined as your system default printer. This means that all printed documents will be sent to this printer unless you select another output device. This is defined using the Printers window previously described. Right click on the printer to be set as the default printer and select Set As Default. Only one printer can be selected as the default printer.

Printer/Plotter Attributes and Customization

The system attributes for a printer can be defined or redefined by modifying the settings for the printer in question. Right click on the printer's icon and select Properties. The printer attributes (i.e., ports, paper size, dithering, and so on) can be set or modified.

Different printers allow for a varying degree of customization. There are literally hundreds of different printers on the market. These range from standard letter-size dot matrix printers to 60-inch wide inkjet plotters and everything in between. Describing all of the settings for various printers and plotters is outside the scope of this book. Refer to the manufacturer's documentation for specific information regarding the settings for your individual printer or plotter.

Page Setup

The Page Setup function allows you to define paper margin values, paper orientation, print scale, and line weights for the active document. The page setup functions should be used to change specific attributes for the current document. The changes made within this section are saved within the current document.

To set default parameters that will affect all documents, the system attributes for a printer can be defined or redefined by the method described earlier in this chapter using the Windows folder. The Page Setup routine in SolidWorks is specific to SolidWorks and is not a part of the Windows operating system. The Page Setup dialog box is shown in the following illustration.

Page Setup dialog box.

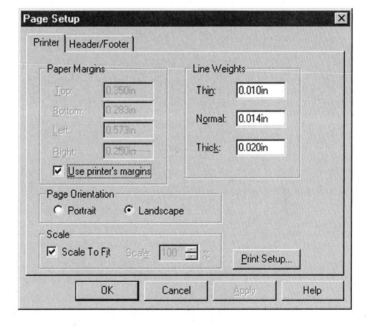

Almost all programs have some sort of page setup utility. AutoCAD's print setup dialog box was a complex array of options somewhat intimidating for new users. The SolidWorks interface is much more basic. There is also a preview function similar to AutoCAD's Full Preview option. The various settings in the Page Setup dialog box are described in material to follow.

Page Setup Options

There are two tabs in the Page Setup dialog box: one labeled Printer, shown in the previous illustration, and another labeled Header/Footer. The Printer tab will be discussed first, starting with setting margins. Discussion of the Header/Footer tab follows this section.

Paper Margins

A margin is the distance from the outside edge of the paper to the available print area. The minimum margin differs for each printer. Think of margins as the white border around the printed image. If the print area exceeds the margin boundaries, portions of the image may be clipped.

Although a printer's margins can be set manually, this is not recommended for most day-to-day printing chores. One example of why you might want to increase the print margins is in the case of publishing. Take, for example, a person that has to run off some prints that are going to be placed in a binder. The default margins are set too small and when the holes are punched in the printed pages, portions of the printed drawing have holes punched through it. In this case, it would benefit you to increase the left margin so this does not happen. However, this is usually not the case; therefore, leave Use Printer's Margins checked.

Page Orientation

The Page Orientation function determines the direction in which output is sent to a printer. The Print Preview function can be used to determine if the page orientation is correct with regard to the data being printed. Portrait is used for most text documents and Landscape for CAD drawings. If the preview function shows the document in an undesirable manner, you can change the page orientation prior to sending the file to the printer. The Portrait option orients the long side of the paper vertically. The Landscape option orients the long side of the paper horizontally.

Scale

The Scale function is used to reduce or enlarge a printed image to fit into a desired paper size or format. The Scale To Fit check box is checked by default. This will automatically scale the output to fit the selected paper size. The scale is defined by a percentage value. Uncheck the Scale To Fit check box to activate the Scale option so that the image can be reduced or enlarged as needed.

It should be noted that the Scale option should not be used to scale drawing views for the sake of fitting the views on a particular sheet size. Drawing views are scaled to fit on a drawing when the views are initially created, as described in Chapter 8. Sheet size is also determined when the drawing is initially created. For this reason, it is best to print or plot drawings at 100 percent. This is extremely critical if individuals will be taking measurements from the print.

If you are printing a part or assembly, or if the drawing scale does not matter, leave Scale To Fit checked and let SolidWorks scale the image to fit the paper.

Line Weights

A line can be defined as having a thin, normal, or thick line thickness. This is defined within the Options dialog box and can be changed. The material that follows explains how this is accomplished, followed by material on modifying line weight settings in the Page Setup dialog box. Perform the following steps to see how line thickness settings are attached to the various line types in SolidWorks. The illustration that follows shows the Line Fonts tab.

1. Select Tools/Options.

2. Select the Line Fonts tab.

3. Click on a line type in the Type of Edge list box to see a preview of that line type.

4. Adjust Line Style and Line Weight as needed.

Line Fonts tab.

SolidWorks breaks up lines into the following nine categories.

- Visible Edges
- Hidden Edges
- Sketch Curves
- Detail Circle
- Section Line
- Dimensions
- Construction Curves
- Crosshatch
- Tangent Edges

Each of these line types can have its line style and weight adjusted. The default settings for these line classifications are acceptable for most users. It is recommended that the default settings not be changed. Take note, however, that in the Line Weight drop-down list box there are three options: Thin, Normal, and Thick. These settings are directly related to the Thin, Normal, and Thick settings in the Page Setup dialog box.

For example, consider visible edges. Visible edges are defined as having a line weight setting of Normal. What this means to you is that all visible edges, when printed, will use the value specified in the Line Weights section of the Page Setup dialog box for normal lines. What that actual value is depends on your printer. Different printers require different settings for line weights. A LaserJet may require a slightly thicker setting because the lines sometimes have a tendency to print so fine that they are almost invisible.

What all this really boils down to is: Do not mess with the line fonts in the Options dialog box. Specify a reasonable line width in the Page Setup dialog box and leave it alone.

How can you determine what is a reasonable line width? Try the following series of steps for the printer in question. You will need to know how to create a part and a drawing because the steps for creating these have been omitted from the following steps. The illustration at left shows a drawing file for testing line weights.

1. Create a simple part. An extruded rectangle will serve the purpose.

2. Create a new drawing. A standard A-size sheet will be fine. Do not bother with a template because you will not need one.

3. Select Tools/Options and select the Line Font tab. Set Construction Curves to Thick. This needs to be done because there are no lines that use the Thick setting by default.

4. Bring in an Isometric view of the part and display the view with Hidden Lines in Gray (or dashed).

5. Draw a centerline (otherwise known as a construction line) and add a dimension if you want.

6. Print the drawing.

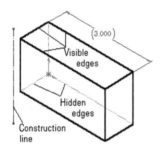

Drawing file for testing line weights.

Now you have a simple drawing containing all three types of line thickness. Experiment with the thin, normal, and thick line weight values in the Page Setup dialog box until good settings are found for your specific printer.

For example, for a Canon bubble jet printer, the line weight settings .010, .015, and .025 inches would probably work well for thin, normal, and thick line weights, respectively. These settings might not be ideal for your printer or plotter, but they might serve as a place to start. Do not forget to set Construction Curves back to thin! If you are wondering why the word *curves* is used, it is because this term refers to arcs, lines, or splines.

Print Setup

This button performs one function, and that is to take you to the Print setup dialog box, shown in the following illustration, for your printers. The information in your Print dialog box may be different, but the layout will be the same.

Print dialog box.

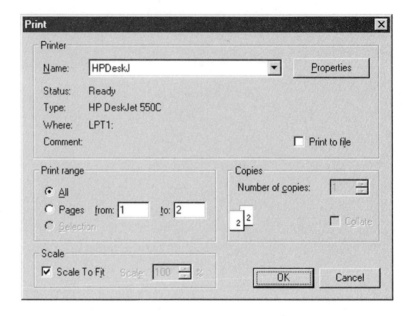

The various options found in the Print Setup dialog box are in the Windows realm. Therefore, they will not be discussed in detail here.

Headers and Footers

Header properties define text added to the top of the printed document, and footer properties define text added to the bottom. The Header/Footer tab is used to add titles, sheet numbers, date, time, and file name to a printed document, to name a few options. Standard header and footer values are available for selection, or custom header and footer values can be defined. The properties defined for a header or footer affect only the active document. The following illustration shows header and footer print locations with regard to a standard drawing document.

Header and footer print location.

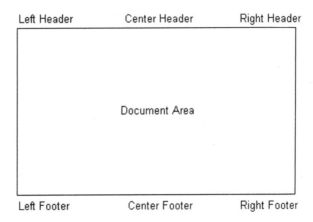

Header/Footer Options

The Header/Footer attributes option is accessed within the Header/Footer tab of the Page Setup dialog box, as shown in the following illustration.

Specifying header and footer properties.

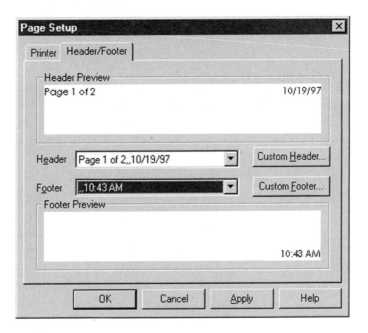

Header and footer areas have three sections each: left, center, and right. Standard header and footer values are available from the drop-down list boxes within the Header/Footer tab. A custom header or footer can also be defined to include the following system attributes. The illustration that follows shows customization of a header or footer.

Page number	Prints the current sheet number being printed.
Number of pages	Prints the total number of sheets in the document.
Date	Prints the current date. Convenient for knowing when the document was printed.
Time	Prints the time the document was printed.
File name	Prints the file name of the active document.

Customizing a header or footer.

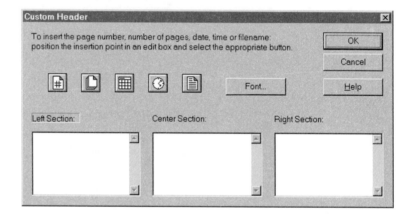

The dialog box for customizing a header or a footer looks exactly the same. The only difference is where SolidWorks places the text, which will either be at the top or bottom of the document being printed. SolidWorks uses its own codes for keeping track of header and footer information. Do not be concerned with this, as it is simply the program's way of doing things.

The system attributes (or codes) SolidWorks uses are prefixed with an "&," followed by the system value enclosed in brackets. Feel free to type spaces or text before, after, or in between the codes to customize the header or footer information even more. For example, you can create a string to display the current page and the total pages with the word *page* before the page number and the word *of* between these values. To see how this is done, perform the following steps.

1. Click on either the Custom Header or Custom Footer button.

2. Click in the section box where the information should be placed, which would be Left Section, Center Section, or Right Section.

3. Type in the word *Page*.

4. Click on the page number icon (see the illustration at left).

5. Type in the word *of.*

6. Click on the Number of Pages icon (see the illustration at left).

7. Click on the Font button to set the font and font size. A small font size, such as 8 point, is common for header or footer data.

8. Click on OK to accept the changes and exit out of the dialog boxes. When you get back to the Page Setup dialog box, you will notice a preview of the header or footer information.

Icons available for customizing a header or footer.

In the header or footer customization dialog box, the code string will appear as follows.

```
Page &[pagenum] of &[pages]
```

In the preview window of the Page Setup dialog box, the same information will appear as follows.

```
Page 1 of 1
```

A Quick AutoCAD Comparison

As mentioned earlier in this chapter, the Page Setup dialog box is specific to SolidWorks. AutoCAD has the ability to manage line weights and pen settings on a much larger scale. Line weights can be associated with pens or with colors. AutoCAD also gives you the ability to draw polylines with varying degrees of thickness, even within the same line segment if you require it. SolidWorks does not have this fine degree of control when it comes to line weights and line types. As of this writing, there are only three line weight settings.

In the area of creating headers or footers, SolidWorks is superior to AutoCAD. In AutoCAD, there are workaround methods to create what looks like a header or footer, but the process is not automated, as in SolidWorks.

Printing a Document

The Print Preview function allows you to view a sample of output prior it to the printer or plotter. This allows for corrections to the header, footer, page orientation, and margin values before spending the time required when creating hard-copy output. Not only does the Print Preview option save time, it can also save money. Paper and ink can be quite expensive. Printer quality has gotten better as the printer prices have fallen, and that trend will more than likely continue. However, ink cartridges are expensive if you have to constantly replace them. The printer or plotter is also tied up when printing the document.

Print Preview

The Print Preview option is found under the File pull-down menu, right where you would expect it. Print Preview is closest in functionality to AutoCAD's Preview feature with the Full option selected. Like AutoCAD's Full Preview, Print Preview provides the ability to zoom in and make some last-minute checks before sending a document to the printer. Changes can be made to Page Setup values if the previewed image is not correct. The following illustration shows the Print Preview toolbar.

Print Preview toolbar.

Almost everything in the Print Preview window is self-explanatory. Included are the Next and Previous buttons, which take you to the next or previous sheet to be printed. This is convenient because drawings can have multiple sheets. Also included is a button named Two Page for viewing two pages at the same time.

Print Preview
Print

Print icon located on the Standard toolbar.

Two Page will place two sheets onto the computer screen at the same time; therefore, do not expect to see a great deal of detail. The data will probably be too small to see. Use Zoom In and Zoom Out if you need to check details at the last minute. Note that the Two Page option is not available if the current preview is zoomed in. The illustration at left shows the Print icon on the Standard toolbar.

Printing a Document

To print a document, select the Print icon or select Print from the File menu. If using the Print Preview function, click on the Print button. An advantage to using the Print command in the File menu as opposed to the Print icon is that the Print dialog box will appear when the document is printed. This allows you to select multiple copies, change the printer, and so on.

The Print icon will print a single document using the currently defined print properties and using the default printer if more than one printer has been defined. If you want to print using another printer (other than the default), use the File/Print command and select the desired printer.

Printing a Selected Area

To print a selected area of a part or assembly, simply get to the desired view orientation and zoom scale. The part or assembly will print using the current display mode (i.e., shaded, wireframe, hidden lines removed, or hidden in gray).

When printing a drawing, the entire drawing is printed no matter what view scale is displayed. To print a portion of a drawing, select File/Print and check the Selection option from the Print Range section. After pressing OK to continue, a dialog will appear, asking you to select a print scale. Select a scale shown or enter a custom scale. After the scale of the print is entered, the print select window will update.

Summary

SolidWorks allows you to access any output device, local or network, available within the Windows operating system. Printers are defined outside SolidWorks, which means that you no longer need to maintain the drivers and configure printers or plotters within the CAD system. The print services (i.e., devices, location, name, spooling, and so on) are handled and maintained at the operating system level.

Printing properties defined outside SolidWorks are those for printer drivers, printer locations, and the default system printer. Printing properties defined within SolidWorks are those for headers and footers, page margins, line weights, print scale, and paper orientation (landscape versus portrait). The SolidWorks Print command uses the standard Windows Print dialog box, which is the same for any Windows application. Use Print Preview to output prior to sending the document to the printer or plotter, thereby saving time and money otherwise wasted on inaccurate prints.

11 Import/Export

Introduction

This chapter discusses how to import and export parts, assemblies, and drawings to and from other CAD systems. The Import/Export function is used to communicate with external sources, to import data, and to reuse legacy data. SolidWorks includes all file import/export formats within the base product and does not require you to purchase add-on modules.

Content

The "DXF and DWG" section describes the import/export capabilities with respect to AutoCAD DXF and DWG formatted files. The section on IGES describes the import/export capabilities with respect to IGES formatted files. The "Stereolithography" section describes the import/export capabilities with respect to stereolithography (STL) formatted files used primarily for rapid prototyping. The section on miscellaneous formats describes the other import/export capabilities within SolidWorks. These formats include STEP, ACIS, Parasolid, and VRML.

Objectives

With completion of the "DXF and DWG" section, you should be able to import and export files using these formats, and understand what type of SolidWorks documents are created when using these file types. After you have finished the section on IGES, you should be able to import and export files using this format, and you should understand what type of SolidWorks document is created when importing this file type.

Upon completing the "Stereolithography" and "Miscellaneous Formats" sections, you should be able to export files using the format descried within this section.

Basics

Files are imported and exported using the Open and Save As commands. This makes saving or opening files of other file types nearly as easy as saving a native SolidWorks file because the commands are the same. The Open and Save As dialog boxes are shown in the following two illustrations. These commands allow you to specify a file type, either for importing (Open) or exporting (Save As).

Open dialog box.

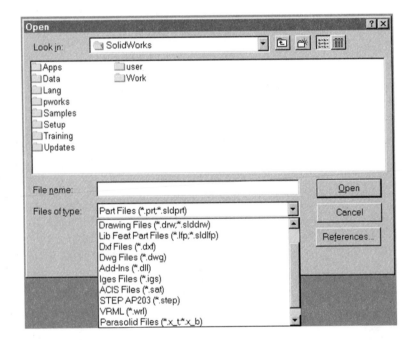

This simplistic and logical approach sometimes catches the new user off guard because they are looking for an Import or Export command when there is none to be found. This is because these commands are built into the Open and Save As dialog boxes.

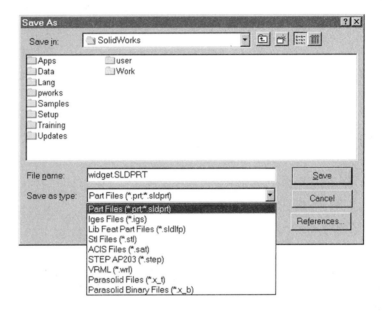

Save As dialog box.

Data imported from an external source can be manipulated within a SolidWorks document (i.e., part, drawing, or assembly) and stored either as a SolidWorks document or in the original file format. One advantage to keeping the file in a SolidWorks format is the ability to add parametric features to the document. Another advantage would be the elimination of exporting the file back into its original format (Save As) when work is complete on the document.

Many factors need to be considered when choosing a file format for import or export. The type of data being translated (i.e., part, assembly, or drawing), the system it is going to (export) or coming from (import). Some systems may not have all the format types. For 2D data, DWG files are usually preferable. For 3D data (surface, curves, and so on) IGES is typically the choice.

Some formats can only be imported as SolidWorks parts, and some as drawings. The reverse is true as well. This is obvious for certain types of formats. For example, it would not be possible to export a solid model as 2D wireframe geometry. Likewise, it would not be possible to take a 2D AutoCAD layout and import it as a SolidWorks 3D part file. The material that follows explores exactly what is possible within SolidWorks and discusses the best methods of translating various file types. First, however, review the types of formats available for importing and exporting SolidWorks part files.

SolidWorks Import/Export Formats for Parts

The following list contains the various file import and export options for SolidWorks parts. Some export options may be better than others for the particular program depending on the target system. It is often a good idea to use a practice file to run some tests and determine the best translator to use. That way, it is also possible to inform clients or customers of the best file type to use when exchanging data, to ensure that important data is not lost or corrupted.

IGES	Initial Graphics Exchange Format file type used to transfer information between dissimilar systems. This file type has an *IGS* extension and is one of the more widely accepted file translation formats.
STL	Stereolithography file type used for rapid prototyping. This file type has an *STL* extension and is available for export only.
VRML	Virtual Reality Markup Language file type. This file type has a *WRL* extension. These file types are usually exported for use on the Internet, but can be imported as a surface.
STEP AP203	ISO STEP (International Standards Organization Standard or Exchange of Product Model Data) formatted file type. This file type has a *STEP* extension.
VDAFS	A German automotive surfacing formatted file can be imported or exported. This file type has a VDA extension.
ACIS	ACIS formatted file type. This file type has an *SAT* extension and is AutoCAD's native solid format.
Parasolid	Unigraphics formatted file type. This file type has an *X_T* extension.
Parasolid Binary	Unigraphics binary formatted file type. This file type has an *X_B* extension.

SolidWorks Import/Export Formats for Assemblies

Import and export options for assemblies are not quite as numerous as for parts. The nature of assemblies makes translating this type of document more difficult. The following are the import and export options for assemblies.

IGES	Initial Graphics Exchange Format file format.
VRML	Virtual Reality Markup Language file type. This file type has a *WRL* extension.
STEP AP203	ISO STEP (International Standards Organization Standard or Exchange of Product Model Data) formatted file type. This file type has a *STEP* extension.
VDAFS	A German automotive surfacing formatted file can be imported or exported. This file type has a VDA extension.
ACIS	ACIS formatted file type. This file type has an *SAT* extension.
Parasolid	Unigraphics formatted file type. This file type has an *X_T* extension.
Parasolid Binary	Unigraphics binary formatted file type. This file type has an *X_B* extension.

The user may choose whether to export the assembly file as separate files or as a single file. The target system will break the single import file back into separate files. It is easier to create, transfer, and import an assembly file as a single file, rather than create and import separate files for each assembly component.

SolidWorks Import/Export Formats for Drawings

When exporting or importing 2D geometry, the formats are logically going to be different from 3D geometry due to the nature of the data. Therefore, the formats for drawings will be different than for parts and assemblies. Also, as you can see, there are not a huge number of options when it comes to translating 2D geometry. Thankfully though, the DXF and DWG translators are very good and widely accepted. The DWG file format is preferable to DXF due to its ability to handle text and dimensions better than a DXF formatted file. The following are the import and export formats for DXF and DWG.

DXF	AutoCAD Data eXchange Format file type. This file type has a *DXF* extension.
DWG	AutoCAD drawing file type. This file type has a *DWG* extension.

Changing and Setting Document Export Preferences

Default settings for import and export characteristics can be defined by reviewing the setting with the Options section of the Save As dialog box. Settings for IGES, Parasolid, DXF/DWG, STL, and ACIS can be defined within the Save As dialog box. These settings are discussed throughout the chapter, as they are applicable to the individual chapter sections.

DXF and DWG

Using the Open and Save As commands, a DXF or DWG formatted file can be used to import or export documents. A DXF or DWG file can be imported into a drawing or part document. Only a drawing can be exported using the DXF or DWG format.

When a DXF or DWG file is opened, the system will prompt the user to select whether it will be imported into a SolidWorks drawing or part. When importing in a part, it is possible to bring portions of a 2D drawing into a part file for use as sketch geometry. AutoCAD versions 12, 13, and 14 DWG or DXF files are supported. The specific version can be defined by selecting the desired version from the Options section with the Save As dialog box (see the following illustration). Centermarks, detail circles, section lines, and note symbols are supported for DXF export.

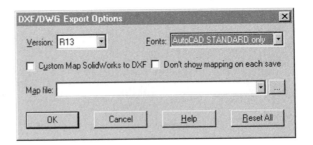

Setting DXF and DWG export options.

Depending on what version of AutoCAD you are exporting to, it may also be desirable to limit the font export to AutoCAD standard fonts only. Not all AutoCAD users have the same Windows TrueType fonts, especially if they are using the DOS version of AutoCAD. Make sure you check the appropriate option for font export, shown in the previous illustration.

DWG or DXF legacy data can also be imported into a Solid-Works sketch using the Sketch From Drawing command. Once geometry is in the sketch, geometric constraints and dimensions can be placed on the geometry. This allows you to add parametric capabilities to static 2D geometry. Multiple sketches can be used from one 2D drawing to define a 3D solid model. You can also break 2D drawings into individual sketches to define multiple features. First, however, step through how to import a DWG or DXF file, and take a look at how to get data from that drawing into a SolidWorks sketch.

Importing a DXF or DWG Formatted File

To import a DXF or DWG formatted file, perform the following steps.

1. Select the Open icon, or select Open from the File menu.

2. Change Files of Type to specify DXF Files (*.dxf) or DWG Files (*.dwg).

3. With the left mouse button, select the file to be opened.

4. Select Open.

SolidWorks will automatically translate the file. If there is geometry you want to use within a SolidWorks sketch, perform the following steps.

1. Begin a new part file, or open the part file to import the data into.

2. Select a plane or planar face to sketch on.

3. Enter sketch mode.

4. From the Insert menu, select Sketch From Drawing. Notice that the cursor changes.

5. Switch to the drawing that has the data to be imported into the sketch.

6. Place a window around the geometry to be imported. Do this by picking with the left mouse button and dragging the opposite corner. The geometry will automatically be imported into the sketch in the part.

Once the geometry is imported into the sketch, there are two things to accomplish right away. The first thing you will notice is that the geometry is in the wrong location and needs to be moved. This is because the AutoCAD UCS *x-y* coordinate of 0,0 directly correlates to the SolidWorks origin point. (Every AutoCAD user knows that the UCS is usually at the bottom left-hand corner of the work area.)

First, use the Constrain All function found in the Relations menu under Tools. This will add as many of the basic constraints to the sketch as SolidWorks can add. Usually, these consist of the simple constraints, such as horizontal, perpendicular, vertical, and parallel. You will have to add others, along with dimensions. The sketch geometry will turn black as it is constrained.

To move the sketch geometry into its desired location, two methods can be used. The first is to use the Modify tool found in Sketch Tools under the Tools menu. Modify will allow you to translate the entire sketch closer to the origin point. The second is the No Solve Move function. This will move the sketch geometry without solving any geometric constraints. Otherwise, the sketch geometry can be difficult to move without adversely effecting the sketch. Once the sketch is moved into the desired location you can then more fully define the geometry as necessary and create the feature.

Exporting a DXF or DWG Formatted File

Only SolidWorks drawing files can be saved as DXF or DWG files for reasons already discussed. The following are the steps required for exporting a DXF or DWG file type.

1. Select Save As from the File menu.

2. Specify the new file name.

3. Change "Save as type" to DXF Files (*.dxf) or DWG Files (*.dwg).

4. Select the Options button.

5. Select the version of the DXF or DWG file (R12, R13, or R14).

6. Click on OK to accept the changes.

7. Select Save to accept the data.

IGES

IGES data files are used to transfer information from one type of CAD system to another. IGES is a neutral file format can be interpreted by other systems. IGES file import or export works with part and assembly documents only. When native file formats are not available as an export type on both systems, an IGES format is typically used to exchange geometric model data.

When importing a closed, contiguous IGES surface file, SolidWorks will sew the surfaces into a solid if possible. If the IGES file contains open or noncontiguous surface data, SolidWorks will import the surfaces without sewing a solid and generate an error log file. This error message means only that the imported data cannot be "knitted" into a solid. This warning message can be ignored if the surfaces are not meant to be a solid feature (i.e., a single surface).

If the IGES file has errors, the Import Diagnostics function, described later in this chapter, can be used to correct the problems. The surface can then be knitted into a solid using the Insert/Boss/Thicken function with the "Create solid from knitted surface" option checked.

Supported Entity Types

If the IGES import was successfully knitted, SolidWorks will treat the new solid model like a base part. Features can be added or material removed from the original model. The original imported surface file can be changed on the original system, and reimported using the Edit Definition function. This will read in the file and create the solid from the new import file. Any changes to import file will be displayed, and any children features created based on the import file may need to be redefined if the related reference to the import file is missing.

When importing IGES files, use trimmed surfaces (IGES entity type 144) whenever possible. Trimmed surfaces are the surface types understood by SolidWorks. All entities supported for IGES translation have numbers associated with them. This is nothing more than a method of identifying the various entities used by the IGES file format. The numbers are not of much meaning to an average user, but they are included here for reference purposes. This is not a complete listing of all entity types. The following entity types are supported for import and export.

IGES Entity Type	Entity Name
144	Trimmed (parametric) surface
142	Curve on a parametric surface
128	Rational B-spline surface
126	Rational B-spline curve
122	Tabulated cylinder
120	Surface of revolution
112*	Parametric spline curve
110	Line
102	Composite curve
100	Circular arc
* Export only.	

Trimmed Surfaces

IGES files can be exported as various "flavors." This means that some programs use a slightly different IGES file type, or that they can read specific entities supported by the IGES translator better than others. For this reason, there are seven export options for the IGES export function. Once again, this setting can be found in the Import/Export section of the Options tab (see the following illustration).

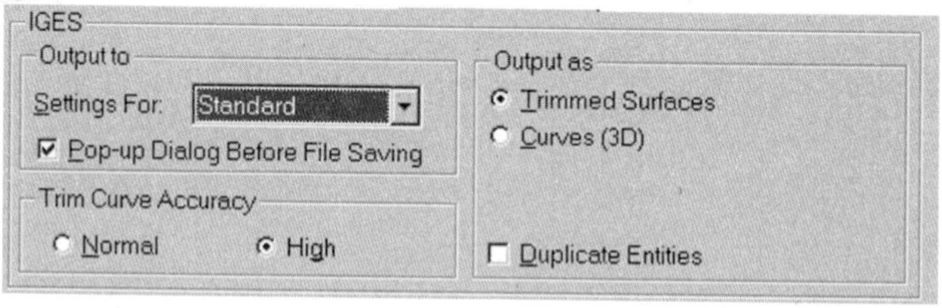

IGES export options.

The following is a listing of the various IGES flavors and the surface types supported by each program listed.

Export Format Types	Exported IGES Entity Types
Standard	144, 142, 128, 126, 122, 120, 110, 102, 100
ANSYS	144, 142, 128, 126, 110, 102, 100
COSMOS	144, 142, 128, 126, 110, 102, 100
MasterCAM	144, 142, 128, 126, 110, 102, 100
SurfCAM	144, 142, 128, 126, 110, 102, 100
SmartCAM	144, 142, 128, 126, 110, 102, 100
TEKSOFT	144, 142, 128, 126, 110, 102, 100

3D Curves and Wireframe Representation

It is not a common practice to export 3D curves if translating files for use in another solid modeling package. Most solid modelers do not need this simple wireframe geometry. This export option is for software that only understands wireframe geometry, or as a reference for surface or solid modeling software. The following are the export IGES entity types for wireframe geometry.

Type of 3D Curve	Exported IGES Entity Types
B-splines (entity 126)	126, 110, 102*, 100
Parametric splines (entity 112)	112, 110, 102*, 100
* Exported only if you select the Duplicate Entities option in IGES Preferences.	

Report Files

SolidWorks creates a report file, *<filename>.rpt*, that lists the entities that were imported. This file can be used to investigate import errors. When errors do occur during the import process, an error log file, *<filename>.err*, is created describing what occurred. You should review this file to ensure that the import function performed properly and that all entities were processed. The report file is an ASCII text document that can be opened with any text editor or word processor (e.g., Notepad or Microsoft Word).

➥ *NOTE: When reporting problems to a vendor, customer, or SolidWorks, include a copy of this report file and the original import file for reference. This can help pinpoint problems.*

Individual surfaces can be imported using Insert/Reference Geometry/Imported Surface. This import technique is different from using the File/Open dialog box in one major way. When using File/Open, SolidWorks attempts to knit the surfaces to form a solid. When using the Import Surface function, no knitting of the surfaces is attempted.

IGES surfaces that do not form a closed solid can be imported, and used in subsequent feature creation, such as by thickening the surface or using the surface to cut geometry already in the model. Individual surfaces can be exported as IGES file types using File/Save As. To save only certain surfaces, hold down the Control key a select the surfaces to be exported. Then select File/Save As and specify the IGES file type. During the export process, SolidWorks will prompt whether to export the selected surface or the entire part.

When Trimmed Surfaces is selected in the Import/Export tab (see the illustration at the beginning of this section), check the Pop-up Dialog Before File Saving option to display a menu with all of the IGES "flavors" prior to exporting. This is a nice option when you frequently want to export IGES files to a variety of programs but do not want to have to enter the Options dialog box prior to every export to verify settings.

When problems occur while exporting IGES files, change the Settings For option under the Import/Export tab for the IGES settings menu to another that may better match the requirements on the target system. For example, if you are exporting to a software program not listed in the IGES Settings For list box, it is usually a good bet to use the default settings. Try changing to an alternate setting if the first try does not work. In addition, the Trim Curve Accuracy option sometimes plays a part when knitting surfaces. Setting the Trim Curve Accuracy to High creates a larger IGES file, but it might be just enough to make a successful IGES translation to an otherwise uncooperative software program. The healing capabilities described later in this chapter will also help fix files that may not transfer correctly.

Importing an IGES Formatted File

You have been through the ins and outs of IGES translation. Perform the following steps when importing an IGES file.

1. Select the Open icon, or select Open from the File menu.

2. Change "Files of type" to IGES Files (*.igs).

3. Select the IGES file to be opened.

4. Click on the Open button. SolidWorks will import the file.

If a large file is being imported, possibly in the multiple megabyte range, the import process may take a few minutes. This depends on your hardware as well. If SolidWorks cannot successfully import the file on its first pass, it will loosen the tolerance and try again. This continues for a few passes until SolidWorks either completes the translation or until the tolerance cannot be loosened any more and SolidWorks abandons the translation with a message to you to this effect.

Exporting an IGES Formatted File

Exporting is a very simple process. To export an IGES formatted file, perform the following steps.

1. Select Save As from the File menu.

2. Change the file name and location, if desired, using the Save As dialog box.

3. Change "Save as type" to IGES Files (*.igs).

4. Select Save to continue. SolidWorks will write the IGES file to the destination selected.

Stereolithography

Stereolithography (STL) formatted files are typically used by rapid prototyping processes. This process produces a triangulated meshed part that other systems can use for viewing, prototype creation, or machining. Setting a display quality, from the STL dialog box, will set the accuracy used to create the STL file. This dialog box is accessed from the Tools/Options menu in the Import/Export section by selecting the STL Options button. Checking the Preview option in the STL export settings dialog box can also preview the STL output file. The following illustration shows the various settings present in this dialog box.

STL dialog box.

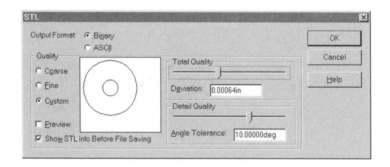

STL File Content

STL files consist of polygons. These polygons can be large or very small. Their size depends on the tolerance required. This is the reason for the preview circles in the STL dialog box, and for the ability to set the deviation. Using a quality (polygon size) that is too large will result in a smaller STL file, but a prototype that would probably not serve its intended purpose. The accuracy of this file is defined by the Total Quality (Deviation) and the Detail Quality (Angle Tolerance) settings with the STL Options settings of the Export tab within the Tools/Options menu.

Prototyping machines will actually build a part using the settings built into in the polygon file for creating the prototype. A good prototype file will have smaller polygons and therefore a smoother surface. The output file can be defined as an ASCII or binary file. The advantage to an ASCII file is that the file can be read with a text editor. However, this is almost never needed. The advantages of a binary file far outweigh an ASCII text file; therefore, unless specifically requested, you should create a binary file. They are smaller and are read by the prototyping machines faster. The following illustration shows an STL polygon file.

STL polygon file.

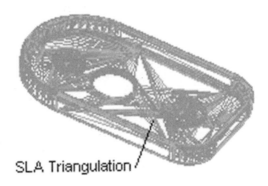

SLA Triangulation

There are a number of STL viewers on the market that allow you to measure features, calculate mass properties, spin a model, and cut cross sections in an STL part. It is not possible to import an STL file into SolidWorks, but the steps for exporting such a file follow.

Exporting a Stereolithography File

To export a stereolithography file, perform the following steps.

1. Select Save As from the File menu.

2. Change the file name and location, if desired, using the Save As dialog box.

3. Change "Files of type" to STL File (*.stl).

4. Select the Options button.

5. Select Show STL Info Before File Saving if you want to see a preview of the STL file before saving it.

6. Specify the Binary file type.

7. Specify the Deviation and Angle Tolerance as required. Higher is better, but results in a much larger file and longer processing times. Experiment with these settings to find an appropriate middle ground.

8. Click on OK twice to exit out of both dialog boxes.

9. Select Save to continue.

10. If Show STL Info Before File Saving has been checked in the Import/Export tab, select Yes to continue or No to abort.

Miscellaneous Formats

A number of other formats can be imported and exported within SolidWorks. The description and use of each format are discussed in the material that follows. Always make it a point to know what software and modeling kernel is being used by the target system. This can help determine logical choices for file formats.

It would be redundant to detail the steps required to import or export a STEP, Parasolid, ACIS, or VDAFS formatted file. The procedure is the same for IGES, Parasolid, STEP, ACIS, and VDAFS formatted files. The only exception is when importing: the correct "Files of type" file extension needs to be selected. When exporting, make sure to select the appropriate "Save as type" file extension. Files that contain IGES, Parasolid, STEP, ACIS, and VDAFS surfaces can also be imported into a Solid-Works part file by running the Import Surface command found in the Insert/Surfaces menu.

STEP

An ISO STEP (International Standards Organization **ST**andard for **E**xchange of **P**roduct Model Data) file format is an attempt to allow for better translation from system to system than an IGES format. A SolidWorks part or assembly can be imported or exported using the STEP file format. SolidWorks currently supports the AP203 STEP protocol. This file format carries a STEP file extension. This file extension may need to be changed to or from STP depending on the system it came from or is going to. Although STEP can offer, in some cases and conditions, the possibility of a more complete and robust transfer format, it remains to be seen whether it offers a consistently better translation than the popular IGES format.

Parasolids

The Parasolids solid modeling kernel (geometry engine) is developed by Electronic Data System (EDS) and is used by numerous CAD vendors, such as SolidWorks and Unigraphics. A SolidWorks part or assembly can be imported or exported using the Parasolid file format. This file format can be directly interpreted by other systems that use the Parasolid solid modeling kernel.

A Parasolid file can carry a v_b (Parasolid Binary) or v_t (Parasolid) file extension. The type of Parasolid file will depend on the target and what type of format that system can accept.

ACIS

The ACIS solid modeling kernel (geometry engine) is used by a number of CAD vendors, including AutoCAD's solid modeling software. A SolidWorks part or assembly can be imported or exported using the ACIS file format. This file format can be directly interpreted by other systems that use the ACIS solid modeling kernel. This file format carries an SAT file extension.

VDAFS

The VDAFS file format is a German automotive surfacing format. A SolidWorks part or assembly can be imported or exported using the VDAFS file format. This file format carries a VDA file extension.

VRML

The VRML viewer allows you to save an active part or assembly as a VRML (Virtual Reality Markup Language) formatted file. This file can be viewed with an Internet browser with a VRML plug-in. The VRML viewer is a plug-in option for web browsers. The VRML formatted file of a part or assembly can be viewed and navigated (panned and zoomed) with an Internet browser or a VRML viewer. Instructions on where to find a VRML viewer are available on the SolidWorks web site. To export a VRML formatted file, perform the following steps.

1. Select Save As from the File menu.

2. Change the file name and location, if desired, using the Save As dialog box.

3. Change "Save as type" to VRML (*.wrl).

4. Select Save to continue.

It is also possible to import VRML files using the Import Surface function in the Tools/Surface menu.

Healing Imported Files

When importing data files, errors can occur. SolidWorks offers a couple of tools that can be used to identify and correct the problem. Check and Import Diagnostics are described within this section. When a file is imported that has problems, SolidWorks will not be able to knit the part as a solid. Feature-Manager will display the object with a red exclamation mark or the imported surface will not be an object that can be knitted. This can be checked by selecting the object and starting the Insert/Boss/Thicken function. If the "Create solid from knitted surface" option is not selectable, the surface may have problems.

Description of Problem	Possible Solution
Duplicate edges or surfaces within the import data file.	Use the Import Diagnostics function to identify and remove the duplicate edges from the import file.
Faces are missing.	Use the surface creation tools (Loft, Planar, Sweep, and so on) to patch in a new face. If the face cannot be created within SolidWorks, either have the surface corrected in the original system or export the required geometry to another system and import the face back into SolidWorks using the Import Surface function.
A face does not translate correctly.	Use the Remove Face option within the Import Diagnostics function to remove the bad face(s) and use the same methodology to add a new face, as described previously.
The accuracy of the import file was greater than SolidWorks could import correctly.	Have the import recreated using a tolerance (1e-8) that is better suited for SolidWorks.
Part did not sew into a solid.	First check to see that the import file was a contiguous solid or surface file. If it was, use the Import Diagnostics function to review and fix the problem with the import file.

Use the Check function under the Tools menu to determine what is wrong with the file. This function will look for bad geometry or short edges that might be causing problems. If the Check function or FeatureManager icon identifies a problem, use the Import Diagnostics right mouse button function to start the healing process.

SolidWorks can correct (heal) import files that contain data that can not be translated properly. Select the imported object from within the FeatureManager design tree, press the right mouse button, and select Import Diagnostics. The Import Diagnostics dialog box shown in the following illustration, can be used to identify and correct errors with the imported data. When the diagnostics are complete, the imported object should knit automatically as a solid feature.

Import Diagnostics dialog box.

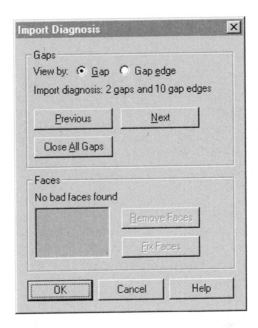

XchangeWorks

XchangeWorks is a free plug-in for AutoCAD and Mechanical Desktop that allows these platforms to import the following data types.

- IGES
- STEPV
- DAFS
- SAT (ACIS)
- Pro/ENGINEER
- SolidWorks®
- Unigraphics

XchangeWorks works with AutoCAD R14.x, AutoCAD Mechanical Rev 14.5, or Mechanical Desktop 2.x or 3.x. This plug-in can be used on existing seats of AutoCAD either within your current company or by vendors and customers. This allows them to read in data in any of these formats.

This plug-in is useful when a using AutoCAD or Mechanical Desktop to communicate with SolidWorks. This plug-in allows native SolidWorks documents to be read in with another data exchange format. XchangeWorks is available from the Solid-Works web site at *www.solidworks.com*.

Summary

When working with vendors or customers to transfer files, it is a good practice to determine the type of system they have, available translators, and preference of file format types. When working with a system unfamiliar to you, it may be useful to provide a couple of different formats. The vendor or customer can then try another format if a problem does occur with the first attempt.

The difficulty with importing and exporting data typically arises due to one of the translators not understanding the format used by the other translator. Both translators must work together for successful results. If one end does not produce or read the data correctly, an error can occur. The log file generated by Solid-Works can be used to determine the cause of the problem. SolidWorks provides a means to heal imported files that do not translate well from another system.

There are inherent difficulties involved with translating. If it is at all possible, translate using the same proprietary kernel format (i.e., Parasolids for moving between Unigraphics and Solid-Works) rather than involving a secondary format.

12 Customizing SolidWorks

Introduction

Aspects of the SolidWorks user interface can be tailored to your preferences. This chapter discusses customizing SolidWorks. The options for doing so are different from AutoCAD's customization options. This chapter also discusses automating repetitive tasks using a programming language (e.g., Visual Basic or C++) or a SolidWorks macro. This chapter is meant to give you an overview of the available programming capabilities and describe how the Visual Basic programming language can be used to expand on the SolidWorks software.

Prerequisite

It is essential that you have a very good understanding of Solid-Works before attempting any customization. Making changes to the SolidWorks interface can result in a very different look and feel for the program and make the learning process much more difficult than it might otherwise be. Learning SolidWorks is not nearly as difficult as learning nearly any other CAD program, but you should be comfortable with SolidWorks before customizing the interface.

Content

Three main ways in which the SolidWorks software can be customized are discussed in this chapter. The first section describes how to use some of the options to change the way SolidWorks operates. The next section goes into detail with regard to using the Customization dialog box to further alter the SolidWorks interface. Finally, this chapter introduces programming languages for automating repetitive tasks. This is intended to offer only an idea of what capabilities are available and where to look for further reference material on programming languages and how they can be used within SolidWorks.

Objectives

After reading this chapter, you will be able to set up the Solid-Works environment based on your preferences. You will also be able to customize toolbars, menu commands, and keyboard shortcuts. Additionally, you will be made aware of the options available for creating user-defined programs and macros and know where to find additional reference material on the subject of programming.

User Preferences

SolidWorks, like many software programs, offers you many ways of customizing its interface. These customization methods are not quite as extravagant as AutoCAD's, wherein you have the ability to modify pull-down menus and define custom Lisp routines within those menus. Nevertheless, the methods used to change the SolidWorks interface are actually quite similar to AutoCAD's customization capabilities. The Customize dialog box allows you to make more drastic changes to certain areas (e.g., pull-down menus) than should sometimes be performed; therefore, you need to be careful with this function.

Before diving into that aspect of the program, you will explore a more tame approach to customizing in SolidWorks. Most of the user preferences can be found under the Tools/Options menu. Some were discussed in Chapter 3. All of the various options will not be covered at this time, as these have already been defined in Chapter 3. Only the options that alter the appearance of SolidWorks are mentioned in this chapter.

The Options dialog box started out being called User Preferences when SolidWorks was first released. The following sections describe some of the option tabs that can be used to alter the way SolidWorks behaves.

Some preference tabs have an "Apply to" scope setting. This is used to tell the system when this change should affect the current document (Active Document), all new documents (System Default), or the current and all new documents (All Possible). Existing documents must be changed as they are opened. Any tab without a scope setting will be a global setting that changes SolidWorks regardless of the document settings.

General Tab

The General tab under Tools/Options has many settings that can be toggled on and off. Others, such as the View Rotation section, require more user input. View Rotation defines how sensitive the mouse will be when rotating a model onscreen. When clicking on the Rotate icon, a smaller movement of the mouse will result in a greater amount of rotation if the slider bar is moved to the right. The angle amount listed in the same section defines how far the model will be rotated when using the keyboard shortcut keys. In this case, the keyboard shortcut keys refer to the arrow keys. In the following illustration, which shows the General tab of the Options dialog box, the model will rotate 15 degrees any time the arrow keys are pressed.

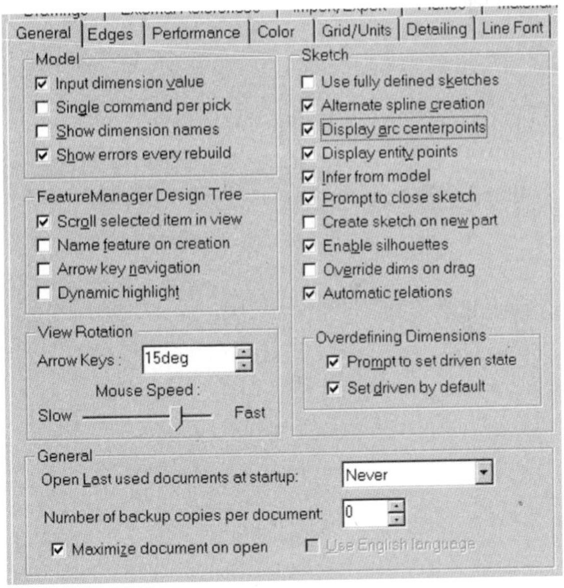

General tab in the Options dialog box.

In the General section of the same tab, checking "Maximize document on open" will result in documents always filling the entire work area when opened. The ability to create backup copies of documents and define a directory for these backups is defined within this section.

The Sketch section of the General tab has some settings relative to how SolidWorks looks to the user. For instance, "Display arc centerpoints" will project all arc centerpoints to the current sketch plane when in sketch mode. This is convenient when trying to sketch to centerpoints. However, when you are working on a part that has many arcs, the arc centerpoints have a tendency to clutter up the display. "Display entity points" will show entity endpoints as small dots. These dots will also display a specific color if they are fully defined. The color scheme follows the standard color scheme used by SolidWorks. Black represents fully defined, blue represents underdefined, and red represents overdefined.

Edges

This tab controls the display of edges within a SolidWorks document. The tab controls the display of hidden edges, ability to select hidden edges within a part or assembly, display of tangent edges, shaded display type, and selection modes.

Performance

This tab is used to define characteristics that affect the graphics performance level of the system. Shaded and wireframe settings can be altered to increase or decrease the display quality. A higher shaded deviation or lower wireframe setting will improve performance, but the display quality will be lower. For most cases, the default settings work well.

For assemblies, the user can determine whether lightweight components are loaded by default. A lightweight assembly component will load only the information required when displaying the assembly component. This can greatly improve the performance for large assemblies. Additional information is loaded when resolving the component.

Color Tab

This option can be used to set default colors for SolidWorks objects. SolidWorks has default colors it uses to display certain entities or conditions. One example is overdefined sketch geometry being displayed in red. This is not something a user will normally want to change, but the capability exists nonetheless. The following illustration shows the Color tab from the Options dialog box.

Color tab in the Options dialog box.

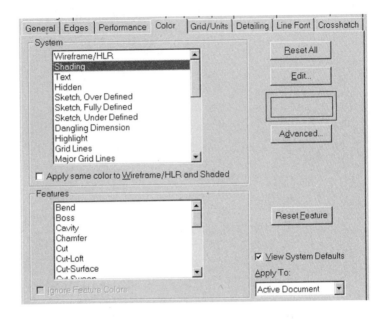

If you click on the View System Defaults button, the System list expands to show all of the options in SolidWorks that can have their color settings altered. Simply highlight the option to be changed, click on the Edit button, and specify a new color. It is almost too easy to make modifications to the system defaults; however, SolidWorks gives you a way out if things become too messed up. All you have to do is click on the Reset All button to reset everything in the Color tab back to default color schemes.

Grid/Units

The Grid/Units tab is used to define the default unit of measure and grid characteristics. The units of measure can be changed at any time. When changing the unit of measure, SolidWorks will ask you whether the default sizes for grid setting and annotations should be changed to correspond with the new unit of measure.

Detailing

The Detailing tab is used to define the dimensioning standard, note characteristics (i.e., font, justification, leader length, and so on), dimension settings, annotations settings, and section and detail font type.

External References

This tab is used to define how externally referenced documents are handled. An external reference occurs when a document uses information from another document. An example would be a drawing that references a part or assembly to create the drawing views and annotations.

The Folders section is used to define additional directory locations for files, blocks (symbols), FeaturePallete parts, features, and forming tools. These settings allow a network drive to be used to store the objects.

Drawing

The Drawing tab defines the default drawing sheet scale, view projection type, display mode for views, and view settings and behavior.

Material Properties

The Material Properties tab is used to establish the part density used for the mass properties function.

Line Font

The Line Font tab defines the line style and weight for different types of lines (i.e., visible edges, hidden edges, crosshatching, dimensions, section lines, and so on). The line style determines whether the line is solid, stitched, or another style. The spacing between the line font cannot be customized.

Crosshatch

The Crosshatch tab defines the default crosshatch type, pattern, spacing, and angle.

Reference Geometry

The Reference Geometry tab defines the default display of reference geometry and the name of the default planes. The status (on or off) of the display filters defined in the View pull-down menu can be changed using these settings.

The name of the default planes should be changed from Plane1, Plane2, and Plane3. Determine which plane is considered the top-front and rename the default plane names. There are differing opinions as to which plane is considered the top-front. The user can use their preferred setting.

Copy Options Wizard

The Copy Options Wizard allows you to back up the default settings defined within the Tool/Options menus on a system. This can be useful if the software ever needs to be reloaded, or to share information and setup configurations with other users. This file can be used to ensure that all users have a consistent setup and configuration.

The Copy Options Wizard copies the Windows registry settings related to SolidWorks into an Options.reg file. The default name can be renamed if so desired. The most typical option is "Create a registry file." This will create a file that when double clicked on will change the default SolidWorks settings based on the content of the registry file.

After a system has been set up, the wizard should be run to store the default SolidWorks configuration. This file can be renamed and stored on a network drive. It is a good idea to store this file in a location that gets backed up, or to store it on disk.

Customize Command

Similar to AutoCAD, SolidWorks gives you the ability to modify toolbars and pull-down menus. Where the difference lies is in the extent to which these changes can be made. For example, AutoCAD allows the creation of new toolbars complete with the ability to create your own icons. SolidWorks allows for adding and deleting icons, but only to the extent of which icons are available to begin with. In SolidWorks, you cannot create your own icons or make new toolbars.

Pull-down menu customization is similar to AutoCAD to the extent that items can be removed from the menus. However, what can be added to SolidWorks pull-down menus is limited by what originally came with the software. The Customize dialog box contains three tabs that allow you to make modifications. The options within these tabs are discussed in the material that follows.

Toolbars

Toolbars contain functionally grouped icons that serve as shortcuts to pull-down menu functions. Each toolbar's content is grouped per various commonly used menu functions. SolidWorks contains 15 standard toolbars (more for solution partner applications), whereas AutoCAD contains over 50. SolidWorks toolbars were listed in Chapter 1, but the names of all the toolbars can be found in the Toolbars dialog box. Click on View/ Toolbars to view the Toolbars dialog box. Notice the check boxes with which to turn the individual toolbars on and off.

The Toolbars dialog box also contains the options Large Buttons and Show Tooltips. Tooltips are the yellow hint boxes that appear when the cursor is held stationary over an icon for longer than a second. These are helpful, and it is difficult to imagine why anyone would want to turn them off. The Large Buttons option displays the icons in a larger size for those who may have a difficult time making out the small icons. These also help if you have a very large monitor running at high resolution. Sometimes at high resolutions, icons can get very small, making work very difficult.

Toolbars can also be customized to your specific requirements so that only commonly used icons are displayed. You do this by adding or removing function buttons using the Customize dialog box. However, a toolbar must be active to be customized. The following illustration shows the toggling of toolbars on and off in the Toolbars dialog box.

Toggling toolbars on and off.

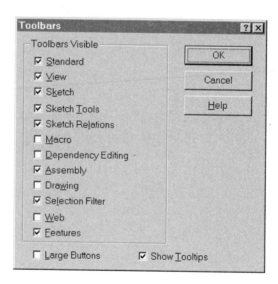

Toolbar Activation

Toolbars automatically appear based on the current type of open document (i.e., sketch, part, or assembly). When a document type different from the current document is opened, the appropriate toolbar is automatically activated. The only toolbars that really should be active all of the time are the Standard, View, and Sketch toolbars, simply because they are used constantly no matter what drawing type is open. The Drawing, Assembly, and Sketch toolbars automatically activate according to document type. Another example would be the Macro toolbar, which turns on automatically if you are playing or recording a macro. To manually activate a toolbar, perform the following steps.

1. Select Toolbars from the View menu.

2. Check the desired toolbars to toggle them on or off.

3. Select OK to close the dialog box.

Toolbar Customization

As stated previously, toolbar icons can be added or removed from toolbars. To customize a toolbar, perform the following steps. The illustration that follows shows the Toolbar customization tab. Currently, the Web toolbar and the Selection Filter toolbar cannot be modified.

1. Activate any toolbars to be customized.

2. Select Customize from the Tools menu.

3. Select the Toolbars tab.

4. Drag the button to be removed from the desired toolbar. Drop it anywhere outside the toolbar to delete it. To add a button, drag an icon from the dialog box to the toolbar. The button should appear on that toolbar.

5. Resize the toolbar if needed by dragging the edge of the toolbar similar to resizing a program window in Windows. Sometimes this must be done before the new icons will show up on the toolbar to which they were dragged.

6. Select OK when finished.

Toolbar customization tab.

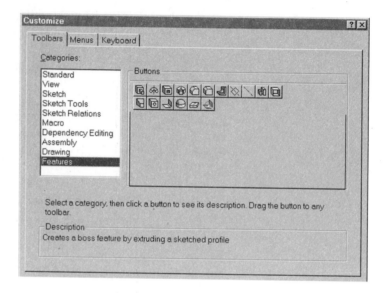

Menus   You may wish to simplify a menu or remove an unwanted command. You can rename or remove commands from pull-down menus by performing the steps that follow. Only commands listed within the Customize dialog box (see the following illustration) may be added or removed. In SolidWorks, user-definable commands and functions cannot be added to a menu, as they can be in AutoCAD. It is possible in SolidWorks to add commands and alter menus using C++ programming language, but this is outside the abilities of most users, and outside the scope of this book.

It is highly recommended that new users do not attempt to customize pull-down menus. This section has been added for reference and for the benefit of those with a good understanding of the SolidWorks program.

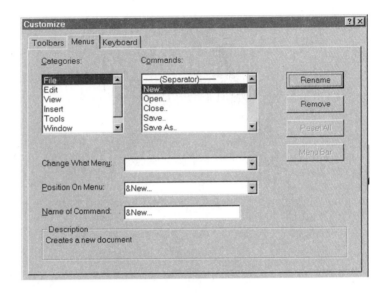

Menu customization tab.

Keyboard Shortcuts

Keyboard shortcuts assign commonly used SolidWorks functions (e.g., Orientation or Measure) to keyboard keys. This can make it easier and quicker to access these functions. Only functions listed within the Tools/Customize/Keyboard menu can be assigned a keyboard shortcut. Additional keyboard shortcuts can be defined by selecting the Keyboard tab in the Tools/Customize menu. The following keyboard shortcuts have already been established within SolidWorks.

Rotate the model	Arrow keys
Rotate the model 90 degrees	Shift + Arrow keys
Rotate clockwise/counterclockwise	Alt + left/right
Pan	Ctrl + Arrow keys
Zoom in	Z
Zoom out	z
Rebuild the model	Ctrl + B
Redraw the model	Ctrl + R

The F2 through F9 function keys are good candidates for a keyboard shortcut. Unlike AutoCAD, these keys are typically not assigned to any functions in SolidWorks. You may want to create some simple labels to place above your defined function keys for easy reference. The keyboard shortcut customizations tab is shown in the illustration that follows.

NOTE: *Do not redefine the F1 function key, as this key is the default system help key.*

Keyboard shortcut customization tab.

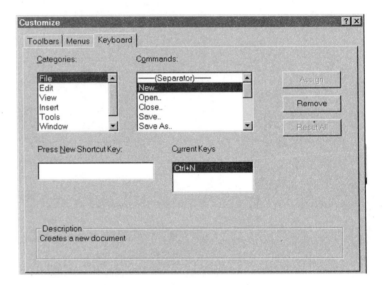

AutoCAD uses the ACAD.PGP text file for controlling which shortcuts (aliases) are assigned to specific commands. There is no need to modify a text file within SolidWorks. You simply select the command from the Customize dialog box in the Keyboard tab and press the desired key. Again, it is not possible to define new commands or macros. You are limited to commands previously defined by SolidWorks. To customize a keyboard shortcut, perform the following steps.

1. Select Customize from the Tools menu.
2. Select the Keyboard tab.
3. Select the desired menu command to which to assign a new shortcut key.
4. Press Assign to set the new shortcut.
5. Select OK to complete this function.

You can also select Remove in this tab to remove a shortcut key, or you can Reset All shortcut keys to their original default state. If a shortcut is already listed in the Current Key list box, it is recommended that a new shortcut not be assigned.

Programming Languages

SolidWorks offers two methods of further customizing and automating design tasks. These are a macro programming language and an application programming interface (API) included with SolidWorks. The macro programming language allows you to automate repetitive tasks. The API allows Visual Basic and C++ programs to access SolidWorks functions.

Macros

The macro programming language can be used to automate repetitive tasks or create features that are geometrically similar but that have different dimension values. If a macro is created, and in the process a dimension is added, when the macro is played back SolidWorks will pause for you to input a dimension value of choice.

Macros can be defined without any programming knowledge. A macro will record the steps taken and allow you to input values during playback. Files created with a macro have an .swb file extension. The macro programming language is described in the SolidWorks help file \sldworks\lang\english\Swxbasic.hlp.

→ **NOTE:** *The English directory may differ depending on the installed language or languages.*

Recording a macro is also a useful tool when creating API programs using Visual Basic or C++. The syntax for a function can many times be recorded using a macro and cut and pasted, using the clipboard, into a program. A macro is a simple API program. Visual Basic or C++ is used to enhance the capabilities and features available to the programmer.

The following procedures describe how to record, edit, and run a macro. To record a macro, perform the following steps.

1. Select Record from the Tools/Macro menu.

2. Perform the steps that define the macro.

3. Select Stop from the Tools/Macro menu.

4. Enter the macro file name (<filename>.swb).

5. Select OK to complete this function.

To edit a macro, perform the following steps.

1. Select Edit from the Tools/Macro menu.

2. Select the macro file name to be edited.

3. Select Open to complete this function.

4. Edit the macro as needed to add or delete macro functions.

5. Select Close from the File menu when finished. There is no Save function. The file will be saved automatically when closed.

When editing a macro, the Macro editor window will be active. The code for the macro itself will be displayed in the window. This can be modified as needed. The Macro editor with a sample macro is shown in the following illustration.

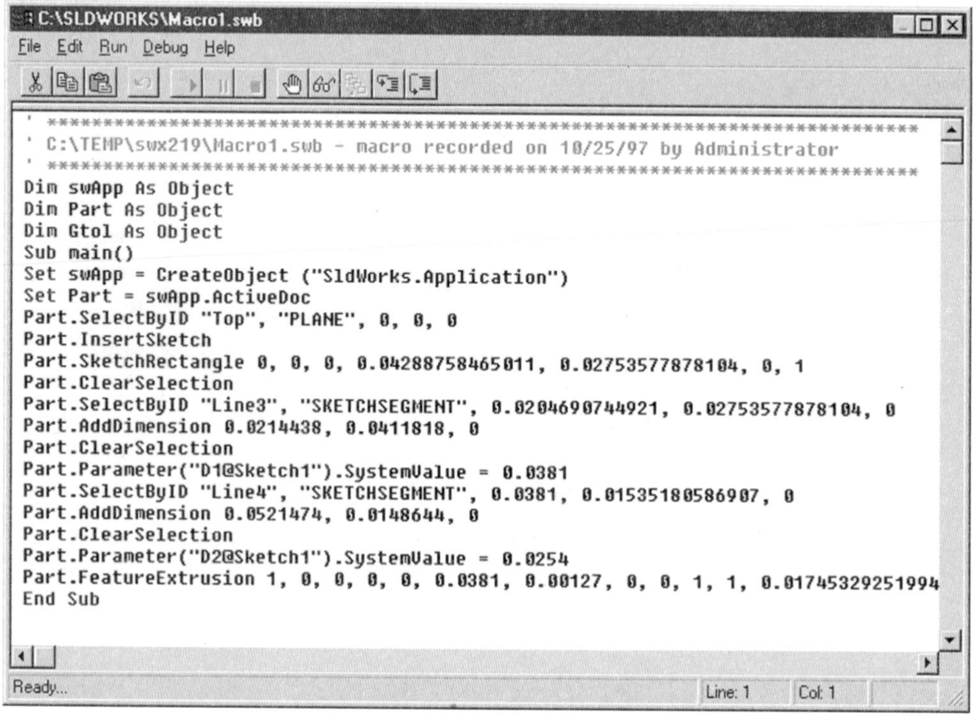

```
C:\SLDWORKS\Macro1.swb                                    _ □ X
File  Edit  Run  Debug  Help

'*****************************************************************
'  C:\TEMP\swx219\Macro1.swb - macro recorded on 10/25/97 by Administrator
'*****************************************************************
Dim swApp As Object
Dim Part As Object
Dim Gtol As Object
Sub main()
Set swApp = CreateObject ("SldWorks.Application")
Set Part = swApp.ActiveDoc
Part.SelectByID "Top", "PLANE", 0, 0, 0
Part.InsertSketch
Part.SketchRectangle 0, 0, 0, 0.04288758465011, 0.02753577878104, 0, 1
Part.ClearSelection
Part.SelectByID "Line3", "SKETCHSEGMENT", 0.0204690744921, 0.02753577878104, 0
Part.AddDimension 0.0214438, 0.0411818, 0
Part.ClearSelection
Part.Parameter("D1@Sketch1").SystemValue = 0.0381
Part.SelectByID "Line4", "SKETCHSEGMENT", 0.0381, 0.01535180586907, 0
Part.AddDimension 0.0521474, 0.0148644, 0
Part.ClearSelection
Part.Parameter("D2@Sketch1").SystemValue = 0.0254
Part.FeatureExtrusion 1, 0, 0, 0, 0, 0.0381, 0.00127, 0, 0, 1, 1, 0.01745329251994
End Sub

Ready...                                    Line: 1    Col: 1
```

Macro editor.

To run a macro, perform the following steps.

1. Select Run from the Tools/Macro menu.

2. Select the macro file name to be run.

3. Select Open to complete this function.

When running a macro, it may be necessary to input dimension values. SolidWorks automatically pauses anytime user input is required.

The macro tab with the Tools/Customize menu offers the ability to define a set of 10 user-defined macros executed with the user macro buttons in the Macro toolbar, shown in the following illustration. The name displayed and the macro executed are definable with the Customize dialog box. When a user icon is grayed out, this means a macro has not been assigned to the button.

Macro toolbar.

The following example shows how a macro can be documented and how a macro can be used to start an executable file. Any line starting with a apostrophe (') will be treated as comment within a macro. Placing a standard header and completing the comments helps document the function for the macro and makes it easier to edit and use. This example will start an application named "app1," located in the Macros directory on the D drive.

```
'***************************************************

'Title: macro_shell

'Date: 12/10/98           Rev:  1.0

'By:   Gregory Jankowski

'Description:  The macro will start an executable (.exe)

'file. This example starts app1 located in the macro directory

'***************************************************
```

```
Sub main()

 ' Start the application

   MyAppID = Shell("D:\Macros\app1.exe", 1)

End Sub
```

The API   A Visual Basic or C++ program can access a wider range of functions and features using the SolidWorks API. Creating an API program requires programming skills. See the API documentation and training guide on how to automatically add functions and menus to SolidWorks. Examples of programs that access the API, which can be accessed through a Visual Basic or C++ program, are given in the training guide. Additional help can be found in the Windows help file *Samples\App-comm\API_help.hlp*.

➣ **NOTE:** *The API documentation and training guide can be accessed through the SolidWorks technical support Web site at* http://www.solidworks.com/html/Contacts/apisup.htm.

To run a Visual Basic program, see the procedure for running a macro presented previously. Visual Basic programs can also be made to run as a standalone (.exe) program. To run a C++ program, select an Add-in (*.dll) file type, accessed with the File/Open command. This will add the program to the current session.

It is also possible to click on Tools/Add-Ins and select the check box for the add-in program you want to activate or deactivate. These third-party programs are purchased separately. Not all programs conform to this function and may not appear in the Add-Ins dialog box. This is similar to AutoCAD's Applications list box, which automatically loads lisp routines when the AutoCAD program is started. The Add-Ins dialog box is shown in the following illustration.

Add-Ins dialog box.

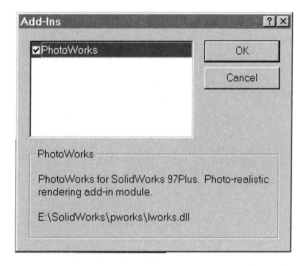

Summary

Adding keyboard shortcuts for frequently used functions that do not have a toolbar icon can reduce the number of menu picks required to access a SolidWorks function, and therefore increase your productivity. Macros can be used to create simple automated tasks in SolidWorks. This function requires little or no programming experience. Visual Basic or C++ programs can be written to access the SolidWorks API to provide access to a wide range of SolidWorks functions and to automate functions specific to your company's needs. Creating programs of this type requires programming experience.

C++ is typically used by professional programmers to produce commercial applications. There are some functions (e.g., adding a pull-down menu) that cannot be added using Visual Basic. Visual Basic is typically used by nonprogrammers, as it is easier to learn and use. Many of the limitations to Visual Basic do not typically affect this type user, as the applications are not as elaborate or commercial in nature.

There are similarities between AutoCAD and SolidWorks with regard to the ability to customize the interface. Unlike AutoCAD, SolidWorks does not allow you to add new commands that do not currently exist within the program. However, with a little Visual Basic programming experience, a SolidWorks user can create customized routines that can accomplish nearly any task. In addition, because Visual Basic is a widely accepted language, many resources are available for the motivated and the curious SolidWorks user.

Appendix

AutoCAD/SolidWorks Functionality Cross-reference

This appendix is divided into two sections: Command Cross-reference and Sketch Entities Cross-reference. It should be noted that many of the cross-referenced commands can only be associated with the listed AutoCAD commands by using a stretch of the imagination. This appendix is not intended as a function-for-function cross-reference, but as a resource for the AutoCAD user moving into the SolidWorks environment. Remember that SolidWorks and AutoCAD are fundamentally different packages, even if many of the commands seem similar. SolidWorks requires a different mind-set than the traditional 2D or wireframe way of thinking.

Command Cross-reference

The following table lists AutoCAD and equivalent SolidWorks functionality, with pertinent notes about individual entries and an indication of the modes in which each function operates. All SolidWorks commands are shown in boldface. SolidWorks pull-down menus are also listed where applicable, in the form <menu name>/<function name>. Any function that has a Solid-Works toolbar icon is denoted with a "*" (i.e., File/Open*).

Not all AutoCAD commands have a direct counterpart in Solid-Works. Many of the commands listed in the following table will show only the closest similar command in SolidWorks, or no command at all, in which case a reference note is typically included.

AutoCAD/SolidWorks Command Cross-references

AutoCAD	SolidWorks	Notes	Mode
About	Help/About SolidWorks 97Plus	—	A D P
ACISIN	File/Open*	Files of type: ACIS files (*.sat).	P
ACISOUT	File/Save As	Files of type: ACIS files (*.sat).	P
ALIGN	Tools/Relations/Add Insert/Mate	—	A S
AMECONVERT	None.	SolidWorks will not read AME files.	—
APERTURE	Dynamic object selection.	No need for an aperture.	All
APPLOAD	File/Open* Tools/Add-ins	Files of type: Add-ins (*.dll).	All All
ARC	Tools/Sketch Entities/Centerpoint Arc* Tools/Sketch Entities/Tangent Arc* Tools/Sketch Entities/3 Point Arc*	— — —	D S D S D S
AREA	Tools/Measure	—	A D P
ARRAY	Linear Pattern, Circular Pattern	—	A P
ARX	None.	—	—
ASEADMIN	Insert/Object	OLE-compliant database application.	A D P
ASEEXPORT	Insert/Object	OLE-compliant database application.	A D P
ASELINKS	Insert/Object	OLE-compliant database application.	A D P
ASEROWS	Insert/Design Table	—	P
ASESELECT	Insert/Design Table	—	P
ASESQLED	Insert/Object	OLE-compliant database application.	A D P
ATTDEF	File/Properties	Entity attributes unavailable.	A D P
ATTDISP	None.	SolidWorks does not use attributes.	—
ATTEDIT	None.	—	—
ATTEXT	None.	—	—
ATTREDEF	None.	—	—
AUDIT	Tools/Check	Checks for invalid geometry.	P
BASE	Insert/Mates/Distance	—	A
BHATCH	Insert/Drawing View/Section Insert/ Drawing View/Aligned Section	Use Properties to change crosshatching attributes.	D
BLIPMODE	None.	Blips do not exist in SolidWorks.	—

AutoCAD	SolidWorks	Notes	Mode
BLOCK	Insert/Component/From File	Parts in an assembly are similar to blocks.	A
BMPOUT	Print Screen key.	Part of Windows OS; copies screen to the clipboard.	All
BOUNDARY	A series of line segments is created when sketching entities using endpoints.	When the option Automatic Relations is selected, a relation constraint keeps endpoints attached to one another.	D S
BOX	Insert/Boss/Extrude	Create the sketch of the wedge.	P
BREAK	None.	Trim/Extend can be used as a workaround.	D S
CAL	Enter the expression when inputting the value.	Input expressions using mathematical operators (+, −, /, *, parentheses).	All
	Tools/Measure	Measure distances, radii, perimeter, area, and so on.	All
CHAMFER	Insert/Features/Chamfer*	—	P
CHANGE (POINT)	Display/Delete Relations.	Select the objects to be changed first.	D S
	Drag endpoints.	Unconstrained entities can be dragged.	D S
CHPROP	Tools/Options/Colors	Select the objects to be changed first.	D P S
	Properties	Right click on the object to be changed.	D P S
CIRCLE	Tools/Sketch Entities/Circle	—	D S
COLOR	Tools/Options/Colors	—	D P S
COMPILE	None.	—	—
CONE	Insert/Boss/Revolve or Loft	Create the sketch first.	P
	File/Open	Use a library part. Files of type: Lib Feature Part files (*.lfp *.sldlfp).	P
CONFIG	None.	Uses standard Windows devices.	—
COPY	Edit/Copy*	—	All
COPYCLIP	Edit/Copy*	—	All
COPYHIST	None.	—	—
COPYLINK	Edit/Copy*	—	All
CUTCLIP	Edit/Cut*	—	All

AutoCAD	SolidWorks	Notes	Mode
CYLINDER	Insert/Boss/Revolve	Create the sketch first.	P
	Insert/Boss/Extrude	Create the sketch first.	P
	File/Open	Use a library part. Files of type: Lib Feature Part files (*.lfp *.sldlfp).	P
DBLIST	None.	Individual objects could be selected to view the Properties information.	—
DDATTDEF	None.	SolidWorks does not use attributes.	—
DDATTE	None.	—	—
DDATTEXT	None.	—	—
DDCHPROP	Tools/Options/Colors	Select the objects to be changed first.	D P S
	Properties	Right click on the object to be changed.	D P S
DDCOLOR	Tools/Options/Colors	Define default colors for objects.	A P
DDEDIT	Double click on the text to change.	—	All
DDEMODES	Tools/Options/Colors	Select the objects to be changed first.	D P S
DDGRIPS	Geometry can be dragged.	Dragged only; no editing while dragging.	D S
DDIM	Tools/Options/Detailing.	Set or change default dimension values.	D S
	Select dimensions, press the right mouse button, and select Properties.	Change selected existing values.	All
DDINSERT	Insert/Library Feature	—	—
DDLMODES	None.	—	—
DDLTYPE	Tools/Options/Line Font	Global setting only.	D P S
DDMODIFY	Right select entity and select Properties.	—	All
DDOSNAP	Dynamic snapping.	Replaces the need to manually change snap modes.	All
DDPTYPE	None.	—	—
DDRENAME	Slow double click the object in FeatureManager.	Same as Windows OS.	All
DDRMODES	Tools/Options/Grid Units*	Tabbed dialog box name.	All
DDSELECT	Uses standard Windows conventions for multiple selections (Ctrl and Shift).	—	All
DDSTYLE	None.	—	—

AutoCAD	SolidWorks	Notes	Mode
DDUCS	None.	Uses the selected plane or face, or the default origin.	—
DDUCSP	None.	Uses the selected plane or face, or the default origin.	—
DDUNITS	Tools/Options/Grid Units*	Tabbed dialog box name.	All
DDVIEW	View/Orientation	—	All
DDVPOINT	View/Orientation	—	All
DELAY	Programming can be done via a SolidWorks macro, Visual Basic, or C++.	—	—
DIM	Tools/Dimensions/Parallel*	Intelligent dimension tool icon will insert any dimension type listed at left. SolidWorks dimension tool knows what dimension to apply, depending on the entities selected.	D S
	Tools/Dimensions/Horizontal		D S
	Tools/Dimensions/Vertical		D S
	Tools/Dimensions/Baseline		D
	Tools/Dimensions/Ordinate		D
	Tools/Dimensions/Horizontal Ordinate		D
	Tools/Dimensions/Vertical Ordinate		D
DIMALIGNED	Tools/Dimensions/Parallel*	Intelligent dimension tool.	D S
DIMANGULAR	Tools/Dimensions/Parallel*	Intelligent dimension tool recognizes an angular dimension versus a linear dimension.	D S
DIMBASELINE	Tools/Dimensions/Baseline	Intelligent dimension tool.	D
DIMCENTER	Insert/Annotation/Center Mark	—	D
DIMCONTINUE	None.	—	—
DIMDIAMETER	Tools/Dimensions/Parallel*	Intelligent dimension tool recognizes a circular feature and places a diameter dimension. A circle will produce a diameter dimension and an arc a radius dimension by default. To change, use the Properties function.	D S
DIMEDIT	Select the dimension. Reposition handles will appear, and the text or extension can be dragged to the new location.	—	D S
	Select the dimension, press the right mouse button, and select Properties.	Pick Display to change extension or dimension line properties.	D S
DIMLINEAR	Tools/Dimensions/Parallel*	Intelligent dimension tool.	D S
DIMORDINATE	Tools/Dimensions/Ordinate	—	D S
	Tools/Dimensions/Horizontal Ordinate	—	D S
	Tools/Dimensions/Vertical Ordinate	Intelligent dimension tool.	D S

AutoCAD	SolidWorks	Notes	Mode
DIMOVERRIDE	Right click dim and select Properties.	—	D S
DIMRADIUS	Tools/Dimensions/Parallel*	Arcs are dimensioned as radial by default.	D S
DIMSTYLE	Right click the dim and select Properties. Tools/Options/Detailing	Global setting.	D S D S
DIMTEDIT	Select the dimension text and drag it.	—	D S
DIST	Tools/Measure	—	All
DIVIDE	Trim/Extend *	Temporary dimensions and entities may be required.	D S
DONUT	Insert/Boss/Revolve	Use a circle to revolve.	P
DRAGMODE	Uses dynamic drag by default.	—	—
DSVIEWER	None.	—	—
DTEXT	Insert/Annotations/Note	The text is entered in the dialog box.	D
DVIEW	View/Modify/Perspective	Camera and target angles can be simulated by rotating and panning the part.	A P
DXBIN	None.	—	—
DXFIN	File/Open*	Files of type: DXF files (*.dxf).	D
DXFOUT	File/Save As	Files of type: DXF files (*.dxf).	D
EDGE	Select the edge, press the right mouse button, and select Properties.	—	D
EDGESURF	None.	—	—
ELEV	Insert/Reference Geometry/Plane	Use an offset plane to set the elevation of a part. Use Extrude to set thickness.	P
ELLIPSE	Tools/Sketch Entity/Ellipse Tools/Sketch Entity/Centerpoint Ellipse	Creates an elliptical arc.	D S D S
END	File/Exit File/Close	—	A D P
ERASE	Edit/Delete Del key.	—	All
EXPLODE	None.	It is possible to explode a solid by exporting it as a wireframe IGES, but it cannot be read back into SolidWorks.	—
EXPORT	File/Save As	Select file type.	A D P
EXTEND	Trim/Extend*	Intelligent tool used for trimming and extending entities.	D S

AutoCAD	SolidWorks	Notes	Mode
EXTRUDE	Insert/Boss/Extrude	Add material to the part.	P
	Insert/Cut/Extrude	Subtract material from the part.	P
FILES	File/Find References	Use to copy files with external references (drawings and assemblies).	A D P
	File/Save As Windows Explorer.	Change name and/or location. Use to rename or delete files.	A D P A D P
FILL	None.	—	—
FILLET	Insert/Features/Fillet Round	—	P
	Tools/Sketch Tools/Fillet	—	S
FILTER	Select an object filter from the pull-down menu.	Filter toolbar can be activated under the View/Toolbars menu.	All
GIFIN	Insert/Object	—	A D P
GRAPHSCR	None.	—	—
GRID	Tools/Options/Grid Units*	Tabbed dialog box name.	All
GROUP	None.	—	—
HATCH	None.	Crosshatching automatically created for section views.	D
HATCHEDIT	Right select crosshatch, and select Crosshatch Properties.	—	D
HELP	Help/SolidWorks Help Topics	General help.	All
	Help	Help button with a dialog box active.	All
HIDE	View/Display/Hidden Lines Removed	Display with hidden line removed. Model can be dynamically rotated.	A P
ID	Tools/Measure	Select both the origin and vertex.	All
IMPORT	File/Open	Select file type.	A D P
INSERT	Insert/Library Feature	Library features are very similar to blocks.	P
INSERTOBJ	Insert/Object	—	A D P
INTERFERE	Tools/Interference Detection	Select the parts to check.	A
INTERSECT	None.	SolidWorks does not use Boolean operations.	—
ISOPLANE	None.	No need to sketch isometric views.	—
LAYER	Layer Properties	Configurations can also be used.	A D P

AutoCAD	SolidWorks	Notes	Mode
LEADER	Insert/Annotation/Note	Hold down the Ctrl key and select the locations prior to inserting the note for multiple leaders.	D
LENGTHEN	Trim/Extend*	Entities can also be dragged.	D S
LIGHT	Lighting Icon	Add directional or spot lights.	A P
LIMITS	Edit/Properties	Edit the drawing sheet size with the sheet selected.	D
LINE	Tools/Sketch Entity/Line	—	D S
LINETYPE	Tools/Options/Line Font	Global setting.	All
LIST	File/Properties	—	A D P
LOAD	None.	—	—
LOGFILEOFF	A log file is created by default.	—	All
LOGFILEON	A log file is created by default.	—	All
LTSCALE	Automatic	Scaling is based on view scale.	D
MAKEPREVIEW	File/Print Preview	—	All
MASSPROP	Tools/Mass Properties	—	A P
MATLIB	PhotoWorks/Edit Material	Optional PhotoWorks module.	A P
MEASURE	Tools/Measure	—	All
MENU	None.	Pull-down menus cannot be modified.	—
MENULOAD	None.	—	—
MENUUNLOAD	None.	—	—
MINSERT	Insert/Assembly Pattern	—	A
MIRROR	Tools/Sketch Tools/Mirror	—	S
MIRROR3D	Insert/Pattern Mirror/Mirror Feature Insert/Pattern Mirror/Mirror All	Mirror single features. Mirror complete part.	P P
MLEDIT	None.	—	—
MLINE	None.	—	—
MLSTYLE	None.	—	—
MOVE	Select objects and drag to the new location. Tools/Component/Move Component*	Fully constrained features will not move. —	D P S A
MSLIDE	None.	—	—
MSPACE	None.	No need for separate paper and model space in a SolidWorks document.	—

AutoCAD	SolidWorks	Notes	Mode
MTEXT	Insert/Annotations/Note	Format per note only.	All
MTPROP	Right click text and select Properties.	—	D
MULTIPLE	Tools/Options/General	Uncheck Single Command per Pick.	All
MVIEW	Insert/Drawing Views	Any number of views in a drawing.	D
MVSETUP	Edit/Properties	Change the sheet's properties.	D
NEW	File/New *	—	A D P
OFFSET	Tools/Sketch Tools/Offset Entities	—	S
OLELINKS	Insert/Object	—	A D P
OOPS	Edit/Undo*	—	All
OPEN	File/Open*	Select type of file to open.	A D P
ORTHO	Tools/Options/Grid Units tab	Set snap angle to 90 degrees.	D S
OSNAP	Tools/Sketch Tools/Automatic Relations	Intelligent cursor snaps to geometric points and adds relations automatically.	D S
PAN	View/Modify/Pan*	More similar to RTPAN in Release 13_c4.	All
PASTECLIP	Edit/Paste*	—	All
PASTESPEC	Insert/Object	—	A D P
PCXIN	None.	OLE can be used to attach PCX files (or any other raster format) to a SolidWorks document.	—
PEDIT	Select the spline and drag the spline handles (points).	Polyline entities with AutoCAD characteristics do not exist in SolidWorks.	D S
PFACE	None.	—	—
PLAN	View/Orientation/Normal To	Select a plane or planar face first.	A P S
PLINE	Draw lines endpoint to endpoint.	Line will not have pline characteristics.	D S
PLOT	File/Page Setup File/Print Preview File/Print*	— — —	All All All
POINT	Tools/Sketch Entity/Point*	—	D S
POLYGON	None.	—	—
PREFERENCES	Tools/Options	Dialog box split into various tabbed sections.	All
PSDRAG	None.	Objects display after drag.	All

AutoCAD	SolidWorks	Notes	Mode
PSFILL	None.	—	—
PSIN	Insert/Object	—	A D P
PSOUT	None.	—	A D P
PSPACE	SolidWorks drawing.	Create a drawing with views.	D
PURGE	None.	Delete unwanted entities.	All
QSAVE	File/Save*	—	A D P
QTEXT	None.	—	—
QUIT	File/Exit	—	A D P
	File/Close	—	A D P
RAY	None.	—	—
RCONFIG	None.	Windows is used to define output.	All
	PhotoWorks/Options/Image Output tab	Specify file type to save as.	A P
RECOVER	None.	A backup file can be created by selecting the backup option in the General tab in the Tools/Options menu.	All
RECTANG	Tools/Sketch Entity/Rectangle*	—	D S
REDEFINE	Tools/Customize/Menus	The Reset All button can be used to return to the default condition.	All
	Tools/Customize/Keyboard	Ibid.	—
REDO	None.	—	—
REDRAW	View/Redraw*	Refreshes graphics.	All
REDRAWALL	View/Redraw*	Refreshes graphics.	All
REGEN	Edit/Rebuild*	Rebuilds geometry.	All
REGENALL	Edit/Rebuild*	Rebuilds geometry.	All
REGENAUTO	None.	SolidWorks will not prompt before rebuilding.	All
REGION	None.	Regions are not necessary.	P A
REINIT	None.	—	—
RENAME	Perform a slow double click on the item in FeatureManager.	Works like renaming files in Windows OS.	All
	Edit the object's Properties.	—	All
RENDER	PhotoWorks/Render	Optional PhotoWorks module.	A P
RENDERUN-LOAD	Tools/Add-Ins	Uncheck PhotoWorks to remove from memory.	A P
REPLAY	PhotoWorks/View Image File	Optional PhotoWorks module.	A P

AutoCAD	SolidWorks	Notes	Mode
RESUME	Tools/Macro/Pause Macro	—	A D P
REVOLVE	Insert/Boss/Revolve	Adds material to the part.	P
	Insert/Cut/Revolve	Subtracts material from the part.	—
REVSURF	Insert/Reference Geometry/Revolved Surface	—	P
RMAT	PhotoWorks/Select Material	Optional PhotoWorks module.	A P
	PhotoWorks/Edit Material	Ibid.	A P
	PhotoWorks/Copy Material	Ibid.	A P
	PhotoWorks/Paste Material	Ibid.	A P
ROTATE	Tools/Sketch	Rotates an entire sketch. Modify	S
	Tools/Modify Dimension objects to rotate.	the dimension to rotate entity.	P S
ROTATE3D	Insert/Reference Geometry/Plane/At Angle	Features can be placed on rotating planes.	P
	Tools/Component/Rotate	Rotates parts in 3D space.	A
RPREF	PhotoWorks/Options	Optional PhotoWorks module.	A P
RSCRIPT	A macro or Visual Basic program can be used to produce a self-running script.	—	All
RTPAN	View/Modify/Pan*	Real-time panning.	All
RTZOOM	View/Modify/Zoom*	Real-time zooming.	All
RULESURF	Insert/Reference Geometry/Extruded Surface	—	P
SAVE	File/Save*	—	A D P
SAVEAS	File/Save As	Select file type.	A D P
SAVEASR12	File/Save As	Select Release 12 from Tools/Options/Import Export tab.	D
SAVEIMG	PhotoWorks/Options/Image Output	Set output to Render to File.	A P
SCALE	Tools/Sketch	Affects a single sketch.	S
	Tools/Modify Insert/Features/Scale	Affects the entire part.	P
SCENE	View/Lighting	Controls the lighting and adds new lights.	A P
	View/Orientation	Controls the view orientation.	All
SCRIPT	Programming can be done via a SolidWorks macro, Visual Basic, or C++.	—	—
SECTION	Insert/Drawing View/Section View/Display/Section View	—	D P
	Insert/Assembly Feature/Cut	For display purposes only. Ibid.	A

AutoCAD	SolidWorks	Notes	Mode
SELECT	Select multiple objects by holding the Ctrl key down.	Objects can be deselected using the same method.	All
	Drag a window around objects to be selected		All
SETVAR	Use Tools/Options to change settings.	SolidWorks has no variables, per se.	All
SHADE	View/Display/Shaded	Image can be dynamically rotated.	A P
SHAPE	None.	—	—
SHELL	None.	Not required for a Windows application.	—
SHOWMAT	PhotoWorks/Select Material	Optional PhotoWorks module.	A P
SKETCH	None.	Not to be confused with SolidWorks Sketch mode.	—
SLICE	Insert/Drawing View/Section	—	D
	View/Display/Section View	For display purposes only.	P
	Insert/Assembly Feature/Cut	Ibid.	A
SNAP	Tools/Options/Grid Units tab*	—	D S
SOLDRAW	Insert/Drawing View	Select the type of view to be created.	D
	View/Orientation	Select or define a new orientation.	A P
SOLID	None.	—	—
SOLPROF	Insert/Drawing View	Select the type of view to be created.	D
	View/Orientation	Select or define a new orientation.	A P
SOLVIEW	Insert/Drawing View	Select the type of view to be created.	D
	View/Orientation	Select or define a new orientation.	A P
SPELL	None.	Spell checker currently not available.	—
SPHERE	Insert/Boss/Revolve	Create the sketch of the wedge.	P
SPLINE	Insert/Sketch Entity/Spline*	—	D S
SPLINEDIT	Select the spline and drag the spline handles (points).	—	D S
STATS	PhotoWorks/Options	Optional PhotoWorks module.	A P
STATUS	File/Properties	View file information.	All
STLOUT	File/Save As	Files of type: Stl files (*.stl).	P

AutoCAD	SolidWorks	Notes	Mode
STRETCH	Select the object's endpoint and drag to the new location.	Stretching is normally done through the use of parametric dimensions.	D S
STYLE	Tools/Options/Detailing	Set the default text and dimension fonts.	D S
	Select the object and select Edit/Properties, or press the right mouse button and select Properties.	Change existing text and dimension fonts.	D S
SUBTRACT	Insert/Cut Insert/Assembly Features/Cut	Create a sketch and select the cut type. Create a sketch to cut through the assembly.	P A
SYSWINDOWS	Window/Cascade	Basic Windows OS functionality.	All
	Window/Tile Horizontal	Ibid.	All
	Window/Tile Vertical	Ibid.	All
TABLET	None.	SolidWorks uses the mouse and keyboard for all functions.	All
TABSURF	Insert/Reference Geometry/Extruded Surface	—	P
TBCONFIG	Tools/Customize/Toolbars	—	All
TEXT	Insert/Annotations/Note	—	A D P
	Insert/Sketch Entity/Text	For creating text features.	S
TEXTSCR	None.	—	—
3D	None.	3D solids created through Insert/Boss.	P
3DARRAY	Create two linear or circular patterns.	—	—
3DFACE	Insert/Sketch*	One plane (surface) at a time.	S
3DMESH	None.	—	—
3DPOLY	Insert/Reference Geometry/Curve though Reference Points	—	S
	Insert/Reference Geometry/Curve though Free Points	—	S
3DSIN	None.	Use a supported file format.	A P
3DSOUT	None.	Use a supported file format.	A P
TIFFIN	Insert/Object	OLE-compliant database application.	A D P
TIME	File/Properties	—	All
TOLERANCE	Insert/Annotation/Geometric Tolerance	—	A D P
TOOLBAR	View/Toolbars	Toggles toolbars on/off.	All

AutoCAD	SolidWorks	Notes	Mode
TORUS	Insert/Boss/Revolve	Create the sketch of the wedge.	P
TRACE	None.	—	—
TREESTAT	None.	Handled automatically.	—
TRIM	Trim/Extend*	Intelligent tool used for trimming and extending an object.	D S
U	Edit/Undo icon	Rebuild wipes undo list clean.	All
UCS	Insert/Sketch	Select plane or planar face first.	A P S
UCSICON	View/Origins	—	All
UNDEFINE	Tools/Customize/Menus Tools/ Customize/Keyboard	The Reset All button can be used to return to the default condition. Ibid.	All
UNDO	Edit/Undo icon list box	Rebuild wipes undo list clean.	All
UNION	None. Insert/Features/Join	All parts must be contiguous. Assembly components can be joined.	P A
UNITS	Tools/Options/Grid Units tab*	—	All
VIEW	View/Orientation	Double click on to select view.	All
VIEWRES	Tools/Options/Performance	Slider bar adjustments.	All
VLCONV	View/Lighting PhotoWorks/Select Material PhotoWorks/Edit Scene	— Optional module. Ibid.	A P P A P
VPLAYER	Configurations can be defined to selectively display objects.	—	A D P
VPOINT	View/Orientation View/Modify/Rotate	Select Isometric, Trimetric, or Dimetric. Dynamically rotate the model.	A P S A P S
VPORTS	Window/New Window Use the split window handles.	— The horizontal and vertical split window handles are located next to the scroll bars.	All All
VSLIDE	None.	—	—
WBLOCK	File/Save As	Save as type Library Feature (*.lfp).	P
WEDGE	Insert/Boss/Extrude File/Open	Create the sketch of the wedge. Use a library part. Files of type: Lib Feature Part files (*.lfp *.sldlfp).	P —
WMFIN	Insert/Object	—	A D P

AutoCAD	SolidWorks	Notes	Mode
WMFOPTS	None.	—	—
WMFOUT	None.	A screen capture program can be used to capture the screen.	All
XBIND	None.	OLE can be used to attach files to a SolidWorks document.	A D P
XLINE	Right select an entity and select Properties.	Entities can be changes to construction lines, but they will not be infinitely long.	D S
XPLODE	Save as an IGES wireframe.	File cannot be imported back into SolidWorks.	A P
XREF	Insert/Base Part	Part will be updated if the original base part is modified.	P
XREFCLIP	None.	—	—
ZOOM	View/Modify/Zoom to fit* View/Modify/Zoom to area* View/Modify/Zoom up down*	Similar to Zoom/All. Similar to Zoom/Window. Similar to RTZOOM.	All — —

Key to abbreviations: Mode or SolidWorks document type: A = Assembly, D = Drawing, P = Part, S = Sketch.

Sketch Entities Cross-reference

The following table contains sketch entities cross-referenced between the AutoCAD and SolidWorks environments. In many cases, commands in SolidWorks function differently than their AutoCAD counterparts, such as is the case with the arc commands. However, through the application of constraints and dimensions, the same objective can be reached. The AutoCAD entity is listed first, followed by the equivalent SolidWorks entity, followed by SolidWorks functionality notes.

AutoCAD/SolidWorks Sketch Entity Cross-references

AutoCAD	SolidWorks	Notes for SolidWorks
Line	Line	Pick start point and drag endpoint.
Construction Line	All sketch entities	Must modify properties. Entities such as lines will not extend infinitely.
Polyline	Line	Lines drawn end to end stay attached to one another but otherwise do not exhibit same characteristics as AutoCAD polylines.

AutoCAD	SolidWorks	Notes for SolidWorks
3D Polyline	Curve Through Free Points	Key in X-Y-Z coordinates.
Multi-Line	Line	Must draw the lines separately. Relation (parallel) placed automatically by system.
Spline	Spline	2D only.
Curve Through Free Points	Curve Through Free Points	2D or 3D.
Arc Three Points	3 Pt Arc	Pick and drag endpoints; pick and drag radius.
Arc Start Center End	Centerpoint Arc	Pick center and drag radius; pick and drag arc segment length.
Arc Start Center Angle	3 Pt Arc	Ibid.
Arc Start Center Length	3 Pt Arc	Ibid.
Arc Start End Angle	3 Pt Arc	Ibid.
Arc Start End Direction	3 Pt Arc	Ibid.
Arc Center Start End	Centerpoint Arc	Ibid.
Arc Center Start Angle	Centerpoint Arc	Ibid.
Arc Center Start Length	Centerpoint Arc	Ibid.
Arc Continue	Tangent Arc	Pick endpoint of a line or arc and drag the arc's length.
Circle Center Radius	Circle	Pick center and drag radius.
Circle Center Diameter	Circle	Drive with a dimension.
Circle 2 Points	Circle	Drive with constraints.
Circle 3 Points	Circle	Drive with constraints.
Circle Tan Radius	Circle	Add tangent relations.
Donut	None	Create as a revolved feature.
Ellipse Center	Ellipse	Pick center and drag first axis radius; pick and drag second axis radius.
Ellipse Axis End	Ellipse	Ibid.
Ellipse Arc	Centerpoint Ellipse	Same as Ellipse, then pick arc start point and drag arc endpoint.
Polygon	Line	Drive with constraints and dimensions.
Rectangle	Rectangle	Pick and drag opposite corners.
Point	Point	Pick to place point.

Illustrated Glossary

Align—To line up objects such as views or vertex points.

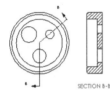

Aligned section view—A section that rotates a portion of the section line normal to the section view, usually for a cylindrical part.

SECTION B-B

Arc—A sketch entity defined by a portion of the circumference of a circle.

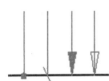

Arrow—A triangular-shaped symbol at the ends of dimension lines.

Assembly—A collection of parts or subassemblies.

Assembly configuration—A configuration created to suppress a group of parts or subassemblies to simplify the assembly or create alternative assemblies.

Auxiliary view—A projected view defined from an angled surface.

Axis1

Axis—Axis of rotation, or the center of a cylinder.

1

Balloon—A circle that encloses a label used to identify objects, often in a bill of material.

Base feature—The first main feature defined in a part.

Base part—Use of an existing part as the basis of a new part.

1.950
1.250
606
.125

Baseline dimension—A set of dimensions that uses a common reference origin and first extension line.

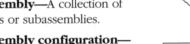
Boss

Boss—A feature that adds volume to a model.

Broken view—A partial view showing both ends of a part, with the center removed.

Cavity—A hollow area created by subtracting a part from a solid area. The cavity can be used to create a mold for the part that created the cavity.

Center mark—A drawing entity that displays the center of an arc or circle.

Centerline—A sketch entity used to define symmetry, rotate a sketch, or show reference.

Chamfer—A feature that cuts along an edge defined by a distance and angle.

Circle—A circular or round sketch entity.

Circular pattern—To copy features around a central point of rotation and specify the included angle and number of copies.

Component—Part in an assembly.

Configuration—A subset of features in a part or assembly that can be used to define an alternative or simplified part or assembly.

Constraint—Geometric relation used to control the behavior of sketch entities.

Construction geometry—Sketch entities converted to reference entities to help define geometric relationships.

Convert edge—To use an existing model edge as sketch geometry.

Cosmetic thread—A cosmetic feature used to define and detail threads.

Cursor—The graphical display of the operating system pointing device (mouse).

Cursor graphics change depending on the action being taken.

Cut—To remove volume from a model.

Datum—A reference plane used to define a geometric tolerance.

Derived part—A new part created from and referenced to another part. The new part is derived from the existing part.

Design table—A Microsoft Excel spreadsheet used to drive model and dimension parameters.

Detail view—A blow-up view of a specific area on an existing view to show greater detail.

Dialog box—A graphical user interface (GUI) menu.

Dimension—A feature that displays the size or length of a feature.

Document—A file that contains information for an application. SolidWorks document types include parts, drawings, and assemblies.

Draft—To tilt or angle a surface, usually for tool ejection for plastic molded parts and castings.

Drawing—Detail drawing of a part or assembly.

Drawing formats—The title block format that contains standard drawing information (i.e., drawn by, title, part number, and so on).

Drawing sheet—A drawing that contains a drawing format, views of the part or assembly, dimensions, and notes. A drawing can contain multiple drawing sheets. Only one sheet is visible at a time.

DWG—AutoCAD drawing file. Used to import information into a SolidWorks drawing or transfer data between programs.

DXF—AutoCAD data exchange file. Used to import information into a Solid-Works drawing or transfer data between programs.

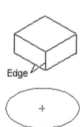

Edge—Single outside boundary of a feature.

Ellipse—An oblong arc/circle defined by a major and a minor axis.

Entity—An object within a SolidWorks document (i.e., line, arc, note, feature, and so on).

Embedded OLE object—A Microsoft object that follows the OLE (Object Linking and Embedding) criteria. Embedded objects no longer reference the host document, and any changes to the host (server) file will not be shown in the target (client) object.

"D5@Sketch3" = "D4@Sketch3"

Equation—A mathematical relation used to define dimensional relationships between feature or sketch dimensions.

Exploded assembly—Displays an assembly in a disassembled (exploded) state. Exploded assemblies can be used as an assembly or customer service aid.

Extrude—To project a profile along a straight line.

Face—A contiguous surface, planar or nonplanar, on a feature or part.

Feature—A solid object element used to create a solid model (i.e., boss or hole).

FeatureManager—A graphical interface for displaying information on parts, drawings, and assemblies.

Fillet/round—A rounded corner that blends together two or more model edges.

Geometric relations—A set of rules added to sketch geometry to control the behavior of sketch entities (same as a constraint).

Geometric tolerance—A set of standard symbols used to specify geometric characteristics and other dimensional requirements.

Grid—A pattern used by the system to aid the user in constructing geometry and dimensioning. The grid can provide a graphical display, as well as optionally allow the user to snap to grid points.

Helix—A curved circular path defined by pitch, rotation, and height (i.e., spring or thread).

Horizontal dimension— The distance horizontally between two selected points or objects.

Icon—Graphical user interface button used to perform functions and commands.

IGES—A data interchange format used to transfer information between different CAD systems.

Inferencing—Intelligent cursor and constraint feature for sketching. Geometric constraints are automatically added by the system. The cursor graphics update to display the current condition.

Interference detection— An assembly function used to determine interference between assembly components.

Layout—A 2D representation of a part or assembly, used to convey dimensions and details about a model.

Leader—A line, usually ending in an arrowhead, used to point to an object.

Leader Anchor

Leader anchor—The point at the end of a leader arrow that anchors its position. This point can be anchored to a face or edge.

Library feature—A part feature (or features) that can be reused for part creation.

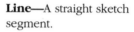

Line—A straight sketch segment.

Pattern features

Linear pattern—Creates a copy of features in a flat plane. Linear patterns can be defined in two directions at one time.

Linked OLE object—A Microsoft object that follows the OLE (Object Linking and Embedding) criteria. Linked objects reference the host document, and any changes to the host (server) file will be shown in the target (client) object.

Linked value—Dimensions that form an equality (i.e., D1 = D2 = D3). Modifying one dimension changes all linked dimensions.

Loft Profiles

Loft—A feature that uses multiple profiles to define a solid feature.

Macro—A program that records selections, menu picks, and keyboard strokes for later playback.

Mass properties—Display the volume, surface area, centroid, and inertia tensor for a part or assembly.

Mate—Similar to a geometric constraint, a relation between entities, usually surfaces, between parts within an assembly.

Model dimensions—The dimensions used to drive the size of feature geometry and to show distances in drawings.

Mold cavity—The hollowed-out area of a block used to create molded parts.

Top
Front   Right

Named view—A defined view using a view orientation and zoom state. These can be system default or user-defined views.

Draft
Neutral Draft Plane

Neutral draft plane—The plane used to define the face pivot location for draft angles.

Normal—A vector perpendicular to a surface, planar or nonplanar, at any point on that surface.

.350

Offset edge—A sketch entity at a given distance from an existing model edge that references that edge.

 **375**

 .000

Ordinate dimension—Dimension in a drawing shown from a common reference point.

Parametric—Dimensions drive the creation of model geometry. Changes to dimension parameters change model geometry.

Part—A single, contiguous solid model.

Part configurations—A subset of features in a part that can be used to define an alternative or simplified part.

Parting line—The edge that defines the mating surfaces between the top and bottom halves of a mold.

Pattern features

Pattern—A part feature copied in a linear or circular fashion.

Right
Front

Plane—Planar entity used for sketching, as well as geometric and dimensional reference.

Point—A reference location defined by a single point entity.

Projection view—A view orthographically projected from an existing view.

Properties—Attributes of an object, such as entities, features, and dimensions.

Window Help
New Window
Cascade
Tile Horizontally
Tile Vertically
Arrange Icons
Close All
✓ 1 Helix.PRT

Pull-down menu—A menu that appears when a title (or menu heading) is selected from the menu bar.

Rebuild—Regenerates a part, drawing, or assembly and incorporates any changes to dimension parameters.

Rectangle—A sketch entity defined by two horizontal and two vertical lines.

Relative to model view— A view relative to planar surfaces on a part or assembly.

Reorder—To change the creation order of selected features. This will affect how a solid model or assembly is created.

Revolve—Creates a solid boss or cut from a closed profile rotated about a centerline.

Rollback—To go back to a specified point in the part creation process. Used to insert features at a specific point in the creation order.

SECTION A-A
Scale 1 : 1

Section view—A cutaway view of a part or assembly that shows interior detail.

Scope—Defines what parts a new assembly feature affects.

Shell—A feature that creates a hollow part with at least one surface open to the outside. Often used for plastic molded and cast parts.

Sheet metal—A thin-walled part that can be unfolded into a flat sheet.

Shrinkage—The percentage a part contracts due to cooling after the molding or casting process.

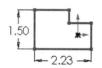

Sketch—A collection of 2D entities used to define features in a solid model.

Standard 3 Views—Three orthographically defined views: top, front, and right.

Stereolithography (STL)— A file format used to create rapid prototyping parts.

Suppress—To temporarily remove a feature or component from a graphics display and from memory. The features or components can be unsuppressed at any time.

Surface finish—A unit of measure that determines the roughness of a surface.

Surface finish symbol— A detailing symbol used to denote the operation, roughness, direction, and special requirements of a surface.

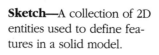

Sweep—Creates a feature by sweeping a closed profile along a trajectory.

Thin feature—A feature of constant wall thickness that can be used to create sheet-metal parts.

Tolerance—The amount of deviation allowable for a given feature, shown in a dimension.

Trim/extend—To shorten or lengthen a sketch entity.

Trimmed surface—A surface defined with a specific boundary used during IGES translation by solid modeling software.

Undo—Reverses previous operations to their original state.

Vertex—Corner or end point of an edge or sketch entity.

Vertical dimension—The vertical distance between two selected points or objects.

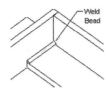

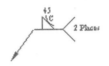

Weld bead—An assembly feature that represents the welded area between assembly components.

Weldment—An assembly of welded components.

Weld symbol—Detail symbol (or symbols) that defines a weldment.

Index

Symbols

! (exclamation point), meaning of 297

Numerics

2D drawings
　See two-dimensional drawings
2D layouts
　See drawings
3 point arc 116
3 Points plane option 228
3D accelerator chips
　See three-dimensional accelerator chips
3D curves, exporting 467
3D wireframes
　See wireframes

A

Accelerated Graphics Port (AGP) 41
accelerator chips 41
accidental dimensions 150–151
ACIS files 235
　about 460
　assembly format 461
　exporting 473
　importing 473
ACIS kernel 34
actions in SolidWorks 312
Add Relations dialog box 161
Add Symbol note option 396
Advanced Face Fillet option 207
Advanced Smoothing loft option 193
Advanced Smoothing sweep option 190
aligned dimensions
　See parallel dimensions
Aligned mate conditions 305
aligned section views 367, 369
aligning grids 92
alignment conditions, mates 304
allowances, bend 278
Alpha chip 36
alternate spline method 118
ambient lighting
　color settings for 427
　options for 427
　source of 426
Ambient lighting option 432
AMD processors 37
anchor points, use with BOM 417–418
Angle chamfer option, Flip Dimensions chamfer option 209
Angle crosshatching option 366
Angle extrusion option 184
Angle helix option 240

Angle mate constraints 304
Angle note option 396
Angle revolved feature option 186
Angle rib option 213
angles, Snap To Angle function and 94
angular dimensions 148
Angular Tolerance option 393
annotations 256, 258
　drawing 343, 379
　types of 257
ANSI geometric tolerance datum 405
anti-aliasing 437
Anti-aligned mate conditions 305
API, SolidWorks 492
application programming interface
　See API, SolidWorks
applied features 100, 181
arcs
　3 point 116
　center marks for 400
　centerpoint 115
　constraints for 160
　elliptical 117
　lengthening 132
　tangent 115
　Trim command for 131
Arrow Style note option 396
Arrowhead finish option 411
Arrowhead Style datum option 407
Arrows dimension option 389
assemblies 281, 340
　AutoCAD, comparison of 284, 288
　bills of materials, adding 413
　bottom-up 289
　cavities, inserting 325, 327
　colors and 435–436
　components, inserting 291, 299
　components, instances 315
　configurations, specifying for 314
　constraints 301, 307
　creating 289–290, 294, 307, 322
　cross sections, showing 320
　cuts in 271
　design layouts for 343
　documents 10, 12
　exploded drawings of 344
　exploding 328, 333
　export formats 460
　external references to 245
　files, locating 313
　files, managing 283, 285
　import formats 460
　in-context editing of 296, 310
　inserting into drawings 346
　large, RAM requirements of 38

　layouts 285, 287
　mates 301, 307
　modifying 307, 316
　patterns 318
　printing characteristics of 443
　quick printing of 441
　references, copying 283
　rotating components in 309
　sample robot 336
　scope of components in 320
　screen shots of 441
　section views of 364
　simplifying 267
　simulating motion in 282, 291, 301, 308
　skeletons 285
　sketches 285
　structure editing 316
　subassemblies 281
　top-down 289
　transparency lighting setting for 432
　weld beads 335
associativity of drawings 344
At Angle plane option 228
Auto Explode option 331
Auto Fillet thin feature option 274
AutoCAD
　comparison of 100
　exporting to 462
　importing files from 463
　limitations on opening documents with 10
　modeling limitations of 17–18
　sketch planes, comparison to UCS 106
AutoCAD comparison 123
auxiliary views 362
axes 230, 232
　mate constraints for 306
　options for 230, 232
axes, examples of 231
axis of rotation, defining 186

B

background color, adjusting 18
backup copies 63
Balloon Style note option 396
balloons 399, 400
　inserting in drawings 394
　notes, option for 398
base features 169
　about 57
　determining 177
Base Part command 269, 272, 313, 325
base parts 269, 272
baseline dimensions 382

beads, weld 335
Bend Allowance default 276
bends
    allowances for 278
    modifying 276
    options for 277
    properties of 274
Bends command 274
Bends feature 272
Bent Leader datum option 407
Bent Leader finish option 411
beveled edges 208
bidirectional associativity 10–11
bidirectional associativity of
    documents, turning off 345
Bill of Materials
    anchoring 417
    editing 414
    moving 418
    properties 419
    splitting 419
Bill of Materials Properties dialog
    box 416
bills of materials 412, 421
    defined 344
    in drawing documents 9
    inserting 413
    options for 416
    updating 417
black, meaning of in sketches 152
blind end condition 183
blue, meaning of in sketches 152
BOM
    See bills of materials
Boolean modelers
    compared to parametric
        modelers 17, 18
bosses
    about 180
    extruded 182
    helixes for 239
    lofted 192
    revolved 185
    solid 192
    solid features for 196
    thin features for 272
Both Directions extrusion
    option 184
bottom-up assembly method
    choosing 289
    defined 285
    using 289
Break command 136
broken views 369
brown, meaning of in sketches 152
Browse component option 315

C

C++, customizing SolidWorks
    with 492
CAD skills, nurturing 44
Cap Ends thin feature option 274

cavities
    inserting 325, 327
    options for 327
Cavity command 271, 325
center marks 400
centerlines 121, 124
    constraints for 160
    in revolved features 185
    mirroring with 125
centerpoint arc 115
centerpoint ellipses 117
chamfers 208, 209
    options for 209
Change Direction of Cut section line
    option 368
child relationships
    See parent/child relationships
circles 116
    center marks for 400
    constraints for 160
    dimensions, creating for 146
circular features, annotating 412
circular holes 201
Circular Pattern command 222
circular patterns
    about 221
    See also patterns
Clockwise helix option 240
Close along Loft Direction loft
    option 193
closed profiles, creating thin
    features for 272
Closest mate conditions 305
Coincident mate constraints 304
collision detection 309
Color component option 316
Color options tab 64
color preferences 480–483
colors
    ambient light, setting 427
    assembly 435–436
    AutoCAD comparison 431
    components, modifying 300
    feature properties, setting 434–
        435
    feature, ignore 434
    feature, setting 433
    meaning of 299
    meanings of in sketches 152
    parts, modifying 299, 430
    preferences for 431
commands
    accessing 114
    Base Part 269, 272, 313, 325
    Bends 274
    Break 136
    Cavity 271, 325
    Chamfer 208
    Circular Pattern 222
    Component Pattern 319
    Component Properties 313
    Copy Files 313
    Curvature 88

Curve Through Free
    Points 242–243
Curve Through Reference
    Points 242–243
Customize 483
Datum Target Symbol 406
Derive Component Part 325
Derived Component Part 271
Design Table 259, 264
Dimension 381
display 75, 90
Dome 254
Draft 215
Drive Component Part 327
Edit Definition 244
Edit Sketch 245
Edit Sketch Plane 245
Edit Template 348
Equation 265
exiting 114
Explode 298, 328, 333
Extend 133
Extrude 182
Feature Scope 320
Fillet 204
Find References 283
Fix 307
Float 307
Grid 90
Helix 239
Horizontal Ordinate
    Dimension 383
Join 333
Knit Surface 237
Light Sources 426
Linear Pattern 221
Link Value 385
List External References 245
Loft 192
Mirror All 218
Mirror Feature 217
Mirror Part 219, 313
Model Annotations 380
Move Component 308
Page Setup 445–454
Pan 79
Plan 57
Point 120
Projected Curves 238
Radiate Surface 236
Rebuild 249
Redraw 89
Revolve 185
Rib 211
Rollback 177, 249
Rotate Component 309
Rotate Component About
    Axis 309
Rotate View 78
Scale 253
Shape 255
Shell 209
sketch 113, 123

Split Line 215
Suppress 251
Sweep 186
Undo 130, 155, 247
Vertical Ordinate
     Dimension 383
view 76, 80
Weld Bead 335
zoom 76, 79
Zoom In/Out 77
Zoom To Area 76
Zoom To Fit 76
See also individual command
     names
Common Properties dialog box 124
complexity of sketches 102
Component Name component
     option 315
Component Pattern command 319
Component Properties
     command 313
component reordering 316
components
     annotations for 380
     colors, changing 316
     colors, modifying 300
     exploded views of 328, 333
     files, locating 313
     fixed 292, 294, 306–307
     floating 292, 306
     inserting 297
     inserting into assemblies 291,
          299
     instances of 315
     interference among 323, 324
     joining 333
     locating 315
     meaning of symbols in
          FeatureManager 306–307
     model dimensions for 380
     motion in 282, 291, 301, 308
     moving 309
     multiple parts for 327
     names, updating 284
     naming conventions 197
     options for 316
     reordering and BOM 317
     restructuring 317
     rotating 309
     scope of 320
     skeletons and 285
     suppressing 315
computer processors
     See CPUs
Concentric mate constraints 304
ConfigurationManager 51, 268
     exploded views,
          controlling 328
configurations 267, 269, 314
     adding comments to 268
     creating 268
     exploding 331
     selecting for components 316

constant radius fillets 205
Constrain All function 166
constraint, simplifications of 105
constraints 103, 105, 108, 160
     adding to unconstrained
          files 166
     assembly 301, 307
     AutoCAD comparison 105, 159
     automatic 105, 111–112, 160
     Coincident 160
     Collinear 160
     Concentric 160
     Coradial 160
     cursor changes and 105
     cursor inferencing and 110
     defined 103
     deleting 162, 164
     dimensional 103
     displaying 162, 164
     Equal 160, 165
     Fix 160
     full 173
     geometric 103
     Horizontal 160
     Intersection 160
     list and description of 160
     list of 106
     maintaining 153
     mate types 304, 307
     Merge Points 160
     Midpoint 160
     Parallel 160
     Perpendicular 160
     Pierce 160
     Symmetric 160
     Vertical 160
construction entities 123–124
consulting for implementation and
     training 31
contents of
     STL files 470
context-sensitive help
     accessing 24, 50
context-sensitive menus
     opening with right mouse
          button 73–74
control key, selecting multiple
     entities with 129
Convert Entities function 133
coordinate system,
     user-defined 232
Coordinate System dialog box 233
coordinate systems 232
Copy Files command 313
Copy Options Wizard 43, 61
copying 167
copying files 312
cosmetic threads 411
Counterclockwise helix option 240
CPUs
     choosing 36
     multiple 37
cross sections, displaying 320

Crosshatch options tab 66
crosshatching
     about 365
     changing 365
     default option 366
     options for 366
cursor
     changes when selecting
          objects 72
     constraints and changes to 105
     symbols, list of 112
cursor inferencing 72–73, 89, 106,
     110–111
Curvature command 88
curvature information 89
curvature radius, showing 88
Curve Through Free Points
     command 242–243
Curve Through Reference Points
     command 242–243
curves 242–243
     AutoCAD comparison of 243
     features for 196
     guide 186
     helixes 239
     projected 238
Custom Properties 352–353
     defining 354
Customize command 483
customizing
     AutoCAD 43
     AutoCAD comparison of 478
     icons 483
     menus 483, 486
     properties 49
     SolidWorks 43, 477, 494
     toolbars 483, 485
cuts
     about 180
     Cavity command for 271
     extruded 183
     helixes for 239
     making holes with 201
     molds, inserting in 325
     section lines and 368
     sheet metal parts and 274
     solid 192
     solid features for 196
cylinders
     aligned section views for 367
     mate constraints for 306
cylindrical features 196
Cyrix processors 37

D

dangling, defined 152
dangling dimensions or
     constraints 164
date, inserting in documents 452
datum point symbols 405
Datum Reference option 407
datum symbols, editing 406

Datum Target Symbol
    command 406
DEC Alpha chip 36
DEC Alpha machines 38
decimal precision, dimensions 389
Default Bend Radius bends
    option 278
Default fillet option 207
defaults
    setting 62, 71
    setting, comparison with
      AutoCAD 71
Defined By helix option 239
definitions, modifying 244
Depth extrusion option 184
Derive Component Part
    command 271, 325
derived sketches 167
Design Component cavity
    option 327
Design Intent 104
design intent 58, 105, 394
design layouts, defined 343
design process, lighting in 425
Design Table command 259, 264
design tables 259–260, 264
    annotating 261
    comparison to AutoCAD 263
    deleting 264
    editing 263
    inserting 262
    use of headings in 260
design trees
    in-context references in 296
    MateGroup object 303
    subassemblies, showing 316
detail drawings, defined 343
detail views 362
    creating 363
Detailing options tab 65
diametric dimensions 146
    creating from linear
      dimensions 146
diamonds, rendering with
    PhotoWorks 437
Diffuse lighting option 432
Dimension command 143, 381
    driven dimensions 151
    settings for 144–145
    starting 143
dimension lines, visibility of 390
dimension properties 388
dimension values, positioning 144
dimensions
    about 142
    accidental 150–151
    adding to drawings 377, 393
    adding to sketches 103, 172
    angular 148
    annotating drawings with 378
    AutoCAD comparison to 378
    baseline 382
    constraints for 103

diametric 146
dragging 386
drawing 343
driven 150–151
    ease of changing 58
    equations and 264–265
    fonts for 390
    horizontal 149
    horizontal, example of 105
    linked 384
    model 378
    modifying 385–388, 393
    modifying values of 153
    modifying, ease of 244
    negative 186
    options for 389, 393
    ordinate 382, 384
    parallel 145
    point-to-point 143, 145
    precision of 389
    preferences for 65, 388
    quick reference drawings
      for 344
    reference 378, 381, 384
    text, adding to 390, 392
    tolerance options for 392
    tolerances for 392
    update option, disabling 386
    vertical 149
dimetric view 57
Direct Edit function 154, 155
Direction Thickness thin feature
    option 274
directional light, options for 427
directional lighting, source of 426
directories
    maintaining 44
    search paths for, creating 67
display
    functions 75, 90
    options 87
    preferences for 480
Display as Dual Dimension
    dimension option 389
Display Only Surface Cut section
    line option 368
display options 80
Display Precision dimension
    option 389
Display with Bent Leader note
    option 396
Display with Leader note
    option 396
Display with Parenthesis dimension
    option 389
Dissolve Subassembly function 317
Distance chamfer option 209
Distance mate constraints 304
Distance mates 302
distances
    See dimensions
documents
    access preferences for 66

assembly 10, 12
closing 15
creating new 12–13
drawing 9
headers and footers for 450,
    454
interdependence of 10
naming conventions for 13
opening 6–8
opening multiple 10
part 9
printing 455
printing, reasons for 441
printing selected areas 455
saving 13, 15
sharing electronically 441
types of 8–9, 12
Dome command 254
Dome dialog box 254
Draft Outward extrusion option 184
Draft Outward rib option 213
Draft While Extruding extrusion
    option 184
drafts 214, 216
    AutoCAD comparison of 216
    creation methods 215, 241
    solid features for 196
    split lines for 240
dragging
    defined 114
    how to 127
dragging sketch geometry 153
drawing formats 10
Drawing options tab 67
drawing sheets 10, 356, 358
    activating 358
    adding 356
    AutoCAD comparison of 358
    changing 357
    deleting 357
    options for 357
drawing symbols 394, 421
    AutoCAD comparison of 394
drawing template, comparison to
    AutoCAD 347
drawing views
    See views
drawings 341, 422
    about 343–344
    adding dimensions to 377, 393
    adding reference dimensions
      to 381, 384
    assemblies, inserting 346
    associativity of 344, 345
    AutoCAD comparison of 347
    creating 345, 358
    documents 9
    exploded 344
    export formats 461
    geometric tolerance symbols,
      inserting 401
    import formats 461
    layers in 375–376

notes, inserting 394
parts, inserting 346
preferences for 62
printing characteristics of 443
printing selected areas 455
sharing electronically 441
templates for 346, 348, 358
types of 343
units of measure 344
views, creating with 347
Drive Component Part
  command 327
Driven dimension option 389
driven dimensions 150
DWG files 462–464
  adding constraints to 166
  exporting 464
  importing 463
  templates, importing with 348
DXF files 462–464
  adding constraints to 166
  exporting 464
  importing 463
  templates, importing with 348
dynamic editing 246

E

Edge options tab 63
edges, beveled 208
Edit Color dialog box 300
Edit Color icon 299
Edit Definition command 244
Edit Sketch command 245
Edit Sketch Plane command 245
Edit Template command 348
editing features 246
ellipses 116
  centerpoint 117
Emission lighting option 432
employees, identifying CAD skills
  of 44
empty views 370
Enable Draft rib option 213
end conditions
  extruded sketches 183
  revolved features 186
engineering layouts
  in drawing documents 9
entities
  centerline 121
  construction 123, 124
  deleting 132
  reference 123
  selecting multiples with control
    key 129
  sketch 113, 123
Entities tab, Display/Delete
  Relations window 164
entity functions 128, 138
entity types, supported for
  IGES 465
Equal Distance
  chamfer option 209

Equal Spacing pattern option 223
Equation command 265
equations 264, 266
  adding 265
  AutoCAD comparison of 265
  basic guidelines 266
  common problems 265–266
  deleting 266
  linked dimensions and 385
error logs 467
error recognition 297
errors
  in importing files 473, 474
  reporting 468
Exit option of File menu 15
exiting SolidWorks 15
Explode command 328, 333
  AutoCAD comparison to 298
  options for 331
Explode Steps option 331
exploded drawings
  defined 344
exploded views 328, 333
  AutoCAD comparison 333
  editing tools for 332
Exponent spot lighting option 429
exporting 457
  3D curves 467
  ACIS files 473
  DWG files 464
  DXF files 464
  formats for 460–461
  IGES files 467, 469
  preferences for 461
  solid models 34
  STL files 471
  surfaces 468
  VDAFS files 472
  VRML files 473
  wireframe geometry 467
exporting files 67, 476
exporting settings 43, 61
Extend command 133
external references
  about 245
  AutoCAD, comparison to base
    parts 270
  See also references
External References options tab 66
extruded features
  end conditions for 183
  options for 184
Extruded Surface function 233
extruded surfaces 234
extrusions 179
  compared to Rib command 211
  direction of 107
extrusions, and the sketch
  plane 108

F

Face Blend fillet option 207
face blends 208

feature names, equations and 265
Feature Scope command 320
feature-based modeling 100
  defined 99
FeatureManager 52
  about 16, 18
  ConfigurationManager, toggling
    between 267
  design tree 62
  external reference symbol
    in 311
  in-context references in 296
  MateGroup design tree
    object 303
  preferences for 62
  symbols, meaning of 306, 307,
    311
FeatureManager MateGroup 305
features
  annotations for 380
  applied 100, 181
  AutoCAD comparison of 179
  base 169, 179
  bends 272
  bosses 180
  chamfers 208, 209
  circular, annotating 412
  classification of 181
  color properties, setting 434,
    435
  colors, setting for 433
  creating 322
  creation order for 178
  cuts 180
  design tables for 259
  determining optimal number
    of 178
  draft 214, 216
  editing 246
  extruded 182
  extrusions 179
  fillets 204, 208
  holes 201
  identifying for parts 168
  in-context 245, 287
  lofted 192, 196
  mirroring 216, 220
  model dimensions for 380
  modeling, defined 99
  modifying 244, 252
  naming 196
  naming conventions 197
  parent/child relationships
    of 198, 200
  pattern 220, 253–255
  renaming 53
  reordering 200
  revolved 185
  ribs 211, 214
  rolling back 177, 250
  rounds 204, 208
  scope of 320
  shells 209, 211

sketched 100, 181
  suppressing 251, 267
  swept 186, 192
  unsuppressing 251
feedback, cursor 72
file extensions
  .dwg 9
  .prt 7
  .reg 43
  .SAT 235
  .sldprt 7
  .WRL 235
  drt 351
  err 467
  finding 7
  IGS 460
  list of 8
  rpt 467
  SAT 460
  STL 460
  v_b 473
  v_t 473
  WRL 460
  X_B 460
  X_T 460
file properties 355
  linking to 351–352
file types
  assemblies, import and
    export 460
  drawings, import and
    export 461
  import and export 460–461
  parts, import and export 460
filenames, inserting in
  documents 452
files
  access preferences for 66
  assembly 10, 12
  assembly, managing 283, 285
  bidirectional associativity
    of 285
  components, locating 313
  copying 312
  drawing 9, 10
  export formats 460–461
  exporting 67, 457,476
  import formats 460, 461
  importing 67, 457, 476
  naming conventions for 13, 44
  part 9
  prototypes 470
  types of 9, 12
files plug-in viewer for VRML 473
fillets 204, 208
  AutoCAD comparison of 207
  options for 206
  sketch 128–129
  thin features and 274
  variable radius 208
Find References command 283, 312
finish symbols
  See surface finish symbols

First Direction pattern option 223
Fix command 307
fixed components 292, 294, 306,
  307
Fixed Edge or Face bends
  option 278
Fixed Radius fillet option 207
flat patterns, sheet metal parts
  and 275
Flip Direction
  chamfer option 209
Flip Side of Material rib option 213
Flip Side To Cut extrusion
  option 184
Float command 307
floating components 306
folded parts 274
folders, creating search paths for
  67
Follow Guide Curves sweep
  option 190
Follow Path and Guide Curves
  sweep option 190
Follow Path sweep option 190
Font dimension option 389
Font Height tolerance option 393
Font Scale tolerance option 393
Font section line option 368
Font toolbar 22
fonts
  dimensions, editing 390
  preferences for 66
footers 450, 454
  comparison to AutoCAD 453
  customizing 451
  inserting date, time, and page
    information in 451
formats
  assemblies, import and
    export 460
  drawing 10
  drawings, import and
    export 461
  export 460–461
  import 460–461
  parts, import and export 460
formats, drawing
  about 348
  customizing 348, 356
  importing 348
  multiple 356
  See also drawing sheets;
    templates, drawing
Full Name dimension option 389
functions
  Base Part 269, 272, 313, 325
  Bends 274
  Cavity 271, 325
  Chamfer 208
  Circular Pattern 222
  Component Pattern 319
  Component Properties 313
  Copy Files 313

Curvature 88
Curve Through Free
  Points 242–243
Curve Through Reference
  Points 242–243
Customize 483
Datum Target Symbol 406
Derive Component Part 325
Design Table 259, 264
Dimension 381
display 75, 90
Draft 215
Drive Component Part 327
Edit Definition 244
Edit Sketch 245
Edit Sketch Plane 245
Edit Template 348
entity 128, 138
Equation 265
Explode 298, 328, 333
Extrude 182
Feature Scope 320
Fillet 204
Find References 283
Fix 307
Float 307
Grid 90
Helix 239
Horizontal Ordinate
  Dimension 383
Join 333
keyboard shortcuts for 487
Light Sources 426
Linear Pattern 221
Link Value 385
List External References 245
Loft 192
Mirror All 218
Mirror Feature 217
Mirror Part 219, 313
Model Annotations 380
Move Component 308
Page Setup 445, 454
Projected Curves 238
Rebuild 249
Redraw 89
Revolve 185
Rib 211
Rollback 177, 249
Rotate Component 309
Rotate Component About
  Axis 309
Shell 209
Snap to Angle 94
Suppress 251
Sweep 186
Undo 155, 247
Vertical Ordinate
  Dimension 383
view 76, 80
Weld Bead 335
zoom 76, 79

G

General options tab 62
general preferences 479
geometric constraints, defined 103
geometric relationships 158
geometric sketching 111
geometric tolerance symbols,
    options for 402, 405
geometry, reference 225, 243
Geometry Pattern option 217, 223
glass, rendering with
    PhotoWorks 437
graphics cards 41
graphics performance 41
gray, meaning of 299
green, meaning of 115
green, meaning of in sketches 115
grid, setting 113
Grid command 90
Grid/Units options tab 64
grids
    aligning to model edges 92
    properties of 90, 93
guide curves 186, 188
    lofts and 192
Guide Curves loft option 193

H

hard drive requirements 39
hardware requirements for solid
    modeling 35, 42
headers 450, 454
    comparison to AutoCAD 453
    customizing 451
    inserting date, time, and page
        information in 451
Height drawing sheet option 357
Height helix option 239
Helix command 239
helixes 239
    AutoCAD comparison 240
    options for 240
Help
    about 23, 25
    accessing 24, 50
    accessing tool tips tool 114
    key 488
Hidden In Gray display option 81
hidden lines
    displaying 81
    removing 83
Hidden Lines Removed display
    option 83
hole callouts 412
Hole Feature window 201
holes 201, 204
    options for 203
    types of 202
    wizard for 202
horizontal dimensions 149
    example of 105
Horizontal Ordinate Dimension
    command 383

I

icon, Sketch 113
icons
    adding to toolbars 485
    customizing 483
    getting help for 6
    removing from toolbars 485
    shortcut, creating 48
    toolbar, standardization of 4
IGES entity type 466
IGES files 465, 469
    about 460
    assembly format 461
    export options 466–467
    exporting 469
    format of 34
    importing 469
IGES translations 33, 34
Ignore Feature Colors option 434
images
    capturing 436
    Print Screen key for 436
implementation plan 29
Import Diagnostics function 474–
    475
imported surfaces 235
importing 457
    DWG files 463
    DXF files 463
    error logs when 467
    errors in 473
    files 67, 476
    formats for 460–461
    IGES files 469
    knitted surfaces and 468
    preferences for 461
    report files 467
    two-dimensional drawings to
        SolidWorks 33
    VDAFS files 472
    VRML files 473
    wireframes to SolidWorks 33
    with XchangeWorks 475
importing ACIS files 473
importing files 67
importing settings 61
in-context editing 296, 310
in-context features 245, 287
Increment dialog box 387, 389
inferencing, cursor 72–73, 110
inferencing cursor, list of
    symbols 111
Initial Graphics Exchange Format
    See IGES files
injection molded parts, draft feature
    and 214
InPlace mates 297–298
Instance ID component option 315
Instances Deleted pattern
    option 223
Intel Xeon processor 37–38
interference, component 323–324
    AutoCAD comparison of 324

internet, obtaining information
    via 31
intranet, disseminating information
    via 31
invalid references, tracing 297
isometric view 57
Items to Copy pattern option 223

J

Join command 333
    AutoCAD, comparison to Union
        command 334

K

Keep Edge fillet option 207
Keep Normal Constant sweep
    option 190
Keep Surface fillet option 207
keyboard shortcut
    AutoCAD comparison 488
    system help key 488
keyboard shortcuts 487, 489
    predefined, list of 487
    view commands 79
    zoom commands 79
K-factor 276–278
K-factor bend allowance value 278
Knit Surface command 237
knitted surfaces 468

L

Label section line option 368
labels, inserting in drawings 394
Landscape orientation 446
Layer dialog box 376
Layer option 374
layers, about 374–375
Layers dialog box 375
layout sketches, minimizing parent/
    child relationships with 287
layouts, assembly 285, 287
leaders open option 398
legacy data, translating 32–35
Light Sources command 426
lighting
    about 424
    ambient 426–427
    directional 426–427
    focusing 429
    intensity of 429
    preferences 432
    sources 426
    sources, adding 429–430
    spot 426, 428
    spread of 429
Line & Point plane option 229
Line Font options tab 66
Line Format toolbar 377
line types 376
line weights, preferences for 66
linear dimensions, creating from
    diametric dimensions 146

Linear Pattern command 221
linear patterns 220–221
    See also patterns
Linear Tolerance tolerance
    option 393
lines
    centerline 121, 124
    constraints for 160
    construction 124
    dimension 390
    drawing 115
    formatting 376
    hidden, displaying 81
    hidden, removing 83
    leader for datum symbols 407
    lengthening 132
    mate constraints for 306
    section 367, 369
    split 240
    Trim command for 131
    types, list of 448
    weights of 447, 449
    weights of, adjusting for
        printing 448
Link to Properties button 354
Link to Properties window 355
Link Value command 385
linked dimensions 384
linking to file properties 351, 352
List External References
    command 245
locked components
    See components, locked in
        space; fixed components
Loft Centerlines 194
loft features
    troubleshooting 195
lofted features
    about 192
    advanced options, examples
        of 194
    AutoCAD comparison 196
    options for 193
lofted surfaces 234

M

macro programming language
    See macros
Macro toolbar 491
macros 489, 491
    creating 489
    editing 490
    running 491
Maintain Tangency loft option 193
Maintain Tangency sweep
    option 190
manufacturing methods,
    documenting 409
margins, setting for printing 446
MateGroup object 303
Material Properties options tab 67
mates
    adding 294

alignment conditions for 304
assembly 301, 307
attribute samples 303
constraint types for 304, 307
deleting 305
Distance 302
parent/child relationships
    in 311
Maximum Variation tolerance
    option 393
MDI 10
memory
    See RAM
mentoring for implementation and
    training 31
menus
    comparison to AutoCAD 20–21,
        23–25, 32–35, 38, 40, 43, 47,
        55–56, 58–59, 6–71, 73, 75–
        81, 83, 85, 87–89, 92–95
    context-sensitive, opening 73–
        74
    customizing 483, 486
    standardization of 3, 4
    View/Display 80
    View/Modify 76
metals, rendering with
    PhotoWorks 437
methodologies for
    assemblies 285
Microsoft Excel spreadsheets,
    inserting as design tables 259,
    264
Mid Plane rib option 213
mid-plane end condition 183, 186
Mid-Plane thin feature option 273
Minimum Variation tolerance
    option 393
Mirror All command 218
Mirror Feature command 217
Mirror Part command 219, 313
mirroring 124, 158, 216, 220
    Geometry Pattern option 217
model annotations 256, 258
    types of 257
Model Annotations command 380
model dimensions
    defined 378
    reference dimensions,
        compared to 378
    update option, disabling 386
Model Document Path component
    option 315
modeling in AutoCAD,
    comparison of 100
Modify dialog box 387
Modify Sketch dialog box 157
Modify Sketch function 156, 158
Modify Sketch icon 157
Modify Text dimension option 389
molds, cuts in 325
monitors 39
motion
    degrees of 302

limitations on 308, 324
simulating in assemblies 282,
    291, 301, 308
Mouse Speed slider bar 80
Move Component command 308
Move Component icon 309
Move/Size Features 246–247
movement
    See motion
Multiple Document Interface
    (MDI) 10
multiple processors 37
multi-thickness shell 209
multithreading 37

N

Name dimension option 389
Name drawing sheet option 357
named views 361
naming conventions 13, 44, 196,
    198
negative dimensions 186
networking software 36
neutral draft planes 214
Neutral Plane draft 215
Next Datum Label drawing sheet
    option 357
Next Detail Label drawing sheet
    option 357
Next Reference rib option 213
Next Section Label drawing sheet
    option 357
No Solve Move 127–128
Normal To view 56
Note dialog box 374
Note Text option 396
notes 394, 399
    adding 398
    AutoCAD comparison of 395
    balloons in 398
    editing 397
    leaders open in 398
    options for 396
    pasting 399
    producing with external
        programs 395

O

objects
    defined 70
    deselecting 70
    dragging 127
    rotating 78
    rotating, adjusting increments
        when 80
    selecting 70–71
    selecting multiple 71
    shaded quality, adjusting 83
    visual feedback when
        selecting 72
Offset Entities function 134–135

offset from surface end condition 183
offset planes 227
Offset Ratio bends option 278
offset surfaces 235
On Surface plane option 229
one direction end condition 186
One Direction thin feature option 273
One Line/Edge axis option 232
One Surface axis option 231
One Temporary Axis option 231
Open command, importing and exporting with 458
open profiles, thin features and 272
opening documents 6–8
operating systems, choosing 36
opposite alignment condition 305

options
    display 80, 87
    Hidden In Gray 81
    Hidden Lines Removed 83
    Perspective 84
    quick reference to 62, 67
    scope of 60
    Section View 86
    Shaded 83
    wireframe display 80
Options dialog box 53
    driven dimensions settings in 151
    renaming planes in 52
    using 59
Options dialog box tabs 62
options settings 61
orange, meaning of 299
ordinate dimensions 382, 384
    aligning 384
    changing 383
    horizontal 383
    vertical 383
orientation, effect on sketching 110
Orientation window 56
    common mistakes with 109, 110
Origin Point, anchoring to 19
overdefined sketches 150
Override properties component option 315

P

page count, inserting in documents 452
page numbers, inserting in documents 452
page orientation, setting 446
Page Setup command 445, 454
Pan command 79
Pan keyboard shortcut 487
paper margins, setting 446
Paper Size drawing sheet option 357

parabolas 118
parallel dimensions 145
Parallel mate constraints 304
Parallel Plane at Point option 228
parallelograms 119
    creating 120
parametric modelers
    compared to Boolean modelers 17–18
Parametrics 5
Parasolid Binary files
    about 460
    assembly format 461
Parasolid files
    about 460
    assembly format 461, 474
    exporting 472
    importing 472
Parasolids kernel 34
parent/child relationships
    about 198
    base parts and 269
    in-context editing and 296
    mates and 311
    minimizing use of 287
    mirrored features and 218, 220
    patterns and 220
    suppressing child features from 251
part documents 9
partial ellipses 117
Partial Section section line option 368
Parting Line draft 215
parts 175, 280
    adding to assemblies 296
    annotations for 256, 258, 379
    base 269, 272
    bends in 274
    color and 425
    color preferences for 431
    colors, modifying 299, 430
    configurations for 267, 314
    constant wall thickness 272, 279
    creating new 50
    curvature of, showing 88
    derived 327
    design tables for 259
    detail drawings for 343
    draft feature for 214
    editing 310, 312
    equations and 264
    exploding 298
    export formats 460
    features, identifying for 168
    folded 274
    holes, adding to 201
    import formats 460
    in-context 298
    in-context editing for 296
    inserting into assemblies 291
    inserting into drawings 346

lighting and 425
mirroring 219
modifying 244, 252
naming conventions 197
new 296, 298
planning 177, 179
printing characteristics of 443
quick printing of 441
rebuilding 249
revolved, creating diametric dimensions for 146
rolling back features of 250
samples for visual feedback on 72
saving 50
scaling 253
screen shots of 441
shaded quality, adjusting 83
sheet metal 272, 274, 279
shells 211
simplifying 267
skeletons and 285
thin feature 272, 279
unfolded 274
welding 335
wireframe view of 80
pattern attributes, in SolidWorks 222
Pattern crosshatching option 366
Pattern Deletion dialog box 224
pattern features 220, 253–255
patterns
    assembly 318
    AutoCAD comparison 223
    axes for 230
    deleting 224
    deleting instances of 320
    flat 275
    instance deletion 224
    options for 223
    retrieving deleted 225
PCI bus, graphics card access to system memory 41
Pentium Pro processor 36
Pentium processors 37
Performance options tab 63
Perpendicular Curve plane option 229
Perpendicular mate constraints 304
perspective, adjusting 84
Perspective option 84
photorealistic renderings, PhotoWorks and 436
PhotoWorks, AutoShade comparison 436
PhotoWorks application 436–437
pick and drag 114
pilot projects 42–45
pink, meaning of in sketches 152
Pitch helix option 239
Plan command 57
Planar surfaces 234–235
Plane names 66

planes
    creating 226, 230
    editing 245
    illustrations of various 227, 229
    mate constraints for 306
    neutral draft 214
    options for 227, 229
    renaming 53
    resizing 230
    See also sketch planes
planning
    implementation 29
    parts 177, 179
    training 29–31
plotters
    about 440
    customizing 444
    defining 443
plug-ins, XchangeWorks 475
Point command 120
points 120
    constraints for 160
    mate constraints for 306
point-to-point dimensions 145
    adding 143
Portrait page orientation 446
precision, dimensions 389
preferences 59, 67
    color 480–483
    determining for SolidWorks 43
    Dimension command 144–145
    display 480
    driven dimensions 151
    exporting 61
    general 479
    importing and exporting 461
    quick reference to 62, 67
    rotation 479
    scope of 60
    user 478
Preview crosshatching option 366
Previous view, comparison to
    AutoCAD 76
Previous view icon 76
Print Screen key 436
Print selection 455
Print Setup command 449
printing
    configuring printer for 444
    customizing printer for 444
    defining a printer or plotter 443
    document characteristics 442
    drivers for 440
    page orientation, setting
        for 446
    page setup for 445
    Print Preview function for 454
    Print Setup command for 449
    quick, of parts and
        assemblies 441
    Scale function for 446
    selected areas 455
    setting margins for 446

processors
    See CPUs
profiles
    multiple, features for 196
    single, features for 196
    thin features for 272
    See also sketch profiles
Profiles loft option 193
Projected Curves command 238
projected views 359
Propagate Along Tangent Edges
    fillet option 207
properties, customizing 49, 415
properties, BOM 419
properties, dimension 388
properties, file
    custom 352–354
    linking to 351–352
    modifying 355–356
PropertyManager 51
prototypes
    assemblies as 285
    STL files as 470

Q

quick reference drawings,
    defined 344

R

radial dimensions
    See diametric dimensions
Radiate Surface command 236
    options 237
RAM requirements for
    SolidWorks 38
ray tracing, full 437
Read Only dimension option 389
Rebuild command 249
Rebuild the model keyboard
    shortcut 487
rectangles 119
    creating 120
red, meaning of in sketches 152
Redraw command 89
Redraw the model keyboard
    shortcut 487
reference axes 230
reference dimensions
    creating 381, 384
    model dimensions, compared
        to 378
reference entity 121
reference geometry 225, 243
Reference Geometry options tab 66
references
    finding 312
    invalid, tracing 297
    lost 314
registry files 43, 61
    merging 44, 62
relations
    deleting 162

    viewing 162–163
    See also constraints
relations, adding 161
relations, viewing 164
relationships, defining 162
relative to model views 360
Remove crosshatch option 366
renaming features 53
rendering 423, 438
    AutoCAD comparison of 425
reordering features 200
report files 467
Reset Feature color option 434
Reverse Direction extrusion
    option 184
Reverse Direction helix option 240
Reverse Direction pattern
    option 223
Reverse revolved feature
    option 186
Reverse rib option 213
Reverse thin feature option 273
Revolution helix option 239
Revolve as, revolved feature
    option 186
Revolve function 185
revolved features
    end conditions for 186
    options for 186
revolved parts, diametric
    dimensions for 146
revolved surfaces 234
ribs 211, 214
    AutoCAD comparison 214
    options for 213
right mouse button 54
    about 73–74
    difference from AutoCAD
        use 73
    use with Edit Definition
        menu 244
robot assembly example 336
Rollback command 177, 249
Rotate clockwise/counterclockwise
    keyboard shortcut 487
Rotate Component About Axis
    command 309
Rotate Component Around
    Centerpoint command
        See Rotate Component
        command
Rotate Component command 309
Rotate Drawing View dialog
    box 373
Rotate the model 90 degrees
    keyboard shortcut 487
Rotate the model keyboard
    shortcut 487
Rotate View command
    about 78
    changing rotation center of 78
Rotate View icon 373
rotating sketches 156

rotation, preferences for 479
rotation motion 302
rotational increments, adjusting 80
Roughness Values finish option 411
rounds 128, 204, 208
    See also fillets

S

SAMPLE.btl bend allowance file 278
Save As command, importing and
    exporting with 458
Save As option of File menu 14
scale
    drawing sheets 357
    drawings 344
    section view 368
    setting for printing 446
Scale command 253
Scale crosshatching option 366
Scale dialog box 253
Scale drawing sheet option 357
Scale with Model Changes section
    line option 368
Scaling Factor cavity option 327
scaling parts 253
scaling sketches, translating 156
Scaling Type cavity option 327
Scan Equal function 165
screen refresh 89
screen shots of parts and
    assemblies, printing 441
scrolling, keyboard shortcut for 79
SCSI bus standard 39
SCSI drives 39
section lines 367, 369
    options for 368–369
Section View option
    about 86
    limitations on 87
section views 364, 366
    creating 86–87, 364
    crosshatching in 365
Seed Face list 238
Selected Items extrusion option 184
selecting objects
    compared to viewing
        objects 54
    planes 54
Settings For extrusion option 184
SGI machines 38
Shaded display option 83
Shaded Display Quality dialog
    box 84
shaded face highlighting 63, 71
Shape command 255
sharing documents
    electronically 441
sheet characteristics,
    preferences for 67
Sheet Metal Attributes 277
sheet metal parts 274, 279
    creating with thin features 272

unrolling 278
sheets
    drawing 10
    See also drawing sheets
Shell Outward option 211
shells 209, 211
    troubleshooting 211
Shininess lighting option 432
shortcut icons, creating 48
shortcuts
    customizing properties 49
    in context-sensitive menus 74
    SolidWorks 43
    See also keyboard shortcuts
shortcuts, SolidWorks 61
shortcuts, Windows 98 49
Show Feature Detail component
    option 316
Show Intermediate Profiles sweep
    option 190, 194
Show Leader finish option 411
Show Model component option 315
Silhouette draft option 241
Silicon Graphics, Inc. 38
single command per pick check
    box 114
Single Side rib option 213
Size note option 396
skeletons 285
sketch 5
sketch area 18
sketch entities 113, 123
    adding dimensions to 381
    changing length of 131
    constraints, maintaining 159
    constraints, removing 162, 164
    converting to sketch
        geometry 133
    deleting 132
    editing 153
    grouping with Scan Equal
        function 165
    lengthening 132
    Offset Entities function for 134
    shortening 131
    turning into construction
        entities 123–124
    See also individual entity names
Sketch Fillet dialog box 128
sketch geometry 167
    creating from sketch
        entities 133
    drafts, incorporating in 216
    dragging 153
Sketch icon 113
sketch mode
    starting 109
sketch planes 52–57
    about 106, 108
    common mistakes with 109
    compared to AutoCAD UCS 55,
        106, 181
    editing 245

initial, identifying 177
    renaming 52
    ribs and 212
    selecting 54
sketch profiles 57
sketch symbols 112
sketch tools 113
    list of 114, 124
    mirroring 124, 126
sketch view orientation 110
sketched features, defined 100, 181
sketches 156
    assembly 285, 287
    automatic constraints in 111,
        160
    basic guidelines for 100–101
    colors, meaning of in 152
    complexity in 102
    derived 167
    dimensioning 103, 172, 381
    dragging objects in 127
    editing 152, 173, 246
    importing data into 462–463
    mirroring in 124, 126, 158
    overdefined 150–151
    parent/child relations in,
        finding 199
    redefining 245
    ribs 211
    rotating 156
    sample, interactive 168
    scaling 156
    text in 121
    translating 156
Smart Mates 294
Smooth Transition fillet option 207
snap behavior 95
    setting 93
snap grids 95
    compared to geometric snap
        functions 94
    unimportance of in
        SolidWorks 94
Snap to Angle function 94
solid features
    See features
solid modeling, computer system
    demands of 6
solid models
    base features of 57
    building from two-dimensional
        imports 33
    difficulty of changing in
        AutoCAD 181
    exporting 34
SolidWorks
    API to 492
    basic concepts of 4, 6
    customizing 477, 494
    99 menu pick 25
    starting 48
    superiority to AutoCAD 244
    user interface of 16, 23

web site, menu for accessing 25
Spacing pattern option 223
Specularity lighting option 432
spin box increments 154
spirals
    creating 240
    Helix command for 239
splines 118
    AutoCAD creation of 243
    constraints for 160
Split Curve function 136
Split Line command 215
Split Lines draft option 240
spot light
    options for 428
    source of 426
springs, helixes for 239
Standard for Exchange of Product
    Model Data files
    See STEP files
Standard Views toolbar 57
standardization of Windows
    interface 3, 4
Start/End Tangency options 190
starting a SolidWorks session 50
Starting Angle helix option 240
starting SolidWorks 48
status bar
    about 23
    comparison to AutoCAD 23
STEP files
    about 460
    assembly format 461
    exporting 472
    importing 472
Stereolithography file type
    See STL files
STL 470
STL files 469, 471
    about 460
    exporting 471
    viewers for 471
Straight Transition fillet option 207
subassemblies
    defined 281
    disolving 317
    inserting 291
    managing 283
    showing in design trees 316
Suppress command 251
surface finish symbols 409
    inserting 409
    options for 410
surface modeling
    compared to solid modeling 6
surfaces 233, 238
    AutoCAD comparison of 234
    exporting 468
    extruded 234
    illustrations of various 234
    imported 235
    importing 468
    knitted 237

lofted 234
offset 235
radiated 236
trimmed 468
Sweep dialog box 189
Sweep function 187
sweep path
    See swept features
sweep section
    See swept features
sweep trajectory
    See swept features
swept features
    about 186
    AutoCAD comparison 191
    options for 189
    troubleshooting 190–191
swept surfaces 234
swOptions.reg 61
symbols
    ANSI geometric tolerance
        datum 405
    cosmetic threads 411
    cursor, list of 112
    datum point 405
    drawing 394, 421
    FeatureManager, meaning
        of 306–307
    geometric tolerance 401
    inferencing cursor, list of 111
    options for 397
    surface finish 409
    weld 407, 409
symbols, list of 111
system feedback 72–73, 112
system help keyboard shortcut 488

T

Tangent 160
tangent arcs 115
    creating with fillets 128
Tangent mate constraints 304
Taper Helix option 240
Taper Outward helix option 240
tapered surfaces, draft feature for
    creating 214
Target Area Size datum option 407
Template drawing sheet option 357
templates, drawing 348, 358
    AutoCAD comparison of 349
    customizing 348, 356
temporary axes 230
text
    computer intensive nature
        of 123
    dimensions, adding 390, 392
    inserting in drawings 394
    sketch 121
Text dialog box 122
texture memory 40
Thickness bends option 278
Thickness rib option 213

thin feature parts 272, 279
thin features
    AutoCAD comparison 272
    options for 273
third-party programs 52
third-party tools, integration with
    SolidWorks 20
thread profiles, helixes for 239
threads, cosmetic 411
Three-dimensional accelerator
    chips 41
three-dimensional solids
    See solid models
three-dimensional wireframes
    See wireframes
through all end condition 183
thumbnails 63
time stamp, inserting in
    documents 452
tip of the day, turning on and off 2
tips, accessing 114
Tolerance dimension option 389
Tolerance Display option 393
tolerances
    dimension options for 392
    geometric symbols for 401
toolbar icons, standardization of 4
toolbars
    about 21, 23
    activating 484
    adding icons 485
    comparison to AutoCAD 21
    customizing 483, 485
    font 22
    Line Format 377
    Macro 491
    removing icons 485
    Standard Views 57
    view 75
tools
    list of sketch 114
    sketch 113, 123
    See also individual tool names
top-down assembly method
    choosing 289
    defined 285
    new parts and 296
    using 289
Total Instances pattern option 223
training
    classroom 31
    schedule 30
training materials 30
training plan 29–31
trajectory, sweep 186
translation motion 302
Transparency lighting option 432
Trim/Extend function 131
trimetric view 57
trimmed surfaces 34, 468
troubleshooting
    imported files 474
tutorials, SolidWorks 30

Two Direction thin feature
option 273
two directions end condition 186
Two Planes axis option 231
Two Points/Vertices axis option 231
two-dimensional drawings,
importing to SolidWorks 33
Type of Projection drawing sheet
option 357

U

UCS
comparison to sketch
planes 106, 181
origin point, comparison to 19
Ultra DMA type hard drive 39
Ultra IDE bus standard 39
Ultra Wide SCSI drives 39
underconstrained geometry,
problems with 173
Undo command 130, 155
about 247
AutoCAD comparison 247
unfolded parts 274
units of measure
drawings 344
preferences for 64
unsuppressing features 251
up to next end condition 183
up to surface end condition 183
up to vertex end condition 183
Up/Down loft option 193
Up/Down sweep option 190
Use Auto Relief bends option 278
Use Bend Allowance bends
option 278
Use Bend Table bends option 278
Use component's in-use
configuration component
option 316
Use Dimension's Font tolerance
option 393
Use Document's Font note
option 396
Use K-Factor bends option 278
Use named configuration
component option 316
Use properties in Configuration
component option 315
User Coordinate System
See UCS
user interface 16, 23
user preferences 478

V

Variable Radius fillet option 207
variable radius fillets 206
Vary Sketch pattern option 223
VDAFS files,
importing/exporting 472

version of SolidWorks,
determining 25
vertex points
constraints for 160
Vertex-Chamfer option 209
vertical dimensions 149
Vertical Ordinate Dimension
command 383
vertices
adding dimensions
between 381
mate constraints for 306
video cards 39
view commands 76, 80
keyboard shortcuts to 79
quick reference to 79
View crosshatching option 366
view orientation,
effect on sketching 110
View Orientation dialog box 55, 57
View Orientation icon 76
View Rotation preference 479
View toolbar, using 75
View/Display menu 80
View/Modify menu 76
View/Toolbars menu 21
viewing objects, compared to
selecting objects 54
views
about 56
activating 359
aligned section 367, 369
aligning 370, 377
alignment of 358
annotations for 380
assembly, creating 347
AutoCAD comparison of 358
auxiliary 362
borders, turning on and off 374
breaking alignment of 371
broken 369
creating 358, 370
detail 362–363
empty 370
exploded 328, 333
hiding 373
model dimensions for 380
modifying 370, 377
named 361
Normal To 56
part, creating 347
projected 359
relative to model 360
rotating 373
scales of 358
section 86–87, 364, 366
standard 358
types of 57
Virtual Reality Markup Language
files
See VRML files

virtual sharps 120, 130
controlling the appearance
of 131
Visual Basic, customizing
SolidWorks with 492
VRML files 235
about 460
assembly format 461
exporting 473
importing 473

W

Wall Thickness thin feature
option 274
water, rendering with
PhotoWorks 437
web site, menu for accessing
SolidWorks 25
Weld Bead command 335
weld beads 335
weld symbols 407, 409
inserting 408
options for 408
Width drawing sheet option 357
Windows 95
pros and cons 36
Windows 98
shortcuts 49
Windows interface 16
Windows NT
pros and cons 36
Windows registry file 43–44
wireframe display option
adjusting quality with 80
wireframe geometry, exporting 467
wireframes, importing to
SolidWorks 33
WWW link note option 396

X

XchangeWorks 475
Xeon 37, 38
xlines, in AutoCAD 121, 123
Xrefs, AutoCAD 270

Y

yellow, meaning of in sketches 152

Z

zoom commands 76, 79
AutoCAD comparison 77
keyboard shortcuts to 79
quick reference to 79
Zoom in keyboard shortcut 487
Zoom In/Out command 77
Zoom out keyboard shortcut 487
Zoom To Area command 76
Zoom To Fit command 76
Zoom to Selection 77

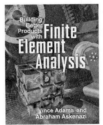